Financial Decision-Making for Engineers

Financial Decision-Making for Engineers

Colin K. Drummond

Yale UNIVERSITY PRESS

New Haven and London

Yale University Press books may be purchased in quantity for educational, business, or promotional use. For information, please e-mail sales.press@yale.edu (U.S. office) or sales@yaleup.co.uk (U.K. office).

Printed in the United States of America.

Library of Congress Control Number: 2017952013
ISBN 978-0-300-19218-6 (pbk.: alk. paper)

A catalogue record for this book is available from the British Library.

This paper meets the requirements of ANSI/NISO Z39.48-1992 (Permanence of Paper).

10 9 8 7 6 5 4 3 2 1

Dedicated to

Lizabeth H. Drummond
1957 – 2017

Contents

Preface

This field guide is intended to help engineers with little or no business background begin the journey of becoming a practitioner in the field of financial decision-making. Many concepts are *introduced* and linkages to textbooks (some of which you may want to buy) and Internet resources (some of which you may want to read) are provided. We hope that you will discover the need to take the initiative to explore further. The chapters to follow introduce key concepts relevant to the accounting, finance, and engineering economy that collectively are the foundation for financial decision-making for engineers. The original use of this book was for the instruction of a intensive 8-week summer session for undergraduate engineers beginning a program of study in a graduate management program.

I hope this book shapes the way you craft responses to situations involving financial decisions, sparks your curiosity, and instills self-confidence in your ability to contribute as a manager. Many of the topics selected and the tone of the narrative reflect my personal opinion and bias from over two decades in industry. Ultimately, your own progress will be marked by development of critical thinking skills reflecting the learning and empowerment we hope that you will achieve in your personal journey to become a '*New Economy*' engineer.

This book is not the result of one author. Many colleagues and students contributed to my knowledge in many different ways over the years I have been teaching at Case Western Reserve University. I thank all of them. I also thank the editors at Yale University Press, especially Joseph Calamia and Mary Pasti; this book would not be the same without their patience, support and sharp editorial eye. Finally, I could not have devoted the time and energy to this project without the support and encouragement of my late wife, Lizabeth. She was a central part of my life for over 40 years and this book is dedicated to her. I will miss sharing the final result with her.

Colin K. Drummond
November 2017

Chapter 1

Decision-Making

Over the next ten chapters, our course of study will improve your competency in financial decision-making. The book spans subjects in accounting, finance, and engineering economics; take a few moments to review Figure 1.1 illustrating the "road ahead" before we forge ahead!

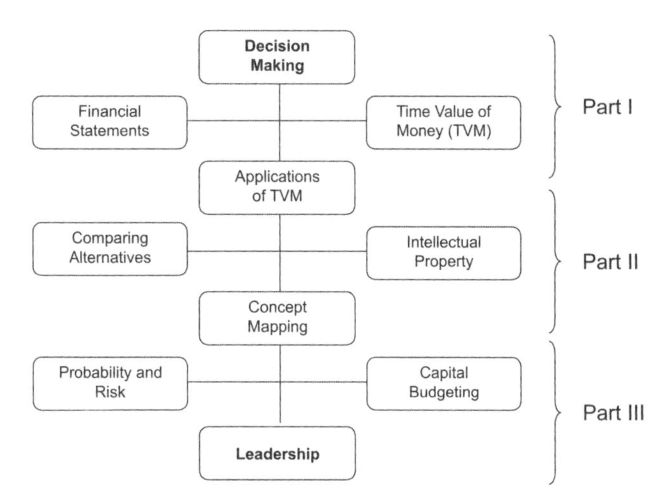

Figure 1.1: The road ahead: Ten learning modules grouped in three parts.

1.1 Three tactical challenges

Much has been written about the subject of accounting, finance, and engineering economics and there are many informative textbooks about these subjects. Three challenges with this abundance of information are that:

- There are a wide variety of topics to introduce, explain, study, and learn that pertain to the everyday life of an engineering manager.

- Readers of this textbook have a wide variety of prior experiences with finance; some may have already taken an accounting class while others may be entirely new to the subject.
- The purchase of several different specialized books (accounting, finance, and engineering economics) results in underutilization when, in practice, we find we only need *pieces and parts* of any one of these books when beginning in this field. Topic variety is a reference selection challenge.

There are so many existing textbooks that we could draw upon – my personal experience, years of work, actually, is with the works of Meigs and Meigs[1], McManus[2], Powers[3], and Wickes[4] – but methodical navigation requires an abundance of time (months!) devoted to each. Later in this introductory chapter we provide some perspectives the reader can adopt to facilitate mastery of so many topics. It is an easy read, so let's go!

1.2 Three perspectives

It is unreasonable to expect that we could simply outline several dozen concepts and expect that, through laborious exercise, the student will somehow recognize the fundamental premise of this course of study. Let's fix that now. If you don't remember *anything else*, the abstraction is to understand a transaction concept map shown in Figure 1.2. Figure 1.2 illustrates a simple, fundamental representation of the daily activities of a typical business; the cycle illustrates the role of accounting, information, and decisions central to the management of business. This diagram is highly simplified (and adopts the so-called *internal perspective*, as we shall soon see), so, of course we could have said "products *or* services" that result from managerial decisions, but hopefully you get the idea. And lots of things are missing from the diagram (the roles of investors, boards, etc.), but, in this simplified diagram, we underscore that at the most fundamental level:

- *Managerial decisions* drive the *conversion or use of assets* into products, that ...
- if are *appealing to a market need* lead to *customer transactions*, that ...
- produce revenue and data to be *converted into information*, that ...
- must be organized and *communicated* to managers to support decision-making.

Pause for several moments to compare the itemized list above with the diagram below. Where might "ethics" issues enter and be a problem?

Now, with Figure 1.2 in mind, is the interconnected nature of activities clear? Can you visualize the activities associated with each step that managers must lead to make things happen? If everything is so "interconnected" then how do we characterize the activities and outcomes in a simple, understandable way? Let's distill the process into three steps:

- Adopt a *transaction-based* view of the business and ways to capture and communicate transaction activity.
- Understand the *engineering managers' view* of asset (cash, equipment,

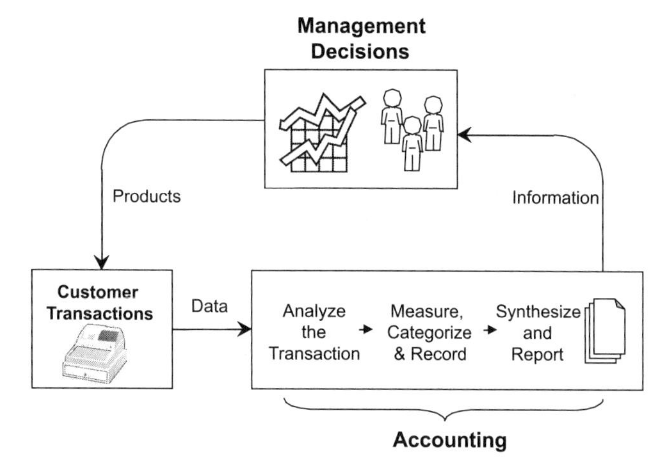

Figure 1.2: Role of accounting to provide support for managerial decision-making.

facilities) management.

- Examine the *decision-making* required to support and direct changes in business activity.

A primary objective of this chapter is to explore these three perspectives with a special emphasis on decision making.

1.2.1 Transaction-based perspective

Let's examine the most fundamental – and hopefully common – business activity: *transactions*! Figure 1.3 helps focus the discussion. You can talk

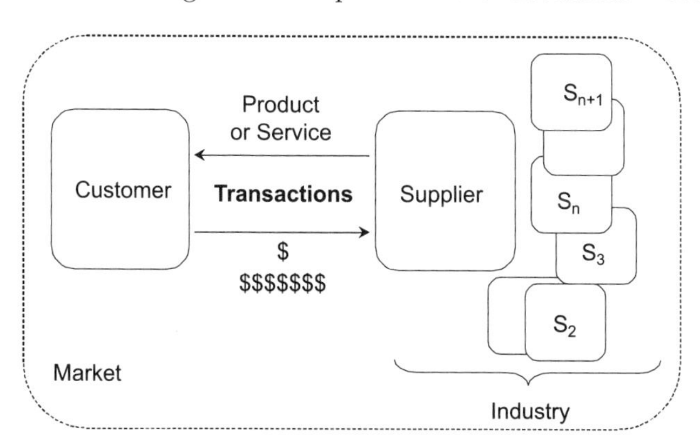

Figure 1.3: Transaction-based view of businesses, industries, and markets.

strategy, dream about products you that think consumers might need, and organize focus groups to determine what end-users say they want, but unless your business is *profitable* (the revenue is greater than expenses) and *solvent* (you can pay bills as they become due), you will never realize those aspirations. Profitability and solvency critically require that customers have a preference for your product and are willing to enter into a transaction with your business in which a product or service is exchanged for payment (i.e., *buyers*, not talkers). Better yet, we look for a series of *ongoing* transactions where a preference (over other suppliers) for your product and perceived value (see Figure 1.4) results in a transaction where the market price is far in excess of the cost of production (more on that later). Very simply, then, the "reality"

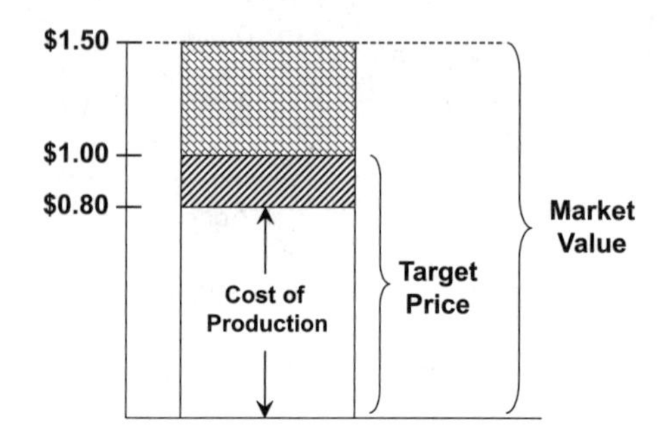

Figure 1.4: Identifying value in a transaction.

of your "day-to-day" business can be characterized by customer transactions. An ability to appropriately capture, analyze, understand, and communicate those transactions is the foundation to knowing your business. Ideally the information provides actionable items and *drives a transaction-based perspective.* Items critical to long-term success (again, profitability and solvency) correlate directly to the magnitude (number of), type (pay now or pay later), and value (market preference) of periodic business transactions. Given that *accounting* is an information management system related to the analysis, measurement, and synthesis of transaction information, it should come as no surprise that the chapter to follow discusses accounting in more detail and the production of financial statements.

To focus on business transactions means, of course, there must be some business transactions to examine, and an implicit assumption is that we are dealing with an *ongoing concern*, one that has been under way for a period of time – say, three years – and can be anticipated to be in operation for sometime in the future (at least five years). Such an assumption is essential for financial statements to have meaning year-to-year. Aspects of the business other than transactions warrant attention, too. These include fund-raising for the start-up, investing for business expansion, and deciding to replace equipment. All involve transactions and play into a vibrant business; these will be identified later as investment activities. For now, our assumption of a transaction-based perspective implies a more simplistic "operational" scope.

1.2.2 Engineering management viewpoint

Many stakeholders have an interest in the activities of a business. Since this *Field Guide* hopes to address those activities most likely to touch engineering students embarking on new *managerial* careers, the present work adopts the *managerial viewpoint*, as distinct from a regulatory, audit, and (to some extent) shareholder viewpoint. Principal stakeholders and the central role of the accounting system are shown below. Note the central role *accounting* plays

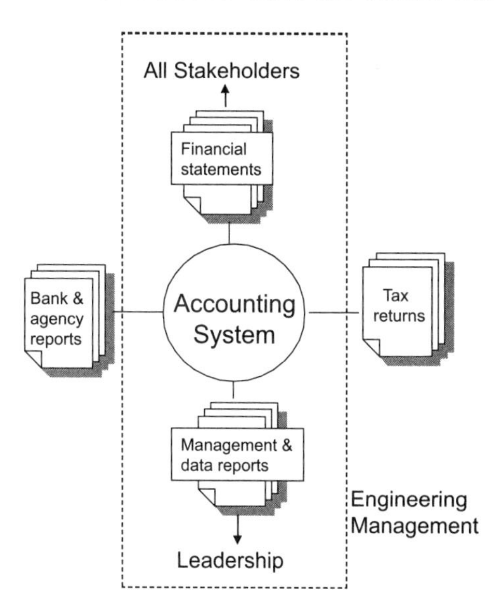

Figure 1.5: Engineering management served by an accounting system (concept adapted from Meigs and Meigs[1], p. 7).

in communication with stakeholders! We'll talk more in Chapter 2 about the critical role of accounting in communicating business results; further, we'll dispel any misunderstanding that accounting is the same as bookkeeping (for fun, try an Internet search of the phrase "accounting versus bookkeeping" – a recent search yields over 750,000 hits on this subject). Section 2.3 on page 23 explains further.

What is the impact of taking a managerial perspective? How does this matter to this course of study? What is gained? Lost? How do *current events* play into the context of solutions? Consider and discuss the following statements in your study group:

- **Taxes** are important and demand attention, but in the current work we do not venture beyond a simple understanding of how decisions can be impacted (or influenced) by tax considerations. The reasoning is that a typical engineering manager relies on the corporate tax department to help frame and present tax issues as a "boundary condition" or a project input parameter, with the idea that managers are not customarily tax accountants themselves. This Field Guide is therefore "light" in the treatment of tax accounting *except* to the extent that taxes pertain to capital investment decisions. Many types of taxes are important, just not right now.

- **Regulatory matters** have a significant bearing on many facets of a modern business, but as with tax accounting, the regulatory and audit function is not explored very deeply in the chapters ahead. Separate from the subject of ethics (which *will* appear to some degree later on in some case studies), this Field Guide lacks depth in the description of the audit function. There is no question that regulatory issues play an important role in decision-making, but, for now, we'll keep these issues on the periphery of our discussions.
- **Stockholders** are quite central to much of managerial thinking, but for the topics, tools, and skills we wish to center on in this book, we relegate stockholders as an environmental variable. Not a good thing to do in a more complete context of financial management, but for the time being we frame our world as having to simply meet established goals and objectives set for us *by* the stockholders *through executive management*, keeping stockholders at an "arm's reach." Do you think this assumption really has much day-to-day impact on the life of an engineering manager?

Another way to distinguish between perspectives is to consider that there are *internal* and *external* users of information created by an organization. *External* users are normally *indirectly* related (auditors, regulators, lawyers, the press, etc.) to business operations and they will have a legitimate need for reliable information about past and planned activities taking the form of regular financial statements, the prospect of and contingency for lawsuits, or the planned purchase or sale of major assets. Returning to the needs of shareholders – external users of information – as investors in a company (and part owners, however small that fraction of ownership may be!) they would need financial information to decide whether they would like to own a bigger part of the company (buy more stock) or if the affairs of the company have become too risky (new products outside the core competencies of the management) for their portfolio and it is time to sell their holdings. Regulatory users are external users, too, as they are monitors of the activities of the company with bearing on managerial decision-making, but the regulators are part of the external context of business operations and are not the managers of the company. Labor unions might also have an interest in the prevailing wages of company employees, even though they are not directly responsible for setting company wages.

In comparison, *internal* users of financial information are those individuals *directly* responsible for the operations and management of the company. Whether it is sales staff, marketing managers, human resources, purchasing agents, supply chain managers, or product development leadership, all have a direct impact on the success of the organization and need reliable financial information to be able to make informed decisions contributing to the efficiency and effectiveness of the organization.

Accountant career paths are another way to observe the difference between internal and external reporting, too. *Management accounting* attends to understanding, analyzing, and reporting information to *internal* decision makers while *financial accounting* has an *external* stakeholder focus.

For the present work, we assume that internal engineering management information needs dominate, thus we set and accept a fairly narrow view of many key stakeholders. Over time, though, managers in practice do have many occasions where the importance of external users takes center stage, and that is not to be trivialized. Here, it is enough to tackle the intertwined, interdisciplinary, multifaceted nature of engineering management by bounding and limiting the environment in which we operate and *make decisions.*

1.2.3 Emphasis on decision-making

We noted at the end of the last section that an important role of engineering management is to *make decisions*; however, up to this point it has been assumed we know what "decision-making" really means! Decision-making is often confused and treated synonymously with *problem solving* and, to some extent, *critical thinking skills.* Does the difference matter? Are we just playing with words? Or is there a structural difference that can help guide development and learning specifically related to the financial affairs of an organization? This section briefly underscores managerial *decision-making* as the impetus for much of the information we collect, analyze, and communicate; further, we identify decision-making as just one *part* of the much larger problem-solving process.

Daily living gives us many examples to observe the gap between "what should be" and "what is." While this sets the stage for spotting a "problem" in many cases, absent any process for *filtering problems by degree of severity*, the domain of problems that confront us would be unbounded! It is therefore useful to narrow thinking (at least in the corporate world) around *those important problems that motivate time and energy to investigate the situation in detail.* Let's clarify with a specific example illustrated in Figure 1.6.

In this example we suggest that a corporate strategic advantage can be gained by creating intellectual capital though the development of product patents, but a problem may be that competitors might already have defensible claims. The process of identifying existing or similar patent claims requires only a simple search of the US Patent and Trademark Office (USTPO) database. Claims retrieved from a keyword search of the USTPO database can be compared against our idea for a patent claim we think is unique. In this situation, let us assume our idea is for a highly fragmented market with a high degree of competition between suppliers. We would want to check if there are patent claims similar to ours to *avoid* the problem of patent infringement. The problem-solving process involves an iterative search of the database, harvesting select information from the database to examine, and then using this information to decide if the product attributes we have in mind are unique. Absent any claims from the database we might consider "similar" to our own ideas, we then decide to proceed with our plans to launch a new product. If we

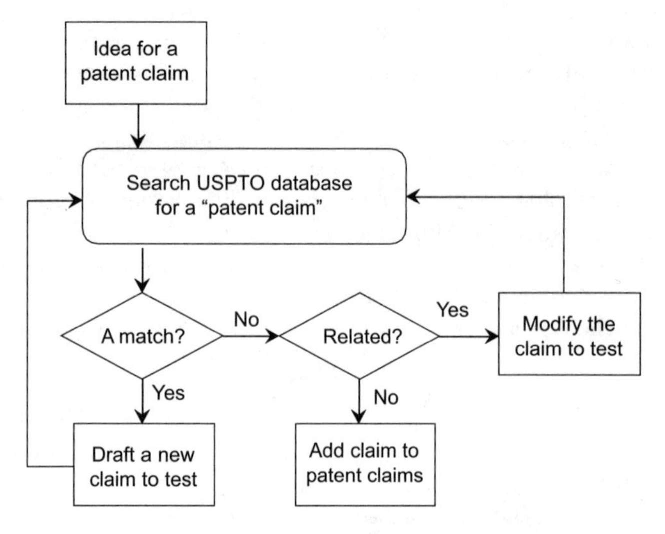

Figure 1.6: A perspective on the patent claim search decision-making process.

are wrong, there is the probability of financial impact on operations through legal expenses, fines, and losses in divestment of a product line. While the idea for the product may have come from, say, marketing, the engineering manager plays a critical risk mitigation role on the management team.

In this corporate growth problem, the choice of core competence to develop (intellectual capital), sources (and legitimacy) of data, the interpretation of that data, and the logic underpinning risk management involve a wide variety of factors contributing to problem definition and decision-making. We can't say that one is more important than the other. Is it better to identify the "right" problem and make poor choices in solving it, or to optimize the solution to the wrong problem? This example is similar to many others one might confront in practice and emphasizes the subtle role decision-making plays in problem-solving:

1. The product launch problem is important and involves an engineering decision among options (in this case, regarding "freedom to operate" in a marketplace).
2. There is a large body of information available and we use analysis tools to identify and refine that subset of information to establish a choice of action.
3. The interdisciplinary nature of the problem (presumably involving lawyers, among other professionals) and an iterative solution helps refine the course of action.
4. Some risk of corporate financial exposure exists and decisions must be made about the probability of that risk.

Problem definition and decision-making are intertwined; for many engineering managers, then, problems present themselves in a variety of ways and must be reduced to a tractable form, a problem-solving strategy crafted, and decision-making undertaken to optimize corporate resources. Accountability follows decision-making and defines managerial success or failure. Leaders must be as good at identifying the right problems as they are good decision-makers. While we attend to both issues in the course of study covered by this book, the abbreviated nature of this course does, however, place an emphasis on engineering decision-making. The successful financial decision-maker is one who embraces and finds delight in the problem-solving process, a person who has some comfort level with ambiguity and interdisciplinary activity. Many times the *cause of the problem* emerges when working through many iterations of the problem definition, in which there is a pattern to relevant changes. And it is most gratifying to find the outcome that explains all the facts. It's sort of like a mystery novel! The reality is that such gratification is hard to come by, so we have to be content with success in the *art of decision-making.* We really do not understand the problem completely, yet still manage to make a decision that satisfies most of our criteria.

Saundra Lipe and Sharon Beasley[5](page 37) emphasize the concept of *decision-making*:

> *"decision making is a purposeful, goal-directed effort applied in a systematic way to make a choice among alternatives. It is action to achieve a foreseen result, which is preceded by reflection and judgment to appraise the situation, and by a thoughtful, deliberate choice of what should be done."*

Decision-making, then, is influenced by many factors – among them "perception" – but we suggest here that many of these skills can be acquired over time. Let's summarize six aspects of decision-making:

1. Outcomes or required outputs must be established.
2. Within the context of an organization or project, the outcomes of required outputs must be prioritized.
3. A set of (actionable) alternatives must be identified or developed.
4. Each of the proposed alternatives must be examined with respect to objectives (and their priorities).
5. A tentative decision is an alternative that appears to provide all the objectives.
6. Decisive actions are taken, and additional actions are taken to prevent any adverse consequences, thus avoiding another round of problem analysis and decision making.

And, it is easy say, for instance, in item 1 that "outputs must be established," but *how* do we go about it? The start of a very long answer begins by looking at the *context* of decision-making, the topic of the next section.

1.3 Financial frame of reference

Among the more frustrating responses to the student inquiry "How do I ...?" is the response "It depends." This fails the test of didactic rigor learned over the students' prior years of college engineering training yet is so common in the practice of management that the credibility of models is undermined and, worse yet, the myth proliferated that much in management is decided simply as a matter of opinion. So, why bother, then, with learning models in the first place? Well, to some extent it *does* depend. The complexity of situations warrants full consideration of the *context* of the decision. This maturity in thinking comes about over time once you realize that there is not a one-to-one relationship between "models" and "problems" but that the manager integrates those models relevant to a problem.

So, the choice of *Present Worth (PW)* or *Equivalent Uniform Annual Cost (EUAC)* models for a specific situation might give the *appearance* of entirely a personal preference for or familiarity with, certain models of analysis; indeed, the process of choosing masks significant *experiential* learning in which what was once a slow, deliberate, conscious process that over time has become a rapid, unconscious selection manifest in the intuition of the decision-maker. Over time, similar situations engaged over and over again prompt a response that seems like and is a *professional* opinion. There is no fault with that strategy, except that both Benner [6] and Quirk [7] would suggest that it becomes harder and harder for the expert to articulate an "if-then-else-do" methodology of decision-making that has largely become intuitive.

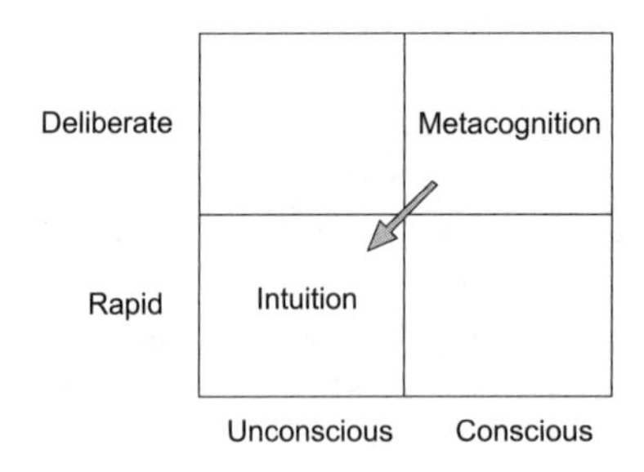

Figure 1.7: Development of expertise in decision-making.

While we can acquire some skills through academic study and practice, there is often no substitute for experience in the field. There seems to be no substitute for college internships enabling juxtaposition of practice and theory for students to explore firsthand the *context of decision making*; context is the key to (a) framing the problem (narrowing scope), (b) prioritizing possible inputs and outputs, and finally (c) identifying actionable alternatives.

Context is so very important, yet escapes those textbook problems designed to make them simple to solve. There is no simple remedy; even as recently as 2011, the workshop report by Koenig[8] on engineering education suggests that increasingly employers demand the type of thinking conveying a deep understanding of a problem domain and those relationships within it; pattern recognition and metacognitive skills are as important as quantitative skills. Wow! We can't solve all *that* here, but Koenig's line of thinking plays a key role in the way that we suggest students approach each of the problems at the end of each chapter. Though you can jump ahead to Section 1.6 for more details, suffice it to say here that your individual effort to supplement the solution to every problem by linking to a current article in the business press – and commenting in a meaningful way – will build competence in framing the context of problems.

In particular, look beyond the obvious in the current press (an article from the Harvard Business Review, Wall Street Journal, or New York Times published within the last two weeks) and *ask questions* – indeed work to *categorize* problems, going beyond the numerical answer and exploring the possible context of the problem. This will make your answer unique (more time for the instructor to grade) but give you experience in shaping new contexts to problems. To be a bit more specific, many users of accounting information are decision-makers. In general, engineering managers and their mid-level counterparts are responsible for the day-to-day operations of a business. Three categories of activities cover most everything in the daily life of a business: (a) operating, (b) investing, and (c) financing activities; these activities are arranged in decreasing order of typical importance to an engineering manager.

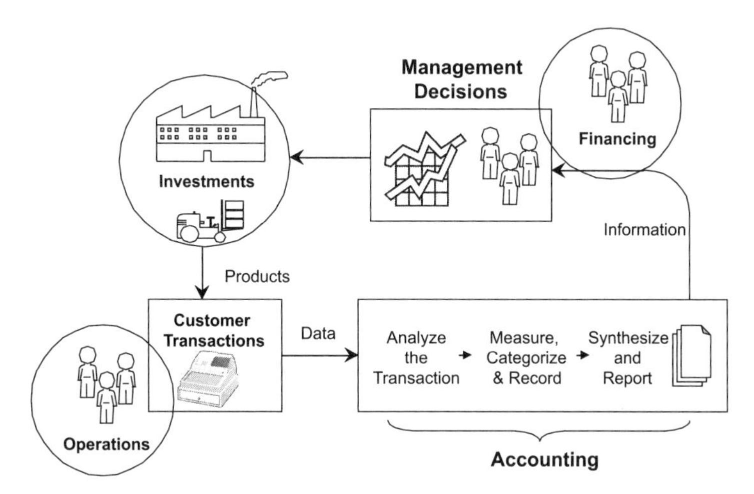

Figure 1.8: Operations, financing, and investing impact on simple day-to-day business activities.

You may have noticed that Figure 1.8 looks much like Figure 1.2, except now expanded on the periphery to include three major activities that managers conduct to help organizations reach their performance goals: operations, financing, and investing activities. Below, we very briefly expand on the nature of these activities relevant to the daily activities of the corporation and for which the engineering manager must be familiar if asset allocation decisions are to be meaningful.

Operating activities

A transaction-based perspective was described in Section 1.2.1. In that context, identifying, categorizing, and communicating the results of transactions are really some of the most fundamental tasks to be mastered as a manager. How will we know if we are profitable? Solvent? Reaching our goals? Did we properly record transactions when they occurred (*accrual accounting*), or simply at the time the payment of cash took place (*cash accounting*)? What are the needs of the internal stakeholders? External? Engineering managers must know how to communicate transaction information in a way that all stakeholders can understand. This is customarily done through the creation of *financial statements*, standardized reports about the analysis of business transactions. Paying workers, buying raw materials, paying taxes, subcontracting services, and selling products are all examples of operating activities.

Investing activities

Managers have an obligation to employ the assets of the company in the most productive way to assist the company reach its goals. Assets – most often in the form of cash or an equivalent – are provided to the management, who then must decide how they should be invested. Through *investment decisions* managers' decision-making activity results in the conversion or reallocation of assets provided by investors or owners of the company into different forms of "capital" that drive the business:

- Human capital: employees of the company who perform numerous purchasing, production, sales, and management activities.
- Intellectual capital: the development of know-how, patents, copyrights, or other intellectual assets that shape organizational uniqueness and core competencies to enable survival in a competitive marketplace.
- Financial capital: the cash needed for day-to-day materials and supplies, investments in new product development, reserves to solve operational problems, or investments in opportunities that arise unexpectedly.

We often think of investing as acquiring assets, but this activity also involves understanding the life cycle of assets and when and how to dispose of them, too.

Financing activities

Finding the initial funding to start a company, securing additional funding to expand operations, or establishing lines of credit for future use are all forms of *financing activities*. Activities such as financing inventory may not be central to an engineering manager's day-to-day concerns, but certainly the decision to introduce a new product in the market in which new inventory is purchased by a third party and bought back over time will impact profitability. Suppose corporate management decided to "free up cash" by selling the land and building holding product X assembly operations and then lease the building back over 15 years. Would this impact your willingness to upgrade (unrecoverable) improvement to the production line? Financing activities are often (dare I say!) of secondary concern to the engineering manager, but most certainly *do* impact the framework of decision-making.

Generally, we distinguish financing activities as those involving *long-term* assets, liabilities, and equity issues such as the issuance or redemption of stocks and bonds. Engineering managers would not normally become intimately involved with, say, the issuance of preferred stock, but to have even a pedestrian understanding of the implications enables one to sit at the boardroom table and to follow such discussions.

Be curious
A well-known research paper by Stuart and Robert Dreyfus [9] on the training of pilots reveals much about how we deal with the context of ambiguous situations, particularly as it relates to decision-making under pressure. Relevant here is the conclusion that while an expert can try to explain a step-by-step methodology for the novice to emulate on the path to development of expertise, there evidently is no substitute for practice and experiential learning. The approach is "study what I have said, then go out and try it on your own."

Concept understanding takes time and effort. We can talk about the difference between, say, an income statement and a cash flow statement, but the two will always be confused unless some "hands-on" activity is part of learning. The problems at the end of the chapter should help with that effort.

In this section we have implied the engineering manager must assume individual responsibility – embrace curiosity – of those other functions of the business, even if no direct responsibility is on the horizon. It is all part of the general context of decision-making. As shown in Figure 1.9, only iterative, self-directed learning can fill the void.

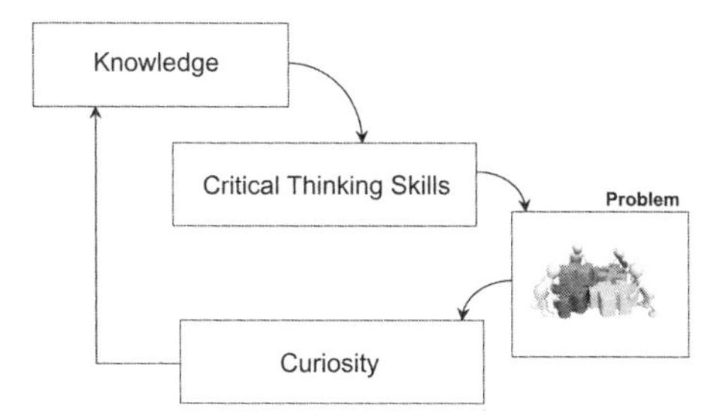

Figure 1.9: A cycle of learning leading to refinement of context for ambiguous problems in engineering.

1.4 Focus on financial decision-making

Section 1.2 outlined three perspectives that are central to this overview: (a) a transaction-based view of business activity, (b) an engineering manager perspective of asset management, and (c) decision-making in support of changes in business activity. We now move forward to link these perspectives to the three major facets of financial decision-making: accounting, finance, and engineering economics. Consider the following statements describing each subject:

1. Accounting: The process of identifying, measuring, and communicating economic information to permit informed judgments and decisions.
2. Finance: The study of assets and the way in which organizations manage assets, resources, and risks over time to accomplish organizational objectives.
3. Engineering economics: Application of engineering or mathematical tools in the quantitative analysis of the economics of engineering alternatives.

At the risk of over-generalizing, it can be suggested that our discussion of accounting will provide tools for developing the transaction-based perspective, finance will expand our thinking about asset management, and engineering economics will provide many ambiguous situations to explore that will sharpen decision-making skills. Recall that Figure 1.1 subdivided each subject area leading to a total of 12 critical "concept blocks" that we will expand upon in the chapters to follow.

Accounting

We often hear that accounting is the "language of business," or an "information management system," or even that it is "the communication tool for the health of a business." While true in all cases and explored in great depth in, say, Meigs and Meigs[1] or Marshall[2], here we must narrow the scope a bit to fit within our strategic intent; that is, in short order, we need to do three things:

1. identify and categorize typical business transactions to produce data about operations,
2. analyze and classify data to produce financial statements about the business, and
3. interpret results by comparing statement summaries and ratios against corporate performance objectives and industry norms or trends.

Consider this list as a basis for formulating engineering management questions. We produce data that are analyzed and classified to produce financial statements which can be used by decision-makers to draw conclusions about the health of the business and to take action as needed to meet corporate performance objectives. This important process turns out to be challenging for those students who have never been exposed to accounting before. Terms like "accounting period" seem academic, yet are critical to knowledge about why *adjusting entries* discussed in Section 2.5.2 are important to, say, capital equipment depreciation. Most important is the ability to look beyond the data presented, and have a sharp eye to identify inconsistencies and *ask perceptive questions.*

In contrast to the other major subject areas, our accounting focus is almost entirely centered around Chapter 2 and the preparation and interpretation of *financial statements*; this leaves a great deal to be desired, but much of the hands-on activity to collect and categorize data does not necessarily prepare the engineering manager for the task of knowing what the data mean. This is a compromise, but we have found that the scope of Chapter 2 is a great way for the engineer without any prior business classes to be successful if that engineers enrolls in (the most basic, anyway) graduate accounting course. This is not a perfect solution, but a compromise we make that should be noted as we steam ahead.

Finance

There are different interpretations of "finance," and one could say that some of what we have chosen to represent could have been under the heading "accounting" or "engineering economics." With an *asset management* perspective, this course of study is separated into Chapter 3 on the *time value of money* and Chapter 4 on the *applications of the time value of money*. Once again, this narrow scope might call into question this being an adequate treatment of finance (especially as MBAs know it!) but we subdivide to create a somewhat "sterile" view of the characterization of assets prior to the section on engineering economics.

Engineering Economics

Most engineers understand the significance of converting units so that the calculations they make are based on equivalent units of measure. In accounting, finance, and engineering economics the same basic principle applies – calculations and comparisons must be made for situations where "equivalent" units of measure are in place, but the situation is slightly different because the basis of measures are time, risk, and the concept of the time value of money. Given that typical investment choices are made over long periods of time, we have to account for the fact that $10 today is simply worth more than $10 three years from now (under typical economic conditions). Risk, inflation, and the cost of money (interest rate) affect estimated values. In this chapter, we describe the tools and techniques for converting time a series of cash or cash flows into a form that permits reasonable assessments of value and decisions to ensue. There is a standard set of formulas for which a complete understanding of the homework problems should leave the student comfortable with this subject.

Previous work in developing an understanding of the basic concepts of the time value of money (TVM) is now applied in the context of decision-making; this chapter frames the basis of most contemporary methods for determining project profitability. To be successful, the student should be fluent in the use of the tables that provide normalized factors as a function of N and i, for single and uniform payments. It may seem peculiar that we have also interjected a discussion of Bonds in this chapter, but this is not only a classic use of TVM, but is sets a financial performance context against which the opportunity cost of pursuing project options can be measured.

1.5 The bottom line: Transactions

We have only briefly looked at the foundations of business analysis, and recited a limited view of businesses at that! Still, this is a foundation for engineering managers who will be decision makers within their company. Let's reiterate a previous position:

1. *Managerial decisions* drive the *conversion of assets* into goods and services, that ...
2. if *appealing to a market need* to lead to *customer transactions*, that ...
3. produce revenue and data to be *converted into information*, that ...
4. must be organized and *communicated* to managers to support decision-making.

Study this very brief list. If you don't remember anything else about the chapter, consider the significance of each and every item. Mastery of these steps can impact your career as an engineering manager. Connect each step to current events – search out examples in the press to underscore the relevance of *failing* to *know* impact on the business; manipulating transactions is at the root of most ethical lapses and fraud in business today.

1.6 Discussion problems

Problem 1.1
Briefly explain the difference between accounting, finance, and engineering economics. Try to put the concepts in your own (or your team's) words and *compare* the concepts where appropriate.

Problem 1.2
What mechanisms are in place to guide the identification, measurement, categorization, and communication of information to stakeholders and users? Discuss differences between users and stakeholders.

Problem 1.3
Among your colleagues in class, identify a term or phrase italicized in this chapter that you think is the most significant from your reading. Absent team consensus, then just provide *your* perspective.

Problem 1.4
What is the purpose of financial statements? Would you want to produce them even if they were *not* required, say, for entity tax reporting?

Problem 1.5
What are the two key financial objectives in the management of a company? How can a focus on these objectives create ethical dilemmas?

Problem 1.6
Distinguish between accounting and bookkeeping – be brief and think in terms of, say, what the differences in training would need to be for employment in those professions.

Problem 1.7
Which of the three components of Figure 1.2 might be a source of concern regarding "ethics"?

Problem 1.8
What are the major forms of a business? Why would there need to be any difference between them? What are some advantages and disadvantages to each form?

Problem 1.9
How would a company know that its product or service was preferred in the marketplace? What could a competitor do to erode this favorable position?

Problem 1.10
This chapter placed emphasis on a "transaction-based perspective" related to customers. What would be different about a transaction involving an investor? What do we mean by transaction?

Problem 1.11
How might the market value of a firm differ from its intrinsic value?

Problem 1.12
With Figure 1.5 in mind, how could changes in government policies affect accounting activities?

Problem 1.13
It is claimed that problem definition and decision-making are intertwined. Explain.

Problem 1.14
Managers must be problem-solvers but are not always decision-makers. Do you agree? Disagree? Provide an example that clarifies your position.

Problem 1.15
Engineers are often accused of having a narrow view of problems, and are claimed to lack skills to deal with ambiguity. In what ways might an engineering manager differ in this regard?

Problem 1.16
Why are most engineers likely to have experience with deterministic risk and not probabilistic risk?

Problem 1.17
Clarify the difference between risk and regret.

Problem 1.18
It was stated that the topic of *tax accounting* would not receive extensive treatment in this textbook. In what way could this be justified? What might be the impact on what can be learned from the text?

Problem 1.19
Why is the concept of an ongoing concern important to the interpretation of financial statements?

Problem 1.20
It is mentioned that for a "decision to be made" there must be more than one alternative under consideration and the possible outcomes must be of unequal value. What are the three general types of problem outcomes?

Problem 1.21
In the decision-making process, what are the ways in which context can impact outcomes?

Problem 1.22
Intuition in decision-making can expedite decision-making (versus cognitive domains that are slow and deliberate), but it can fail or mislead us, too. How can that be?

Problem 1.23
We live in a global economy, and many different external activities have an impact on the business. On what basis would we narrow the scope of this book to de-emphasize tax law, currency exchange, inflation, and financial accounting for the engineering manager?

Problem 1.24
You are in a meeting to decide on a new product development investment. The VP in charge of the meeting seems to be asking a lot of questions of the design engineers. And it also seems like the VP changes position on controversial issues, which is frustrating the engineers, too. If the VP just a curious person? Possibly incompetent? How might this behavior be helpful (rather than frustrating)?

References

[1] R.F. Meigs and W.B. Meigs. *Financial accounting.* McGraw-Hill, 6th edition, 1989.
[2] D.H. Marshall, W.W. McManus, and D.F. Viele. *Accounting: What the numbers mean.* McGraw-Hill, 7th edition, 2007.
[3] B.E. Needles, M. Powers, and S.V. Crosson. *Principles of accounting.* Houghton Mifflin, 10th edition, 2008.
[4] W.G. Sullivan, E.M. Wicks, and C.P. Koelling. *Engineering economy.* Prentice Hall, 15th edition, 2011.
[5] Saundra K. Lipe and Sharon Beasley. *Critical thinking in nursing: A cognitive skills workbook.* Lippincott Williams and Wilkins, 2004.
[6] Patricia Benner. *From novice to expert: Excellence and power in clinical nursing practice.* Prentice Hall, 2004.
[7] Mark Quirk. *Intuition and metacognition in medical education: Keys to developing expertise.* Springer Press, 2006.
[8] Judith Koenig. *Assessing 21st-century skills.* National Academies, 2011.
[9] Stuart E. Dreyfus and Robert L. Dreyfus. *A five-stage model of the mental activities involved in directed skill acquisition.* University of California Berkeley, 1980.

Chapter 2

Financial Statements

2.1 Learning objectives

The world of accounting, finance, and engineering economics is at the core of business operations and is central to the process of gathering, classifying, analyzing, and reporting of financial and economic information. Although the novice might initially sense accounting is more of an "art than a science," recent industry scandals have led to an increase in the provision of guidelines and recommended practices. Overall, much actually goes "according to plan," and the latitude in systems and processes provides the needed flexibility for the broad range of entities they must serve.

There is a new vocabulary for engineers that goes with this general subject; knowledge of the basic terms is essential. This chapter explores this fascinating industry and sets the discussion framework for the chapters ahead. It is imperative to perform the preparatory work outlined in the following section. If you fail to grasp the foundations of financial statements, your ability to ask questions as a manager is significantly compromised.

After reading and discussion sessions, the student should be able to:

1. Identify the difference between the "inside view" and "outside view."
2. Understand who uses accounting information and why this information is useful.
3. Identify the accounting tools and principles that shape financial statements.
4. Understand the ethical issues in accounting and the challenges accountants face.
5. Know the different types of financial statements and the relationship between them.
6. Explain what a company's annual report is and why it is used.

2.2 Jumpstarting our understanding

Explore the following links and download the documents to read off-line prior to or in parallel with this chapter. Might as well jump right in and see some real-life examples!

1. Merrill Lynch's "How to read a financial report" available at `http://www.ml.com/media/14069.pdf` is a classic. Many MBA programs use this concise guide as a way to jumpstart an understanding of financial statements and their use. There were several versions of this report floating around cyberspace, especially prior to the acquisition of Merrill Lynch by Bank of America.
2. Some people seem to think the Merrill Lynch document is too long (it isn't) and hope to find shortcuts by typing "How to read an annual report" into their web browser. This will produce a variety of sites such as `www.investopedia.com/articles/basics/10/efficiently-read-annual-report.asp`. Some will be biased, and this is where using critical thinking skills will help decide what websites are the most useful.
3. Once you have read an introductory guide to financial reports, now download and review a corporate annual report. Many are available from `http://www.sec.gov/edgar.shtml` or other finance websites.
4. The annual reports from the IBM websites are well organized and easily available from `http://www.ibm.com/annualreport/`; you can download entire reports (letters to the shareholders are interesting) or just parts.
5. You can tell if you are getting the hang of finance if you find the Berkshire-Hathaway site `http://www.berkshirehathaway.com` interesting reading. If you immediately sense the tone of Berkshire-Hathaway is quite unlike IBM, then it is safe to say your ability to gain insight from financial reports has improved.

A few things to think about while you read an annual report:

1. Is the business easily categorized as producing a product or service? Why or why not?
2. Is there any hint as to why this company's product or service is preferred over others?
3. Are there indications this is a commodity-based business? Is it capital intensive?
4. What information is being provided in footnotes?
5. Is there a "business outlook" section? Is it upbeat or cautionary?
6. Does the company provide any comparables between its business and others?

I'm not suggesting these questions are relevant to the plan you read or that you'll get satisfactory answers; but I do believe that if you *read with these types of questions in mind*, your experience will be more enriching and memorable. You'll begin the process of creating questions of your own, too.

2.3 Foundational topics

2.3.1 The outside view

MBAs frequently learn "the story is in the numbers." This perspective is annoying to some engineering project managers who are frustrated that "the bean-counters make all the decisions with little understanding of the *real* business at hand." A few engineers even declare, "A good MBA would have taken the time to immerse themselves in technology to fully appreciate what is unique about the business." Is wallowing in technology an essential rite of passage? By themselves, are the sterile numbers adequate to characterize the affairs of a company? We take a closer look at these and other questions to build a greater appreciation for the value provided by financial statements. Briefly, though, differences in perspective have roots that have been examined before. Kahneman and Loval [1] describes the "inside view" versus the "outside view." In the "inside view" a highly optimistic view of minimal project risk is based on detailed insider information while the latter "outside view" more often extends from professional assessment of comparables. From Kahneman and Loval (p. 25):

> *Two distinct modes of forecasting were applied to the same problem ... The inside view of the problem is the one that all participants adopted. An inside view forecast is generated by focusing on the case at hand, by considering the plan and the obstacles to its completion, by constructing scenarios of future progress, and by extrapolating current trends. The outside view is the one that the curriculum expert was encouraged to adopt. It essentially ignores the details of the case at hand, and involves no attempt at detailed forecasting of the future history of the project. Instead, it focuses on the statistics of a class of cases chosen to be similar in relevant respects to the present one.*

Perspective matters. A lot. In this chapter we will promote the "outside view" as a way to avoid the trap of internal optimism and a "can do" attitude. While these are noble traits, Mauboussin[2] explains such tendencies to fall prey to unusually optimistic views and give credibility to "anecdotal evidence and fallacious perceptions" (p. 4). In comparison, an outside view seeks a statistical basis for decision-making based on comparable situations. Mauboussin has special suggestions that are worth exploring on your own (pp. 13-16):

1. Select a reference class - find statistically similar situations to yours.
2. Assess the distributions of outcomes - look at the rates of success *and* failure.
3. Make a prediction - estimate chances of success *and* failure.
4. Assess the reliability of your prediction and fine-tune.

The *outside view* dominates thinking in the present work just as a seasoned manager would.

2.3.2 What matters most?

Earlier, in Chapter 1, we discussed that a business must be both *profitable* and *solvent*:

1. *Profitability* of a business results when total revenue exceeds all expenses.
2. *Solvency* is that capability of a business to pay bills as they become due.

Profitability and solvency seem similar, but are quite different. Two scenarios illustrate:

1. Sven operates a florist shop and during the month of June discovered how easy and profitable it was to import tulips from Zwolle Exporters in the Netherlands. Electronic banking made international purchases easier than he thought possible. Customers pay $1.95 per tulip that only costs Sven $0.75 each. Zwolle is giving Sven credit terms of 45 days to pay for the tulips after they arrive in Cleveland. However, rather than pay Zwolle, Sven decides to use his newfound profits to make a down-payment on an additional delivery truck, hoping expanded, faster delivery service will improve business. Unfortunately, faster deliveries with the additional truck have not increased revenue. He soon finds the truck purchase used all his free cash, created additional business expenses, and Sven does not know how he will pay his bills.
   (a) *Sven makes a profit of $1.20 for each tulip.*
   (b) *Although profitable, Sven's poor business decisions have reduced solvency.*
   (c) *Continued decrease in solvency leads to insolvency and bankruptcy.*
2. Pryanka's proficiency in enterprise resource plnaning (ERP) software installation led her to start a consultancy practice with two friends five years ago. Over the past three years the downturn in the economy exacerbated the glut of consultants and has driven consulting rates to under $100/hour. As a labor-intensive business, rates have to be at least $195/hour to cover costs. This unfavorable situation has been tolerated since the owners are friends and they want to "wait out the storm" and to not have to deal with the unpleasant task of laying off employees. Undercutting the competition has not subsided, and the partners believe the best they can do is raise rates to $145/hour.
   (a) *Pryanka's firm is selling services cheaper than they cost to provide.*
   (b) *Her costs remain high because she has not raised rates.*
   (c) *She has too much staff compared to what the business can sustain.*
   (d) *They may have money in the bank now, but the business model is fundamentally unprofitable and will eventually lead to insolvency.*

Note the hazard of the "inside view": Sven forecasts growth by extrapolating current trends or constructing optimistic scenarios of future progress. Sven is speculative and fails to factor in the "external view" in decision-making by considering trends that other florists may have already entertained. Sven needs to improve solvency by trimming costs and improving earnings; he might be turning a nice profit in the short run but is headed towards bankruptcy. Consider the following when thinking about solvency:

1. Is the trend of company assets rising or falling? Is owners' equity rising?
2. Are they continuing to put money into the business?
3. Do debts (all liabilities, actually) seem to be "under control"?

Pryanka has to become more realistic about pricing strategy and adjust the cost of providing service so every hour charged to the client is profitable. Her firm might be financially solvent at the moment but in the long run remains unprofitable. When exploring profitability we might ask:

1. Is this company *really* making money? As much as comparable firms?
2. In comparison to last quarter and last year, is profit higher or lower?
3. Is sales revenue rising?

This is not to say a business *has* to have *high* profitability. There may be good reasons profit margins are low yet overall cash flow is attractive (think about "big-box" or other discount stores). Still, the situation might exist where unacceptably low profitability is a consequence of poor management and is *masked* to the outsider if cash flow *appears* to be healthy. The company Board of Directors should serve an independent role, asking probing questions based on an *external view*. Just a few examples are:

1. Are salaries lower than market rates and is turnover creating quality problems?
2. Could certain employee costs be hidden or subsidized by grants?
3. Might loans intended for capital projects actually be funding operations expenses?
4. Is the firm simply drawing down on inventory? Creating excessive depreciation?
5. Are payments to vendors being delayed?

Investors use numerous mathematical ratios to calculate the financial health of a business; once there is a satisfactory story for the "big two" (profitability and solvency), investors might also wonder about:

1. Efficiency - the relative productivity of a firm's use of assets.
2. Stability - the relative financial health or position of the company.
3. Job creation - the ability to contribute to the community workforce.
4. Social impact - the "triple bottom line"

These additional factors are important to society, but it is proposed that failing the "big two" a firm is financially unable to be of value or service to anyone.

2.3.3 Naming of financial statements and aliases

Financial statements have been around for centuries, so naming of financial documents has varied over time. The first level of each of the itemized lists below will be the name we'll tend to use in this book.

1. **Balance Sheet**
    (a) Statement of Financial Condition
    (b) Statement of Financial Position
    (c) Statement of Financial Situation
2. **Income Statement**
    (a) Statement of Earnings

   (b) Profit and Loss Statement (P&L Statement)
   (c) Operating Statement (Statement of Operations)
3. **Statement of Cash Flow**
   (a) Statement of Changes in Financial Condition
   (b) Statement of Changes in Cash Position
   (c) Cash Flow Statement
4. **Statement of Owners' Equity**
   (a) Statement of Stockholders' Equity
   (b) Statement of Changes in Capital Accounts
   (c) Statement of Changes in Stockholders' Equity

Once the function of each of the statements is understood, we see that the different names simply reflect common usage and context to satisfy user needs. CitiBank's annual reports are hundreds of pages in length (`http://www.citigroup.com/citi/investor/quarterly/2012/ar11c_en.pdf`) to address millions of stakeholders' interest in their activities. Public firms file a Form 10-K (Annual report pursuant to the Securities Exchange Act of 1934) and the following might typically be found:

1. Consolidated Balance Sheets
2. Consolidated Statements of Income
3. Consolidated Statements of Cash Flows
4. Consolidated Statements of Shareholders' Equity
5. Notes to Consolidated Financial Statements

Let's begin by examining the purpose of each of the statements; once understood, the specific name is of less concern than recognizing the function served in communication with stakeholders.

2.3.4 Overview of financial statements

In Chapter 1 we underscored the transactions associated with ongoing operations of a business, and in particular the need to communicate with a variety of stakeholders shown in Figure 1.5 (page 5). For most companies a set of standard financial statements is produced to communicate with the stakeholders. Statements are different in two very important ways. Differences impact presentation, format of document titles, and source of information:

1. Financial position at a *specific point* in time
   (a) Balance sheet - to convey financial position at the beginning or end of a period.
2. Financial performance *over a period* of time.
   (a) Income statement - describing the earnings, profit, or loss over a period of time.
   (b) Statement of cash flow - depicting the increase or decrease in cash that has occurred.
   (c) Statement of owners' equity - activity related to owners' equity holding and distributions.

Figure 2.1 illustrates a few of the relationships between statements and helps you in developing *your* "elevator pitch" that describes financial statements

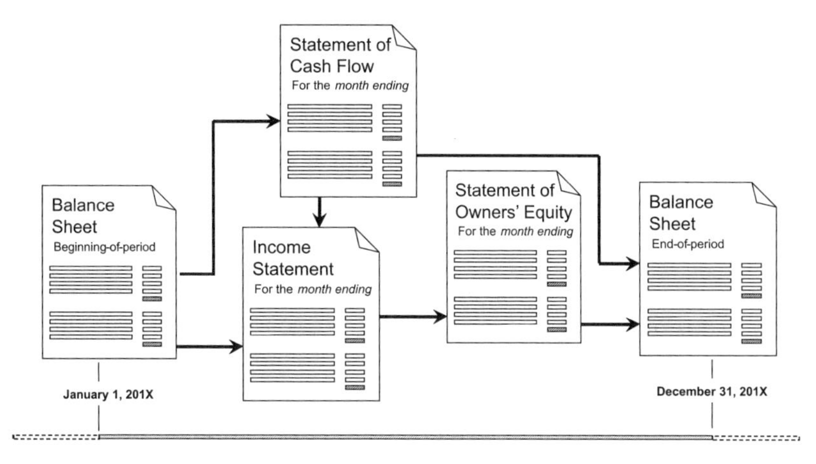

Figure 2.1: Sample set of financial statements and their interrelationship along a time-line.

and their usage. Here's mine:

> "Beginning at the far left-hand side of Figure 2.1, assets available (and the claims on those assets in the form of liabilities) to managers for business activities are presented on a *balance sheet.* Choices are made about how to use those assets, and the *statement of cash flow* provides how cash was used for investing, operations, and financing business activity. With a focus on transactions, the *income statement* informs about how profitable the operations were over the year. Profit creates wealth for the owners (losses erode wealth!), and the *statement of owners' equity* informs how beneficial it is for owners to have invested in the company. Cash flow and owners' wealth changes over the course of the year are then used to update the balance sheet of assets and liabilities at the end of the period."

Let's continue by briefly describing each financial statement. If we jump too soon into expanded descriptions with detailed examples we can avoid getting lost in the numbers too early. Easing into this subject is especially important if you've never had an accounting course before! Engineers drive a lot of technology projects that frequently impact capital expenditures – executives tend to be delighted to work with those engineering managers who can properly identify, categorize, and analyze activities and expenses in the spirit of problem-solving challenges introduced in Section 1.5.

Balance sheet

The balance sheet is simply a listing of the resources (assets) and sources (liabilities and owners' equity) that are involved in the operations of the organization. Assuming the business is "alive and well" (the so-called "on-going concern" assumption) then this financial statement represents a snapshot of the solvency of the organization at any given point in time. The balance sheet is based on what is known as the *accounting equation.* Let's use the example of home ownership to clarify. Suppose you have a home worth $100,000 (asset). If the outstanding mortgage is $60,000 (liability) then you have a stake of $40,000 (equity) in the home.

It is customary for the assets to be on the left-hand side and the combination of liabilities and equity to be on the right. The balance sheet is a view of assets available to run the operation and also the claims against those assets held by the owners and non-owners of the business. Later we will see how all the data needed to create a balance sheet emanates from the other three financial statements.

Assets	=	Liabilities	+	Owners' Equity
$100,000	=	$60,000	+	$40,000

Income statement

Some investors consider the *income statement* to be the most important financial statement of an enterprise, evident by its alternate names such as *profit and loss statement* or *statement of earnings.* Recall that the profitability of an organization is one of the two key performance measures (the other being solvency) managers *must* accomplish in the long run, and the income statement is central to that management task. Typical income statements have many line items needed to fully characterize profit (or loss), grouped in the category of either *revenue* or *expense* (sometimes, certain *gains* are included, too). Net income is simply the difference of revenue and expense. Let's assume that the home described earlier is used as a rental property. If rent is $775 month ($9,300 annually) and expenses (property tax, maintenance, utilities) are $5,200 a year, then the annual income statement would read:

Revenue		$9,300
Expenses	-	$5,200
Net Income	=	$4,100

This is a highly oversimplified situation, but the basic idea should not be lost as details unfold. A very subtle point about the income statement is that the revenue and expenses are (a) associated with the ongoing operations *and* (b) that the revenue recited is "matched" by those expenses required to produce the revenue. We will return to this idea of the "matching principle" in Section 2.5 when discussing how long-term assets are apportioned to revenue. The matching principle ensures an accurate depiction of operations activity in the income statement but does not include those *investment* activities (such as the expense of purchasing capital equipment, or buildings such as a house). Given the association of the income statement with operations, it is no surprise that the income statement reflects the outcome of activities that occur over a period of time (and is not a snapshot in time, as in the case of the balance sheet).

Statement of cash flow

Often it is hard for the novice to differentiate between the *statement of cash flow* and the income statement. That is easy to do since line items and mathematics will often look similar. But while the income statement points to the profitability of an enterprise, the cash flow reveals its *solvency*. The cash flow reflects *all* of the cash flowing into and out of the business over the course of the year - not just operations-related activity. *All* cash flows related to *investing*, *financing*, and *operating* the business are included (the income statement only focuses on operating activity). The income statement contains speculative items (like sales based on credit), while the cash flow statement represents "real" cash transactions. Comparing cash flow and income statements underscores the emphasis made back in Section 1.3 on the difference between operating, financing, and investing activities. Suppose for the rental property we installed a new heating system that cost $14,500, and we financed $10,500 of the cost. The statement of cash flow would look something like this:

Cash Receipts	
Rent	$9,300
HVAC Loan	$10,500
Cash Disbursements	
Operations Expenses	$5,200
HVAC Investment	$14,500
Net Cash Flow	$100

Statement of owners' equity

Owning and operating a successful business means the value you create for your customers translates into a profitable and solvent enterprise. Value creation for the owner occurs, too. The *accounting equation* shows that assets available to drive the business are provided by non-owner financing (liabilities) and owner financing (equity). If the value of the company is rising, the value of owners' equity (OE) normally rises, too. The details of changes in equity positions are provided in the *statement of owners' equity*. Like the income statement, the statement of owners' equity reflects activity that has occurred over a particular period in time. This includes new investment money that has been raised and the annual rise in owners' equity resulting from operational profitability. Suppose our rental property were part of a holding company that was composed of several external investors. Then:

While this table tends to oversimplify many details of a complete statement of OE, it is representative of the basic concepts involved. Changes in stock

Beginning Balance	$8,500
New Stock Sale Investments	$1,500
Retained Earnings:	
Net Income	$100
Dividends	-$25
Ending Balance	$10,075

and the price paid for stock by new investors is discussed in detail later in this chapter and also in Section 6.4.1.

Supplemental statements

Annual reports to shareholders vary widely in length and depth. In the US, reporting requirements vary with the type, size, and composition of investors, and there may be several different agencies that might need to be provided reports on an annual basis. Financial statements audited by professional accounting firms are the norm. Most businesses now have an Internet site with investor pages, and the posted information helps a great deal with public access to information and transparency of activity. A quick review of even two-three annual reports will provide the reader insight on the nature of reporting.

For example, the CitiBank report for 2011 is over 300 pages long and has a wide variety of financial statements that reflect the size and complexity of its business. In comparison, many non-profits have modest assets to work with and the annual report might (only need to) be a very terse document highlighting just one or two essential financial facts such as income and revenue. Over time, the demand for transparency in the non-profit world has resulted in much more (easily available) financial information to be reported. The role of supplemental documents in financial reporting continues to expand as external stakeholders demand (a) more information, (b) improved readability of reports, and (c) simplified accessibility (transparency).

2.3.5 Accounting to support financial statements

Accounting is the process whereby financial information is gathered, organized, categorized, and reported. Like any interdisciplinary subject there are subspecialties that focus on specific facets of the profession. Many people trivialize the field, confusing the task of bookkeeping (data entry and record-making) with the much broader and more sophisticated spectrum of accounting activities. For instance, it would not be unusual for accountants to be setting up reporting processes, involved in audits, presenting to key internal and external stakeholders, creating budget forecasts and working extensively on corporate tax issues. Figure 2.2 illustrates the various tasks you might expect accounting and financial management professionals to conduct.

While this book is not intended to offer a comprehensive treatment of accounting, understanding the accounting processes for recording changes in

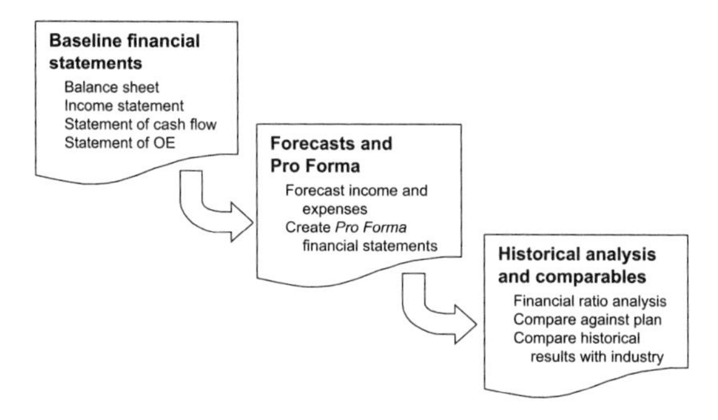

Figure 2.2: Typical accounting and financial management activities.

financial position are important, even if your organization has a computerized accounting system. Over time, it has become clear that some readers may have had a basic bookkeeping or accounting course at some point in their engineering education, but for others the topic is entirely new. As a supplement to this chapter Appendix B (p.357) on "Mechanics of Accounting" has been included for completeness.

As indicated above and in the opening chapter of this book, Section 1.2.2, the field of accounting is quite broad and it helps to briefly note some important distinctions in "type" as this will simplify the task of seeking references to supplement this book. Entire books are dedicated to the specialties of managerial accounting and financial accounting.

Managerial accounting

A variety of accounting activities *internal* to a company are important for the support of ongoing operations. For instance, a "dashboard" of weekly performance indicators may be important during weekly meetings of senior management. Or the preparation of financial forecasts may be important in projecting the timing of a certain threshold of free cash that could be used for capital equipment upgrades or purchases. Related to managerial accounting is *cost accounting*, the process of collecting, analyzing, and reporting information about product and process costs.

Financial accounting

There are a variety of external stakeholders that need (or want) a lot of information to support investments and other interests, and it falls primarily upon the financial accountants to develop and communicate financial information. Often, the task is marginalized as simply "preparing financial statements" but in fact there are many behind-the-scenes activities to ensure the information (**not** just data) is reliable, auditable, and provides transparency to the activities of the company.

Tax accounting
The field of tax accounting is quite complex, as tax laws change in an effort to influence the behavior and decisions of companies. Efforts to improve the transparency and accountability of senior management have resulted in laws like SARBOX, that has had a dramatic impact on firms, both large and small (see Section 6.5.3, p.201). A firm may very well have on staff a person well-versed in tax accounting, even if the firm regularly works with external auditors in the preparation of annual reports.

This book offers a mix of financial and managerial accounting, but with more of an emphasis on managerial accounting activities that would pertain to the typical engineering manager; minimal treatment of tax accounting is included in this book.

2.3.6 Brief note on ethics

Much could be (and has been) written about ethics. If we think of all the steps involved and points where interpretation of data can vary or be influenced without easily being detected, it is clear why ethics plays a central role in the development of information communicated to stakeholders. Consider just a few aspects of financial statement development that could be open to manipulation:

- On the first business day of a new year we have assets and claims on those assets described by the balance sheet, assumed to be correct from the prior year. If the "initial condition" is in error (prior-year data) then current-year reporting is compromised.
- Over the course of the year, management uses assets to produce *cash flow*, *income*, and *wealth for owners.* Identification, analysis, and recording of *accounting entries* associated with such business transactions enable this information to be translated into periodic financial statements. Often errors are spotted during routine checking and audit, and adjusting entries are made to correct. Checks and balances as to *who* makes the changes and *why* are crucial.
- At the end of the year a new balance sheet is constructed from periodic report information and ultimately tells us, for instance, if our asset base has grown and how claims on the assets may have shifted (between external and internal owners). Items such as depreciation schedule changes and onetime charges can influence results and seem justified though footnotes.

A moderately sized business may have as many as 10,000 transactions to track with many different people and points of entry for the data into an accounting system. The opportunities for abuse abound. Ethical intent is essential for us to have any faith at all in what is reported. We proceed assuming the data we are given is accurate and auditable.

2.3.7 Guidelines evolve

Guidance for analysts is available from the Securities and Exchange Commission (www.sec.gov), the Internal Revenue Service (www.irs.gov), the Financial Accounting Standards Board (www.fasb.org), and state and local organizations, among others. Interestingly, US accounting standards are set by common law (not by statute), though regulatory agencies *can* enforce certain recommended practices. For instance, the SEC can require publicly traded companies to follow the Generally Accepted Accounting Principles (GAAP) set forth by FASB (www.fasab.gov/standards.html). And on the horizon are the recommendations of the International Financial Reporting Standards (IFRS) Foundation, an independent, not-for-profit private-sector organization, whose standards are expected to supersede GAAP in the next few years. In short, with so much guidance and assistance, how can there be ethics issues in financial accounting? Don't managers know what they are doing?

The reality is that oversimplified views of accounting and finance lead to the assumption that the field is an "exact science" when in fact there are many assumptions involved in preparing financial statements. For instance, while the term "asset" might seem well-defined to an engineer working with production machinery purchases, the term in its more general accounting form can mean other things (www.fasb.org/project/cf_phase-b.shtml). An asset might be

1. a *probable* future benefit involving a capacity to contribute to future cash inflows, or
2. a transaction giving rise to a *right to* a benefit that can be used to settle a liability.

Suppose we have a customer who has acknowledged the receipt of merchandise and has promised to pay within 60 days. Cash is a tangible asset and for this situation the *cash flow* is zero. But on an income statement the customer's promise to pay (an *account receivable*) is listed as income despite the intangible nature of the entry. Even if we meet the asset criterion, as a manager we may have to also make a judgment call about our ability to collect. How this is performed varies from company to company, and GAAP allows companies latitude in assessment. Variations in estimates of accounts receivable that will eventually be collected result in financial statement variability. It is difficult to set standards that precisely fit every business model, yet at the same time too much latitude can result in abuse.

The news is filled with stories whereby financial statements have been crafted and influenced by managers who believed they could profit by manipulating outcomes. The stories of Enron, WorldCom, Tyco, Adelphia, and others are well known and easily explored in press reports and case studies on ethics. In short, it is very difficult to "legislate" ethical behavior, and instituting "red flags" only partially helps[3]; it must be an intrinsic part of the character of managers entrusted with financial reporting. We only briefly discuss the Sarbanes-Oxley Act of 2002 in Section 6.5.3 that was a legislative response to ethical lapses, with much more available from www.sec.gov/

about/laws.shtml and numerous interpretations available from nearly every large legal services firm website.

Summary of relationships

Figure 2.1 suggested interrelationships between the financial statements. We expanded upon the relationships in the narrative above which is now handy to summarize for clarity:

1. The balance sheet and the income statement are linked through the statement of owners' equity.
2. Several categories on the statement of owners' equity are *also* shown on the balance sheet.
3. The income statement is linked to the statement of cash flows through the *operations portion* of the cash flow statement.
4. The balance sheet *cash asset* is linked to the cash flow statement though the net change in cash.

2.4 Financial statement ratios

Each of the brief financial statement descriptions and examples presented in Section 2.3.4 provided business results on an *absolute* basis. For instance, revenue was simply given as $9,300. As well, the net income for our simplified situation was $4,100. While it is informative to know the rental operation produces a profit, it is hard to infer organizational performance or operational efficiency with just a single number; some frame of reference or context is essential. This is often accomplished with the use of financial ratios. Financial ratios provide the "power of insight" and can be used to compare year-to-year (historical) performance of the company relative to its competitors (as well as how the company fares relative to its general industry).

It is very easy to calculate a large number of ratios from financial statements.[1] Traditionally, classification categories align with key financial considerations: (1) profitability, (2) liquidity, (3) leverage, and (4) efficiency. These categories answer some of the usual question about a business: Are we making money? Are we profitable? Can we pay our debts? Is our investment productive? Industry and investment service organizations can provide data needed for comparison calculations. Many services are subscription-based, though internet sites such as Yahoo Finance (http://biz.yahoo.com/ic/index.html) generally have most of the information needed to get started.

Financial ratios also provide clarity to *investment trending*. Indeed, it may not be the absolute value of a financial ratio that has as much meaning as

[1]Wikipedia lists 122 (http://en.wikipedia.org/wiki/Category:Financial_ratios)! Investopedia defines a more manageable number of 30 ratios distributed among 6 categories (http://www.investopedia.com/university/ratios/), while HBS provides 16 of the most commonly used ratios in 6 categories (http://www.alumni.hbs.edu/new_alumni/toolkit/forecasting/glossary.html).

year-to-year variations in value or patterns relative to the industry norm. Investment trending is an important part of investment analysis and is primarily intended to provide *insight* about a company and the industry within which it operates. Trends can provide a broader context for crafting strategic options and decision-making.

Despite the apparent informative nature of ratios, they can be misleading, irrelevant, or manipulated, so care must be taken to explore full meanings prior to use. There may be (intentional or unintentional) ambiguity to financial data, whether it is inaccurate averaging of a parameter, the use of market versus book value, values biased by date changes (to enhance revenue), write-downs to mislead asset investment, or the hiding of debt to avoid accounting for depreciation expenses. It is often said that financial statements "reveal as much as they conceal," so we can't expect ratios to be any more accurate than the data from which they are computed!

To keep things simple and make the analysis process as straight-forward as possible our attention for the moment is going to focus on the "big three" financial ratios: (A) return on investment, (B) return on equity, and (C) current ratio.

Return on investment

Consider again the time-line shown in Figure 2.1, page 27. As mentioned, our assessment of operations is improved if we compute *relative* as well as absolute numbers, since this enables us to make meaningful comparisons between our operations and those of comparable or competitive entities. Many managers and investors believe that one of the most important parameters is *return on investment*, also known as *return on assets*, as this indicates how well management has used the assets of the company to produce revenue or returns to the owner of the company. In its simplest form,

$$\texttt{Return on investment} = \texttt{ROI} = \frac{\texttt{Net income}}{\texttt{Average assets}} \tag{2.1}$$

This ratio compares the profits with the assets that generate profits, but sometimes the value of assets may fluctuate over the course of a year, so an *average value* must be estimated. For example, the balance sheet showed the asset underpinning the rental operation is a house worth $100,000; here, *return on assets* is simply,

$$\texttt{ROI} = \frac{\$4,100}{\$100,000} = 0.041 = 4.1\% \tag{2.2}$$

Recall, though, that a new HVAC system was installed and thus the book value of the house has actually increased by $14,500 over the course of the year, so the yearly average asset value has increased by 7.24%. At the end of the year the average asset value is $107,250 and therefore

$$\texttt{ROI} = \frac{\$4,100}{\$107,250} = 0.038 = 3.8\% \tag{2.3}$$

Clearly, adjustments affecting averages can make a difference; in this case it might seem moderately small, but sometimes small changes can have unintended consequences and result in a big impact. For instance, suppose the minimally acceptable rate of return[2] for company assets is firmly set at 4.0%. Were we to "accidentally" forget about the HVAC investment, then it is entirely possible a strategic decision to *not* divest an asset was ill-informed; the asset underperformed and was below strategic goals.

Even with a base asset value of $100,000, it is encouraging to make $9,300 from an investment; whether the `ROI` is 4.1% or 3.8% we might conclude investing in the property was the "right" thing to do from the perspective that this return is certainly higher than the alternative investment in, say, a municipal bond with a return of 2.5%. This choice and our assessment of alternatives are explored more fully in Chapter 9.

A slight modification of Equation 2.1 can produce further insight on operations. Consider the calculation of *profit margin* (also known simply as "margin") that is the generation of income fraction for each $1 of sales:

$$\texttt{Margin} = \frac{\texttt{Net income}}{\texttt{Sales}} \tag{2.4}$$

Consider also another measure of asset productivity expressed in the form of *turnover*:

$$\texttt{Turnover} = \frac{\texttt{Sales}}{\texttt{Average assets}} \tag{2.5}$$

From this we see that `ROI` can be computed from the product of `margin` and `turnover`:

$$\texttt{ROI} = \texttt{Margin} \times \texttt{Turnover} = \frac{\texttt{Net income}}{\texttt{Sales}} \times \frac{\texttt{Sales}}{\texttt{Average assets}} \tag{2.6}$$

So the utility of `ROI` as a measure of profitability, productivity, and management use of assets is remarkable when used in conjunction with other supporting information. Some interesting outcomes are possible. Consider that we want to look at the performance of a high-volume commodity business like consumer electronics. In this case, profit margins might be in the vicinity of 5% and turnover at a level of 3.0. For this business,

$$\texttt{ROI} = (5\%)(3.0) = 15\%$$

A medical device company might have higher margins (say, 30%) but a lower turnover (0.5) but with a similar overall return on investment:

$$\texttt{ROI} = (30\%)(0.5) = 15\%$$

Note carefully how different sets of parameters can produce the same computed `ROI` outcome. In such a situation we begin to understand the importance of

[2]We expand on the meaning of the minimally acceptable rate of return later in Section 4.4.1.

the *context* of investments; supplementary information such as seasonal cycles or other factors might contribute to business risk assessment and thus play a part in establishing the probability of actual returns.

An interesting aspect of the ROI calculation is the way it bridges income statement revenue to assets from the balance sheet. This unique role is what makes ratios like ROI worthwhile understanding; they integrate information from statements that might otherwise be viewed as somewhat independent.

A cautionary note on ratio definitions is in order. For instance, turnover in Equation 2.5 has been described as an *asset turnover*, meaning that we are examining sales (i.e., revenue gained from a business selling its product) relative to the assets needed to produce those sales. Another version of "turnover" could reasonably be interpreted as *inventory turnover*, which indicates the liquidity of the inventory and is calculated by dividing the cost of producing the goods (say, production and shipping costs) by the average inventory. Speaking in "shorthand" can be convenient but subsequent ratios can be quite different, and this remark is simply a call for care in usage of terms.

Return on equity

The accounting equation Assets = Liabilities + OE makes clear that the assets available to generate revenue are composed of two general types, the non-owner-borrowed assets (liabilities) and owners' equity (OE). In the same way that ROI measures revenue generated by assets, the *return on equity* (ROE) is a measure of revenue generated by asset investments, too, but for the specific instance where only the productivity of the owners' equity is being examined. This focus is easy to understand: it is entirely reasonable that the owners of a company know how much income is created by their investment. We define

$$\texttt{Return on equity} = \texttt{ROE} = \frac{\texttt{Net income}}{\texttt{Average owners' equity}} \tag{2.7}$$

Variations on ROE definitions derive from the different types of stock that might be held by different owners (e.g., the different portion of *preferred* stock versus *common* stock); as discussed earlier, there might also be differences in the way averages for the denominator in Equation 2.7 are computed as well as minor nuances in the way net income might be calculated, too.

For the general case of non-zero liabilities, the accounting equation yields Assets > OE and thus as a general rule we would expect ROE > ROI.

Returning to our simple rental property situation, the net income is \$4,100 and the OE is \$40,000. In this case:

$$\texttt{ROE} = \frac{\texttt{Net income}}{\texttt{Average owners' equity}} \frac{\$4,100}{\$40,000} = 0.1025 = 10.25\%$$

and, of course ROE = 10.25% >> ROI = 3.8%. This is of keen interest to prospective investors. All things being equal, project returns are a significant component of the assessment of *risk versus reward.*

Liabilities often involve contracts that may have legally enforceable covenants; in such cases owners' equity is more "at risk" - terms and conditions of liabilities incurred are important to understand (yes, you should read contracts). In the case of bankruptcy, legal obligations will have a preferential position making claims against (liquidated) assets, thus representing lower risk to the creditor. When the business is profitable and running efficiently, the owners tend to be in a favorable position, and the astute manager must be on the watch for complacency.

Current ratio

At this point `ROI` and `ROE` ratios have been presented as measures of profitability and productivity. In this section we change gears a bit to explain how the *current ratio* is a measure of the current bill-paying capability of the firm. It is necessary to now introduce a few financial terms to help characterize assets in more detail, and plan to revisit these terms again when discussing the balance sheet in more detail in Section 2.6.

1. **Liquidity** is the ease with which non-cash assets can be converted into cash.
2. **Current assets** are cash and those assets expected to be converted to cash within a year.
3. **Current liabilities** are those loans, accounts payable, wages, rent, and other liabilities that are expected to be paid within the year.
4. **Working capital** is the difference (excess) between current assets and current liabilities, and is a measure of "liquid" assets available to pay short-term (less than 1 year) obligations.

Assets on a balance sheet are listed in order of *decreasing* liquidity. Our fairly trivial example of the rental property does not help in clarifying asset hierarchy, but for the `current ratio` calculation we need to know the most "liquid" of our assets, namely cash and items that we expect can be converted to cash in less than a year. Such assets are known as `current assets`. Cash and working capital are at the heart of a business' ability to drive the cash flow cycle, shown in Figure 2.3.

The current ratio is

$$\texttt{Current ratio} = \frac{\texttt{Current assets}}{\texttt{Current liabilities}} \tag{2.8}$$

To meet obligations, a `current ratio` must be ≥ 1.0. Our very simple example of the rental business does not provide enough data to compute the current ratio, but as a general rule you would expect the `current ratio to be` > 1.5.

This is a good point to reflect again on the "inside" versus "outside" view. In looking at the balance sheet, the internal view might be that it does not matter too much if your assets profile is long-term or short-term; as a practical matter owners' equity remains the same! But the "outside view" can make a difference. If you have a healthy balance of cash and short-term assets then

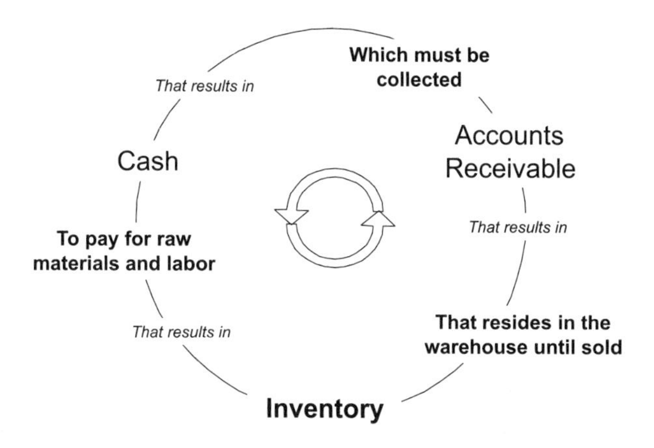

Figure 2.3: General cash cycle for a business.

it may not be a challenge to meet your liabilities, assuming of course `current ratio` is > 1.5. If, however, your cash cycle is experiencing delays, days sales outstanding (DSO) is rising, several large liabilities are on the horizon (payroll within 30 days), and `current ratio` is < 1.0 it may become quite important to monitor the difference between current assets and the current liabilities, also known as *working capital*. The ability to work though unexpected business cycles will depend a lot on working capital and a weak structure could impact the ability to obtain or maintain lines of credit for a "rainy day."

2.4.1 Ratios provide performance insight

The return on assets (`ROA`) computed earlier in Equation 2.1 is one of many business performance measures that can be extracted from financial statements. There are four general areas that management would generally be inquisitive about:

1. Activity ratios (tests of efficiency)
2. Profitability ratios (productivity; ability to translate sales into profits)
3. Liquidity ratios (tests of solvency)
4. Leverage ratios (ability to meet debt obligations)

For instance, assume routine business activity can be characterized as shown in Figure 2.3 and further assume the "velocity" (moving product faster and getting paid quicker) of doing business is of concern. Business activity ratios might have the following priority:

1. Profitability ratios (ability to translate sales into profits)
   (a) Net profit margin: What is reasonable for your industry?
   (b) Return on investment: again, compare to others.

2. Activity ratios (tests of efficiency)
   (a) Days' Sales Outstanding (DSO): How quickly AR to cash?
   (b) Inventory turnover: Can you sell inventory quickly?
3. Liquidity ratios (test of solvency)
   (a) Current ratio
   (b) Acid test ratio
4. Leverage ratios (ability to meet debt obligations)
   (a) Debt ratio: Can you meet your debt?
   (b) Debt to equity: Are you a risk to creditors?

Ratios and parameters enable organizational performance comparisons against industry norms. This can be extraordinarily useful in the process of diagnosing where process improvements can be made. For instance, if there is concern about efficiency of operations, *activity ratios* can provide insight on what your competitors might be doing better than you. For instance, let's suppose you are in a competitive market and you attract customers by selling product on credit. This could be as easy as simply providing generous "terms" of sale and offering a customer a maximum of 90 days to pay for goods received rather than the conventional 30-day period. By deferring the receipt of cash the company is acting like a bank for the customer, doing so because they feel it may result in stronger customer loyalty. The calculation for DSO is simply

$$\texttt{Days' Sales Outstanding (DSO)} = \frac{\texttt{Accounts Receivable, (AR)}}{\texttt{Net Sales}} \times 365 \tag{2.9}$$

Suppose your company tends to average `AR` of $446,000 on net sales of $1,875,000, with the result that `DSO` = 87 days. This is under the corporate threshold of 90 days, but suppose your next largest competitor in the industry has that `DSO` = 45 days based on `AR` of $142,000 with net sales of approximately $1,150,000. It might be true that your sales are higher because you are essentially funding the inventory for your customers, but what is the opportunity cost of having $215,000 ($446,000 - $231,000) unavailable to drive operations? No easy answer, but you can see how a few simple ratios and parameters can help define questions that need to be asked when managing assets to monitor performance, identify operational strengths and weaknesses, and serve as the basis for strategic initiatives.

Table 2.1 provides a sample of ratios that might be used when comparing company performance to the industry norms and commonly watched indices such as the Standard & Poor's 500.

Interestingly, let's suppose corporate policy remains firm at `DSO` = 90 days even though the industry standard is 32 days. There may be historical or other reasons a manager may not view this as a problem. Your business may have unique aspects for which other ratios might be more important to track. You would still want to track *trends* for `DSO`, since a "creeping" `DSO` value might suggest growing problem such as filling a distributor warehouse to record sales

Financial condition	Company	Industry	S&P 500
Debt/equity ratio	0.91	0.47	1.22
Current ratio	2.10	2.60	1.20
Quick ratio	1.30	2.00	1.30
Collections (DSO)	87	32	20
Profit margin	9.5%	5.2%	4.70%
Inventory turnover	2.6	1.7	4.0
Return on Investment	24.7%	8.8%	18.8%

Table 2.1: Key ratios in comparison to comparable industry ratios and the S&P 500

when the customer is unable to move the added inventory. This might be cause for management to consider stricter credit policies or even to launch a new program where favorable product discounts are offered with cash purchases.

Dictionary of terms

Dictionaries of terms can be useful reference guides. Handbooks such as the Vest Pocket CPA (Dauber [4]) or Barron's Accounting Handbook (Siegel and Shim [5]) were traditionally quite popular, but starting about 2011 when teaching this course of study it was clear many students found internet sources much more convenient. Knowing that there are *many* more websites than we have time to explore, the different flavors of on-line dictionaries are represented by the following:

1. WSJ: `interactive.wsj.com/documents/glossary.htm`
2. HBS: `www.alumni.hbs.edu/new_alumni/toolkit/glossary.html#a`
3. Dictionary.com: `www.dictionary.com/`

Experience has shown that on-line sources are (a) often limited in scope and (b) sometimes biased if not derived from refereed sources. Although the handbooks are delightful references and more comprehensive than websites, the internet is worth a look!

2.4.2 Normalization of line items

We've seen that quite a wide variety of ratios are possible to compute (and easy to create with spreadsheets); there are, however, two general types of ratio calculations:

1. Ratios providing insight on business performance measures (we just covered this!)
2. Ratios that normalize the line items in a financial statement (we'll cover this now!).

At the risk of oversimplifying the situation, Table 2.2 is the balance sheet for our rental operation (arranged in tabular form) with each line item expressed as a fraction of the total of each major category on the balance sheet.

This particular example states the obvious for what are very trivial normalized ratios, but the concept behind the format should be clear. Table 2.3 is a normalized OE financial statement that provides slightly more insight on business activity.

Assets		
Real estate	$100,000	100.0%
Total Assets	$100,000	100.0%
Liabilities and OE		
Liabilities		
Mortgage	$60,000	60.0%
Owners' Equity		
Stock	$40,000	40.0%
Total L&OE	$100,000	100.0%

Table 2.2: Simple example of a normalized balance sheet.

It is easy to imagine how a very detailed balance sheet, income statement, or cash flow statement, could produce many more normalized line items than shown in Table 2.3. Here, our focus is to highlight two points:

1. The ratios are expressed as a percentage of the ending balance; the sum of the numbers in the percentage column is not 100%, though certainly the sum of the beginning balance fraction (84.37%) and the net increase (15.63%) equals the ending balance fraction of 100%. The key is that each line item is expressed as a fraction of a key reference number, in this case the ending OE balance.
2. The ratios illustrate that the increase in OE was not a result of profitable operations, but from an increase is stock sales, possibly to cover the cost of the new HVAC system.

Beginning balance	$8,500	84.37%
New stock sale investments	$1,500	14.89%
Retained earnings:		
Net Income	$100	0.10%
Dividends	-$25	0.02%
Net increase (decrease) in OE	$1,575	15.63%
Ending balance	$10,075	100.0%

Table 2.3: Simple example of a normalized statement of owners' equity (OE).

2.5 Special considerations for long-term assets

When introducing the income statement in Section 2.3.4, a point briefly made was that the income statement provides a statement of revenue over a specific period and only those expenses incurred in producing that revenue for that period. The idea that revenue is recorded at the time the goods or services are delivered to the customer is an accounting practice known as the *realization principle.* The process by which revenue earned is matched to expenses incurred (to earn that revenue) and further adjusted to the reporting period is a feature of *accrual accounting.* The idea of "cause and effect" to match revenue and expense is critical to understand: we only recite *this* month's expenses incurred to produce *this* month's revenue, and not to include expenses from a prior month. Further, the expenses are only those costs incurred to produce the revenue, not the entire cost of the asset itself. Of course, matching does *not* imply that revenue and expenses are equal; it is about the timing and allocation of revenue and costs.

A discussion of long-term assets underscores the fundamental differences between cash transactions and revenue, and further differentiates operations expenses from investment expenses. For instance, a building is an asset we anticipate using for a long period of time in service to the business. In our simple example we purchased rental property for $100,000 that is an asset for which tenants pay $775 monthly to use. The first month's income statement would not show $775 in revenue matched by $100,000 in expenses since the house is an asset we estimate will be in service for 30 years, we must apportion the cost of the asset of the 30-year period (say, $600 per month) and use the apportioned cost as the expense incurred that matches the $775 in revenue. Although this general topic is explored later in Section 2.5.1 when we examine the role and use of *adjusting entries* for financial statements, it helps at this point to introduce a few management issues with long-term assets since the engineering manager will often play a role in the acquisition, use, and disposal of long-term assets. Overall, long-term assets are similar to pre-paid expenses, and awareness of this general topic is essential to a clear understanding of realizing how investments create value for the owners of an enterprise.

2.5.1 Some general features of long-term assets

Several features of long-term assets are useful to address now as this will facilitate a more detailed discussion of financial statements later. We mentioned earlier that current assets (inventory) were those assets which could reasonably be expected to be converted into cash with a period of a year. Long-term assets are those assets acquired for the purpose of supporting the business, but are not immediately available for resale to a customer. We often hear the term "plant, property, and equipment" when discussing long-term assets. These tangible assets have useful lives that may last for decades and can be thought of as providing an ongoing stream of services to the organization. Intangible assets (patents, copyrights, and trademarks) are also considered long-term assets and are of increasing importance to modern business; some estimates are that intangible long-term assets account for 60% of the value of a business today; the topic of intangible asset wealth creation is detailed in Chapter 6. Long-term assets can be categorized in three ways:

1. **Tangible assets** are long-term assets that have physical substance, exemplified as production equipment, office equipment, land, buildings, and plant expenses required to make the asset ready for use in the business. Depreciation is the accounting tool used to allocate tangible asset investment as an expense.
2. **Intangible assets** are long-term assets used in the operation of the business but that have no physical substance, exemplified as patents, copyrights, trademarks, and goodwill. These assets represent value by virtue of the legal *rights* they provide to the owner. Amortization is the accounting tool used to allocate intangible asset investments as an expense.
3. **Natural resources** are long-term assets based on value generally derived by extraction from land. Natural gas, iron ore, oil, and lumber taken from the land become exhausted over time. Depletion is the accounting tool used to allocate natural resource investments as an expense.

Aspects of long-term assets that set them apart from short-term ("current") assets are:

1. Long-term assets are expected to have a useful life of *greater than one year* and are often repeatedly used in operations (such as production machinery), unlike current assets (like inventory) that are expected to be converted into cash in less than a year. We can think of long-term assets as *supporting* operations and current assets as *part of* operations.
2. Long-term assets are not immediately available for sale to the customer because they support operations, unlike a current asset such as inventory that is immediately available and sold directly to the customer.
3. Long-term assets must be relevant and actively used in the operations of a business, and not being held for speculative reasons or reasons that are unrelated to the business.

Long-term assets are often referred to as a capitalized assets and are recognized in financial statements within the company's balance sheet, not the income statement. Capitalized costs make their way from the balance sheet to the income statement over a period of time, in which a "fair share" of the capitalized cost is "charged" as an expense through accounting methods of (a) amortization, (b) depreciation, or (c) depletion. These techniques enforce the matching principle whereby we match expenses with revenues over the period in which the asset is used. This is not the same as matching revenue to the period in which the long-term asset cost was actually *incurred.*

Comparison of amortization, depreciation, and depletion

Subtle differences exist between the concepts of amortization, depreciation, and depletion, with "amortization" and "depreciation" sometimes used interchangeably.

Amortization can take on several meanings, depending on the context of usage, but usually is used to suggest a process of decreasing the value of some type of intangible asset over a period of time. This use is consistent with accrual accounting where we wish to prorate the cost of an asset over a period of time. Table 2.4 summarizes the fundamental difference between amortization, depreciation, and depletion:

Proration Type	Description
Amortization	Expensing the cost of an *intangible* asset over the asset's useful life.
Depreciation	Expensing the cost of a *tangible* asset over the asset's useful life.
Depletion	Allocation of a natural resource over a period of time.

Table 2.4: Comparison of amortization, depreciation, and depletion.

Engineers who have been through an engineering economics course might immediately think of amortizing a loan when they hear the term "amortization." This is a correct context for usage of the term. Loans, of course, are often used to buy assets, and the loan represents a contractual obligation to pay a future sum of money (an asset!). In this way the loan might feel like a tangible entity, but is actually a somewhat intangible concept, just like a patent or a trademark. These items are valued primarily for what they evoke in the mind of the customer and are therefore intangible.

Unlike a simple interest uniform principal payment (see Section3.6, p. 79), payment of the loan principal gradually decreases over time. Amortization often uses the time value of money (TVM) equation framework to set the structure of the payment or expense values. For loans, a standard *term* and *risk* of the loan are negotiated, and the payments then determined from TVM calculations based on the number of years for the loan (term) and interest rate (risk); similar methods are used for amortization of intangibles; we explore TVM calculations in more detail in Chapter 3.

Depreciation and amortization might seem like very similar concepts with the difference being that a tangible (versus intangible) asset is being prorated. As a practical matter, though, the expensing schedule is largely governed by *type* of tangible asset. Furthermore, in the US, there are tax implications and very specific GAAP depreciation schedules customarily used to specify exactly how to distribute the asset cost over that asset's life. We work through several details of depreciation in the next section.

The cost of the acquisition of a coal mine is a capital investment, and the asset has a finite life. There may be reasonable and necessary set-up costs involved to position a well to deliver oil for sale to customers. But since the coal mine is a natural resource there are schedules for *depletion* that apply. Two methods are commonly used to account for depletion.

1. **Percentage depletion**. This very simple methodology involves a prescribed asset-dependent percentage value that is multiplied by the current tax year gross income (from the asset).
2. **Cost depletion**. Cost depletion requires an estimate of (a) how long it will take to deplete the natural resource and (b) the total amount of the natural resource. The expensed amount is a prorated fraction of the cost of the asset and the estimated consumption.

The more common tasks confronting an engineering manager would be amortization and depreciation. Amortization is a broad enough topic and fundamental financial matter that warrants the extensive discussion of TVM in Chapter 3. For the present chapter it will simplify the financial statement discussion to come in Section 2.6 by taking a few moments here to discuss the concept of depreciation in a bit more detail.

2.5.2 Depreciation

Distinguishing between long-term assets and expenses (assets purchased but not depreciated) is important in understanding the flow of data and appreciating why certain items are present or absent on the various financial statements. For instance, let's take a closer look at Figure 2.1, centering in on the relation between transactions, the balance sheet, and the income statement.

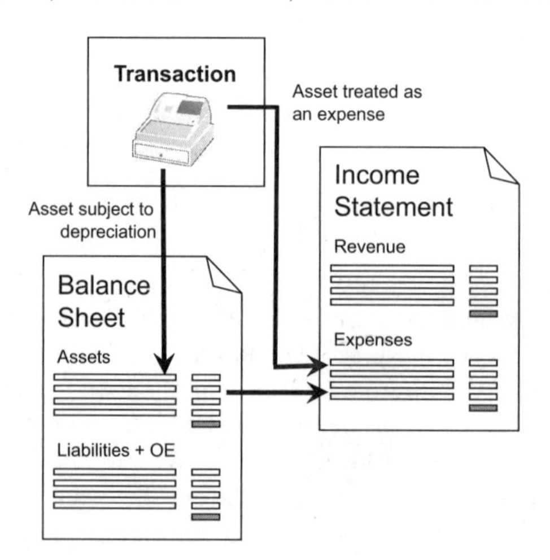

Figure 2.4: Difference in data flow for depreciable assets and expense items.

With Figure 2.4 in mind, items purchased in support of the business are either long-term assets that provide immediate benefit or a long-term asset whose cost is prorated over time according to the matching principle. Items used to directly support the business and not considered long-term assets appear directly on the income statement as an expense. In contrast, those items subject to depreciation must first be recorded on the balance sheet, and through depreciation calculations, the prorated matching cost finds its way over to the income statement. These two pathways are necessary to ensure that revenues are matched by relevant expenses whether the asset purchased is a long-term asset or a legitimate direct expense. Parking the capital expense on the balance sheet simply allows the proper deferral of expense on the income statement. Depreciation calculations are actually quite straightforward, and we will see in Section 2.6 that once the depreciation amount has been calculated, it is a matter of making an *adjusting entry* to adjust the book value of the asset on the balance sheet and record the equivalent incremental change as an expense on the income statement.

Three key factors are involved in depreciation calculations:

1. **Unadjusted cost basis** (or just *cost basis*, or *cost*). The initial asset acquisition cost plus any of the "reasonable and necessary" improvements required to enable the asset to be placed into service. The cost basis may be adjusted for any capital improvements during the life of the asset.
2. **Salvage value** (or *residual value*) is the estimated value of the asset at the end of its useful life, and involves an estimate of what the market value of the asset is at the time the asset is to be disposed. If there is a cost of disposal, that would be deducted from the estimated salvage value.
3. **Useful life**. The estimated useful life is the period - typically expressed in units of years - that the asset is estimated to be in active service to the business (for some assets we might use service units, such as "miles driven"). The estimated useful life is not the same as how long the asset will last, as it could provide another owner many years of service after we sell it at the salvage value.

The two most common methods for computing depreciation are the straight-line (SL) method and the Modified Accelerated Cost Recovery System (MACRS). It is highly recommended that the reader visit the National Council of Examiners for Engineering and Surveying website (`http://ncees.org`) as there are several guides you can download that are quite concise and provide *many* depreciation calculation examples to follow and questions to test your understanding. In particular, Section 17 of the Study Materials section of the NCEES website has a well-written reference guide for typical engineering depreciation calculations. Of course, spreadsheets are increasingly popular for many types of engineering economics calculations and for completeness, Appendix A of this book illustrates a 5-year MACRS calculation example for a company pro forma.

Once the SL or MACRS method is selected, the basic depreciation computation details are quite simple, though there are specialized textbooks and websites that can assist with some of the more nuanced aspects of computing depreciation. For instance, in a capital equipment decision (see Chapter 9, p.267), the decision might be to upgrade the existing asset (the "defender") instead of replacing it if there has been a change in a production plant process. The upgrade might be significant, implying that the cost of improvement is greater than one year and the investment qualifies for an adjusted cost basis. Some circumstances might warrant a change in depreciation calculations or method. For the present work we assume that once a method has been selected it will suffice for the duration of the project and the following three steps then follow to establish the *book value* of the asset at any point in time:

1. Establish the *initial cost* or *cost basis* for the asset, usually found in the prior period balance sheet financial statement (or buried in the footnotes).
2. Compute the amount of asset depreciation deduction that has occurred for the period.
3. Compute the so-called *book value* of the asset, which is the difference between the asset's cost basis and the net depreciation deduction.

Again, in our simplified book value calculation approach, the only fundamental difference in calculation sequence is that SL involves a simple equation and MACRS involves the use of a look-up table. For use in financial statements, we need to know the cost basis and asset depreciation adjustment for the period. Recording the adjusting entry would be as shown later in Figure B.9, p.371; essentially the depreciation amount will be recorded as an expense on the income statement that offsets OE and long-term assets on the balance sheet. Let's walk though some simple calculations.

Straight-line depreciation

The computational strategy for straight-line depreciation is actually fairly simple. For simplicity, we follow the nomenclature of the NCEES and define:

$$\begin{aligned} D_j &= \text{Depreciation in year } j \\ C &= \text{Initial cost} \\ S_n &= \text{Salvage value in year } n \\ N &= \text{expected life of the asset} \\ BV &= \text{Book value of the asset} \end{aligned}$$

Now, the (uniform) depreciation for any accounting time period is given by the straight-line equation,

$$D_j = \frac{C - S_n}{n} \tag{2.10}$$

Knowing the amount of asset depreciation from Equation 2.10, book value of the asset for any period is simply

$$BV = C - \sum_{j=1}^{N} D_j \tag{2.11}$$

Consider the example of a rapid prototyping machine for polymer materials that was purchased for \$28,000 and a 7-year depreciable life with an approximate salvage value of \$4,500 at the end of 7 years. We want to know what the book value of this asset is after 4 years of service. First, we compute the unit depreciation for each year of service

$$D_j = \frac{\$28,000 - \$4,500}{7} = \$3,357$$

And so for the three-year period under consideration, the book value is

$$BV = \$28,000 - \sum_{j=1}^{4} \$3,357 = \$28,000 - \$13,428 = \$14,572$$

Depreciation calculations are an area where the difference between tax accounting, engineering analysis, and GAAP financial statement recommendations differ. From an engineering management perspective MACRS is very popular for engineering project analysis as this method recognizes asset depreciation more quickly than standard accounting practice. But the MACRS calculation strikes some analysts as odd (in comparison to the SL method) since properties are assigned a class for which the salvage value and useful life are not needed. Situations exist where MACRS might be included in tax documents but not viewed as acceptable for preparation of financial statements. Appendix A walks through the MACRS method for an engineering project pro forma.

Impairment

Amortization, depletion, and depreciation are accounting techniques that enable the periodic allocation of the cost of an asset to be distributed over the life of the asset (not immediately expensed in the period it was purchased). Once purchased, the asset is reported on the balance sheet with a value known as the so-called *book value*. This is different than the *market value* of the asset that may fluctuate in time. This is an important distinction, since the price paid for the asset in, say, January may differ from the market value of the asset if a better, faster, and cheaper version of the equipment becomes available 6 months later. Despite the change in market conditions, the book value of the asset continues to be recited at the original purchase price since the transaction materially affecting the business was conducted at the price we *did* pay, not at the price we *could* have paid at another time. There are, of course, situations in which deterioration or obsolescence of an asset might reduce the useful life of an asset, and some firms make annual assessments of "impairment" of assets that might warrant adjustment to the asset's book value.

2.5.3 Opportunities for fraud and abuse

Accounting for the amortization, depletion, and depreciation of long-term assets involves a process of properly apportioning a capital investment to impact profits over multiple future reporting periods. Although the novice might assume all "large" expenses are capital investments, the two are not synonymous and misinterpretations offer many opportunities for error or even fraud. For example:

1. **Capitalizing marketing and advertising costs.** A business might spend millions on Super Bowl marketing and advertising to develop a product "brand" that is believed to develop a future customer base. The branding expense cannot be distinguished from the normal operating expense of business development.
2. **Capitalizing start-up costs.** Losses associated with business start-up must be expensed. Many times an investment will not result in immediate profitability; consider the launch of a new resort that will most likely run under capacity during start-up. WorldCom is a good example of the way in which communication network expenses that should have been shown on the income statement were instead listed as a capital expense – the obvious impact being that the company appeared profitable when in fact there were significant losses being hidden by transferring the expense to the balance sheet.

Some costs associated with placing an asset in use are allowable. For example, costs for the installation and trial runs of a new production line may be allowable if they are necessary to make the asset "ready for use." Because of the opportunity for misrepresentation, these type of costs must be carefully scrutinized, but many "reasonable and necessary" costs are admissible to be added to the value of the long-term asset on the balance sheet. As with other long-term assets, such capitalized costs will be subsequently recognized in future periods after the system "go-live" date is reached.

2.6 Balance sheet

We have introduced pieces and parts of the Balance Sheet in Section 2.3.4 and other sections of this chapter. In this section we center our attention on the balance sheet shown in Table 2.6. Although the table format is for a real start-up company and has been part of actual fund-raising efforts, the reader should appreciate that there are many variations to statement formats; what is provided here is characteristic of what you might find in other textbooks, company 10(K) reports, annual reports or business plans.

Thus far, comments about the balance sheet have portrayed this financial statement as simply a statement of the resources (assets) and sources (liabilities and owner's equity) that are involved in the operations of the organization. We also demonstrated that the balance sheet is based on the *accounting equation.* We now wish to expand on our prior comments and describe in more detail the key line items of a balance sheet.

Let's begin with a fairly standard set of classification categories for balance sheets that has emerged over the year; at a high level many schema have some or all of the following:

- Assets
  - Current assets
    - Cash
    - Accounts receivable
    - Inventory
    - Pre-paid expenses
  - Long-term assets
    - Notes receivable
    - Long-term investments
    - Plant, land, and equipment
    - Intangible assets
- Liabilities and OE
  - Current liabilities
    - Accounts payable
    - Notes payable
  - Long-term liabilities
  - Owners' equity
    - Common and preferred stock
    - Retained earnings

Table 2.5: Typical categories and subcategories for a balance sheet.

Table 2.6 is a balance sheet for a start-up, so a variety of entries are absent from what you might see in other statement presentations, most notably in the area of long-term investments, intangible assets, current liabilities, and long-term liabilities. Assets are the engine of economic growth for the firm. To be listed in the balance sheet, assets cannot be "fictitious" and must be

1. Owned by the company, *and*

2. Positioned to provide economic benefit in the operations of the business. It is inappropriate to list assets that provide speculative gain (unless, of course, that is the business you are in!). It can be startling the number of times a proforma for a technology venture includes assets (say, intellectual property patents) that have not been awarded or that are compromised due to contingencies; the latter does not render the organization to be in complete control of the asset. For instance, were a patent (an intangible) to have been valued at $1,200,000 based on a five-year discounted cash flow projection (more on that in Section 4.2) yet the patent is contingent on a predicate patent that requires a license to be executed with a third party, it is safe to say the actual value of the patent is closer to $0 until there is a document confirming "future economic benefit" (recall Section 2.3.7). Absent meaningful transactions

and related documents it is unrealistic that the company can expect revenue related to the patent.

Consider again the idea of a transaction-based perspective presented in Section 1.2. If there is no documented transaction to validate ownership, control, and relevance to the business, then for the purposes of creating a balance sheet that asset is essentially non-existent. As they say, "hope is not a strategy."

It is customary to list assets in order of decreasing liquidity. Liquidity is the ability to convert an asset into cash within one year (or at least within the normal business cycle of the company). Suggesting an asset is liquid can be actually be quite speculative and the ability to have an "outside view" (Section 2.3.1) is critical to obtaining an objective view of liquidity. For instance, the company may claim inventory is highly liquid, but if a last-in-first-out (LIFO) inventory system is used and DSO is greater than 90 days, then some inventory might "age" on the shelf, become obsolete over time and be equivalent to scrap. Table 2.5 contains the line item "pre-paid" expenses that is often listed last in the list of current assets. This might seem odd given that a concrete transaction took place and this would seem to be just like "money in the bank" in terms of balance sheet assets. Even if a cash transaction was involved, since the asset is anticipated to be *used* by the organization and not *sold* to a customer, it is unlikely the prepaid expense can be liquidated to help solve, say, a payroll problem (unless the prepaid expense can be refunded by the vendor or employees don't mind being paid with magazine subscriptions instead of cash).

Long-term assets – with a focus on depreciation – were discussed earlier in Section 2.5.1. We see from Table 2.6 the treatment of depreciation is as described in Section 2.5.1; note, though, from Appendix A that since Table 2.6 is a pro forma statement (not for tax purposes) we have used the MACRS schedules to compute machinery asset depreciation.

Prior discussions have focused more on assets than liabilities. We should know by now that the liabilities represent owner and non-owner claims on the assets used in the production of goods and services. Further, we understand that non-owners will tend to have a preferential position on claims to assets in the case of insolvency and bankruptcy proceedings.

Description of the liabilities shown in Table 2.5 follows a familiar pattern of priorities, with current liabilities listed first since their maturity date is shorter than for long-term liabilities *and* reiterating that non-owner claims ("creditors") have a priority claim on assets. It is very common for a balance sheet to include the following liabilities:

1. **Accounts payable**. In the course of business many items are purchased from vendors for which the company will pay vendor invoices within 30 to60 days. A sign that a company might be in trouble is when they attempt to relax tight cash flow by not paying items purchased on credit.

2. **Unearned income**. Although many start-ups may not have significant unearned income to report, a high-technology firm may require "up

Balance Sheet
Five-Year Projection ($000) **31 December 20xx**

	Bethesda Imaging, Inc. Year Ending December 31				
ASSETS	**Y1**	**Y2**	**Y3**	**Y4**	**Y5**
Current					
Cash	$232.63	$861.51	$1,301.67	$2,431.38	$4,971.30
Accounts Receivable	$19.53	$71.40	$169.76	$329.65	$479.34
Doubtful Accounts	($0.68)	($3.18)	($9.12)	($20.66)	($37.44)
Ending Inventory	$2.47	$12.87	$30.34	$59.55	$75.16
	$253.95	**$942.59**	**$1,492.64**	**$2,799.92**	**$5,488.36**
Fixed Assets					
Equipment	$15.00	$25.00	$30.00	$35.00	$40.00
Capital Equipment	$25.00	$25.00	$125.00	$125.00	$125.00
Accumulated Depreciation	($5.00)	($13.00)	($20.30)	($27.68)	($34.15)
	$35.00	**$37.00**	**$134.70**	**$132.33**	**$130.85**
TOTAL ASSETS:	**$288.95**	**$979.59**	**$1,627.34**	**$2,932.25**	**$5,619.21**
LIABILITIES					
Current					
Accounts Payable	$3.74	$23.13	$53.55	$104.91	$120.89
Taxes Payable	$10.83	$0.00	$67.44	$226.61	$492.59
Short-Term Notes	$17.00	$76.00	$12.00	$0.00	$0.00
	$31.57	**$99.13**	**$132.99**	**$331.52**	**$613.48**
Long-Term Liabilities					
Notes Payable	**$0.00**	**$0.00**	**$0.00**	**$0.00**	**$0.00**
Owners' Equity					
Stock and Paid-in Capital					
Common Stock	$4.50	$1,004.50	$1,289.11	$1,289.11	$1,289.11
Paid-in Capital	$200.00	$200.00	$200.00	$200.00	$200.00
Total Stock and Paid-in Capital	**$204.50**	**$1,204.50**	**$1,489.11**	**$1,489.11**	**$1,489.11**
Retained Earnings					
Beginning Balance	$0.00	$52.88	($324.04)	$5.25	$1,111.62
Net Income	$52.88	($376.91)	$329.28	$1,106.38	$2,405.00
Dividends	$0.00	$0.00	$0.00	$0.00	$0.00
Retained Earnings	$52.88	($376.91)	$329.28	$1,106.38	$2,405.00
Ending Balance	**$52.88**	**($324.04)**	**$5.25**	**$1,111.62**	**$3,516.62**
Total Owners' Equity	**$257.38**	**$880.46**	**$1,494.35**	**$2,600.73**	**$5,005.73**
LIABILITIES and OE:	**$288.95**	**$979.59**	**$1,627.34**	**$2,932.25**	**$5,619.21**

Table 2.6: Sample balance sheet.

front, good faith" payment on capital equipment projects to reduce the financial exposure of the company and also to "lock in" the commitment of the customer. In the case where cash has been received but there is no expense to match this cash payment, then the balance sheet would report this as unearned income.

3. **Short-term notes payable.** Not all accounts payable are short-term transactions with suppliers. Companies may have a short-term line of credit with a bank to help with unexpected cash flow issues. For instance, there may be a delay in capital equipment manufacture so payment from the customer has not been invoiced, yet the company still has to pay workers at the end of the month for work performed to date. Unpaid salaries fall into the category of *incurred expenses* (Quadrant AE, Figure B.8) that are unrecorded. The line of credit from a bank can help a company meet their payroll obligation.
4. **Long-term notes payable.** The reader can hopefully infer by now that long-term liabilities are those obligations where the debt will *not* be settled within a year. Long-term loans (equipment, vehicles, etc.) are in this category, along with liabilities such as deferred tax or pension obligations.

Balance sheets are financial statements that portray the resources available to operate and grow a business, as well as help to understand the sources and claims to those resources. The structure of the sheets should provide insight on the solvency of the business and the outcomes of resource management and operations. We have seen that ratios provide insight, too. Consider that equity stakes in the company are listed at *book value* and can differ from the *market value* of the company, the latter being an outside view of the perception of company performance. In the next section we examine the profitability of a company as expressed through the income statement.

2.7 Income statement

A pressing question for many firms is, "Are we making money?" Growth in *net income* is critical to business owners as this is a source of owners' equity (wealth). No matter how committed to quality and service employees may be, the company will not survive if it is not profitable. And, if it *is* viable, the next logical question is whether revenue is growing and shrinking. How are assets being used, and how do we compare with our industry counterparts? These and many other questions can be answered in full or in part with the *income statement*, also known at the *profit and loss statement*. We introduced the concept of an income statement in Section 2.3.4 as representing the outcome of operations over a period of time (and is not a snapshot in time as in the case of the balance sheet), in which financial outcomes reflected the ability of the management team to favorably leverage assets provided by the owners and creditors of the company. As a minimum, income statements are prepared annually for the annual report, and (even for a private company) it would be very unusual if the income statement was not produced quarterly.

As in the case of balance sheet statements, income statements come in a variety of formats, too. In the case of the income statement, this has a lot to do with the nature of the business the company conducts. Though the income statement reflects the outcome of management decision-making skills, the statement itself will reflect differences between services, manufacturing, and merchandising companies. Still, some commonalities exist and some generally accepted practices in reporting on the income statement are expected by third parties. When statements are prepared in a multi-year format it is easy to spot expense and revenue trends. An abstract of key categories is provided in Table 2.7; a sample pro forma multi-year income statement is shown in Table 2.8 and has the majority of line items one would expect.

Revenue
Net sales
Cost of goods sold
Gross margin
Operating expenses
Selling expenses
General and administrative expenses
Income from operations
Other revenue and expenses
Net income

Table 2.7: Typical categories and subcategories for an income statement.

The income statement simply seeks to report on the *revenue* of the firm and subtract from it the day-to-day *operating expenses* that were incurred in producing the goods and services that produced the revenue in the first place. Non-operational items in the form of gains and losses are also reported on the income statement.

Measuring income is actually quite challenging, and frequently there is confusion about the difference between cash flow and income statements. The novice might imagine business income is similar to their own personal income clearly defined by a salary. Unfortunately it is just not that simple, and it surprises many when they fully grasp how many *speculative* elements there are to the statement of income. Some key differences are:

- A cash flow statement records cash expenses and income and therefore reflects actual, historical transactions. Further, the cash flow activity could be for the purposes of **investing, financing, and operations** aspects of the business, not just operations. A cash flow statement would also include the sale of an asset (used equipment) that would not show up on the income statement.
- Income statement records reveal an organization's profit or loss result-

Income Statement
Five-Year Projection ($000) **31 December 20xx**

	Y1	Y2	Y3	Y4	Y5
REVENUE					
Gross Sales	$310.0	$1,040.8	$2,474.7	$4,756.9	$6,916.8
Less: Returns/Allowances	10.0%	2.0%	2.0%	1.0%	1.0%
Sales Discounts	0.0%	0.0%	0.0%	0.0%	0.0%
Net Sales	**$279.0**	**$1,019.9**	**$2,425.2**	**$4,709.3**	**$6,847.6**
	$259.5	$964.3	$2,303.7	$4,495.9	$6,593.1
	$19.5	$71.4	$169.8	$329.7	$479.3
	$0.7	$2.5	$5.9	$11.5	$16.8
COST OF PRODUCTS SOLD					
Starting Inventory	$0.0	$2.5	$12.9	$30.3	$59.5
Direct Manufacturing Cost	$46.3	$273.7	$635.1	$1,243.9	$1,461.7
Add: Royalties	$9.1	$33.8	$80.6	$157.4	$230.8
Add: Transportation	$2.2	$9.6	$23.0	$45.0	$65.9
Add: Miscellaneous	$13.0	$48.2	$115.2	$224.8	$329.7
Cost of Goods Available for Sale	$70.6	$367.8	$866.8	$1,701.4	$2,147.5
Less: Ending Inventory	$2.5	$12.9	$30.3	$59.5	$75.2
Cost of Products Sold	**$68.1**	**$354.9**	**$836.5**	**$1,641.8**	**$2,072.4**
As % of Net Sales	24.4%	34.8%	34.5%	34.9%	30.3%
Gross Profit on Sales	**$210.9**	**$665.0**	**$1,588.7**	**$3,067.5**	**$4,775.3**
As % of Net Sales	75.6%	65.2%	65.5%	65.1%	69.7%
OPERATING EXPENSES					
General and Administrative	$78.7	$344.8	$394.7	$570.9	$624.4
Sales and Marketing	$5.0	$291.7	$325.6	$449.5	$500.0
Product Development	$28.1	$117.8	$123.7	$129.9	$136.4
Depreciation	$5.0	$8.0	$7.3	$7.4	$6.5
Operations (Salary only)	$29.7	$277.2	$334.7	$565.4	$593.7
Total Operating Expenses	**$146.5**	**$1,039.4**	**$1,186.1**	**$1,723.0**	**$1,860.9**
As % of Net Sales	52.5%	101.9%	48.9%	36.6%	27.2%
NET INCOME					
Income from Operations	$63.7	-$376.9	$396.7	$1,333.0	$2,897.6
Provision for taxes 17%	$10.8	$0.0	$67.4	$226.6	$492.6
Net Income After Taxes	**$52.9**	**-$376.9**	**$329.3**	**$1,106.4**	**$2,405.0**
As % of Net Sales	19.0%	-37.0%	13.6%	23.5%	35.1%

Table 2.8: Sample income statement.

ing **primarily from operations activity**; records track *earned income*, whether the sales were cash or credit (income, but unpaid). Other non-cash expenses such as depreciation are also part of the income statement, as well as credit items. Since statements include *expected* (not guaranteed) future cash payments, it should be clear the income statement has a very speculative flavor in comparison to the cash flow statement.

Knowledge of just a few terms is quite helpful in understanding the uniqueness and importance of the income statement:

- **Net sales**. Describing revenue from the sale of products would seem simple enough, but we have to be a bit patient when trying to identify the *net sum of sales*, as some initial purchases (or part of a purchase) could be voided for often good reasons. There may be re-work, warranty and discounts applied to a capital purchase, or with consumer merchandise the store might allow returns simply because the consumer may not like the color! In Table 2.8, we have separate line items for initial sales and discounts and returns, and the ratio can be indicative of operational performance.
- **Cost of goods sold (COGS)**. We've noted on several occasions that revenue is matched by expenses in the creation of goods and services. It is informative to know the cost of merchandise – a "direct cost" to manufacture or produce a product. Items like raw material, direct labor, and consumables needed to produce inventory items constitute the *cost of goods sold*. COGS is significant in several ways. For instance, knowing COGS helps when benchmarking performance against comparable companies.
- **Gross margin**. From a mathematical perspective *gross margin* is simply the difference between net sales and COGS. But the significance of gross margin cannot be overstated. Its importance is that it is the *starting point for all other expenses*. In short, without adequate gross margin you are unable to absorb all your other operational costs. Gross margin impacts all other aspects of your business. It sets the ceiling of the profit level and frames enterprise risk and return.
- **Operating expenses**. Expenses incurred in day-to-day operations (payroll, utilities, rent, phone, copier, R&D, etc.) are the firm's *operating expenses*. Operating expenses are also known as operational costs, fixed expenses, indirect costs, and SG&A. Operating expenses are closely watched and questioned since the so-called "bottom line" is inversely proportional to the level of operating expenses. Sometimes companies will try to improve profitability by cutting operating expenses when they find they are unable to raise prices or penetrate new markets; though it might seem easy to lay off employees, managers have to watch that quality or competitive position is not damaged or compromised along the way.
- **Income taxes**. Like any legal entity, a corporation must pay taxes.

Taxes are based on *net income* and therefore are one of the last items to be listed on the income statement.

- **Research and development (R&D)**. A company's activities that are directed at developing new products or procedures.
- **Unusual or one-time charges**. A variety of situations may exist where there might be a gain or loss on sale of assets (as would be the case with the disposal of depreciated assets), impairment of assets (p. 49), restructuring charges, capital write-offs (obsolescent inventory), or loss due to discontinued operations. These one-time charges would be the last items to be recorded on an income statement.
- **Depreciation**. As explored in Section 2.5.2, depreciation is the accounting tool used to allocate tangible asset investment as an expense.
- **EBITDA**. Earnings before interest, taxes, depreciation and amortization (EBITDA) is a popular measure of a company's financial performance since it suggests the profitability of operations outside "unusual or one-time" charges. This has been the source of some misrepresentation, particularly for companies that have a habit of shifting "expenses" (losses) to a one-time charge in order to enhance EBITDA. It is important to note that EBITDA is *not* the same as cash flow.
- **Net income**. This is the "bottom line" for earnings, net of taxes ("net of" meaning that this is the *net* value representing a sum for which taxes have already been deducted). Care must be taken in the use of the term "net" since lots of games can be played with the way taxes are accounted for, and whether the item is net of *all* taxes may not be clear.

Taxes can present difficulty in interpretation of "income," and so it helps to divide tax into categories of "permanent" and "temporary." Sometimes a difference in tax might occur between financial income and tax income arising from the use of different depreciation schedules. If straight-line depreciation is used for financial accounting, but MACRS is used for tax accounting, then there will exist a *temporary difference* in income in the form of a deferred tax (assuming the MARCRS is more aggressive than the straight-line) that will be corrected on a future adjusting entry as, say, an income tax payable. A so-called "permanent difference" would be derived from tax exemptions such as fines, tax-free bonds, and tax incentives for investment. We would not see tax for these items on an income tax statement but they would possibly be present on the internal income statement.

Some companies report additional items after income tax expense on their income statements. These items represent special items outside of normal business operations. They are shown separately to ensure users can identify what income from continuing business results will be. If any special items are included on the income statement, the income tax expense or savings related to each item is net against the special item to report it after taxes. These additional special items may be one of three types: discontinued operations, extraordinary items, and changes in accounting principles. Table 2.9 illustrates an income statement in which we have *normal and recurring* activities listed at the top and the so-called *unusual and extraordinary* items listed below.

Bethesda Imaging, Inc Condensed Income Statement for the period ending June 1, 20xx		
Net sales	....	$278,500
Costs and expenses		
COGS	$201,250	
Selling, G&A	$42,000	
Inçome taxes	$7,500	$250,750
Income from continuing operations	....	$27,750
Extraordinary items:		
Salvage gain on equipment sold	$15,000	
Inventory loss due to flooding .	($900)	$14,100
Net Income	....	**$41,850**

Table 2.9: Simple example of an income statement with extraordinary items.

Accounting statements are notorious for being abused, and the use of "extraordinary" is no exception. Management may try to claim that a *unique* or *significant* event can be classified as *extraordinary, unusual,* or*infrequent.* Inventory is a target since obsolescence and perishability occur, and it might be tempting to claim that the write-down of inventory is an "extraordinary" event simply to hide a uniquely poor sales cycle or significantly poor inventory planning. If EBITDA is a popular measure, then to have inventory written off as an extraordinary expense would result in a more favorable income statement.

2.8 Statement of cash flows

Companies will not only fail when business activity is unprofitable, but also if the business lacks *liquidity*, meaning that the company does not have the cash (is not liquid) to pay obligations to creditors, employees, and other accounts payable. It is entirely possible that the income statement shows "profitability" yet there is no cash to pay bills. The primary purpose of the *statement of cash flows* is to identify cash receipts and cash payments for a given reporting period and to understand how cash flow is parsed between operating, investing, and financing activities.

It has been recommended that the balance sheet and income statement should be based on *accrual accounting* to comply with the matching principle; recall from Section 2.3.4 that the matching principle recognizes revenue when it is earned and expenses when they are incurred, *regardless of whether cash is received.* This is important from a cash flow perspective since the income statement can be viewed by the pessimist as simply a barometer of the company's ability to sell a product or service at a price that covers cost of production. The income statement is speculative in that we assume - but don't know - that the cash flow cycle will keep pace with the sales cycle, as illustrated earlier in Figure 2.3 when discussing the cash flow cycle and current ratio. In other words, you can report healthy income even if you've not received any cash at all! In the end, "cash is king," and the statement of cash flows summarizes the ability to create cash from transactions.

To be clear, the term "cash" refers to *currency* as well as *cash equivalents.* Cash equivalents are those very short-term investments that can be converted to cash within a quarter-year period (90 days). This precludes 6-month and 1-year securities, as the speed of conversion to cash is inadequate to help alleviate economic hardships. Creditworthiness will enable short-term loans (financing) to help with cash-flow issues, and this underscores how the balance sheet, income statement, and statement of cash flows integrate to communicate with outsiders about the financial health of the organization.

To summarize, there are three main uses of the statement of cash flows:

1. Managers assess liquidity and make decisions on policies to pursue to address issues (e.g., short-term financing)
2. Managers and investors assess the ability of the firm to generate cash from business transactions.
3. Investors develop a better understanding of management's ability to manage current cash, generate future cash, and options available when liquidity is an issue.

Stakeholders use the statement of cash flows to examine three key facets of business activities affecting cash flow: operations, investing, and financing.

Like other financial statements, the statement of cash flows comes in many forms; further, some confusion can exist since the document integrates activities from many different business functions *and* other financial statements, often including reversing entries! Table 2.10 highlights the general classification, sources, and uses of funds. Table 2.11 illustrates the statement of cash flows for the product manufacturing start-up described in Appendix B, and reflects the minimal role of investing in the business.

It is not always the case that a decrease in cash is a bad thing. It may be the case the company has just closed on the past quarter's predictably high seasonal revenue, and want to "park" the revenue in a so-called "trading security" as a near-term investment. The point is to look at the reasons underpinning changes in cash flow.

Cash flow statements can sometimes be confusing since they do represent the crossroads of many of a company's activities; it is not unusual to read them a few times to fully comprehend the information being presented. One source of confusion is the presence of reversing entries when describing cash flow associated with operational activity. This comes from the use of accrual accounting for the income statement; when the net income is reported on the cash flow statement, the non-cash components that went into net income need to be reversed. An interesting aspect of this is that the "bottom line" of a

1. **Operations** - purchase and sale of products and services.
   (a) Cash inflows ...
      i. Sale of products or services to customers
      ii. Loans to the company
      iii. Equity investments thought the sale of stock
   (b) Cash outflows ...
      i. Pay operations expenses (wages, inventory)
      ii. Payment of taxes
      iii. Payment of interest on loans to the company
2. **Investing** - buying or selling long-term assets.
   (a) Cash inflows ...
      i. Sale of long-term assets (equipment, plant, property)
      ii. Collecting on any loans (not AR, though)
      iii. Sale of long-term investments
   (b) Cash outflows ...
      i. Purchase of long-term assets (equipment, plant, property)
      ii. Providing loans
      iii. Purchase of long-term investments
3. **Financing** - borrowing funds and selling securities (stock).
   (a) Cash inflows ...
      i. Sale of stock
      ii. Issuing debt
   (b) Cash outflows ...
      i. Payment of dividends
      ii. Payment of debt or stock re-purchasing activity

Table 2.10: Statement of cash flow: categories, sources, and uses.

net increase or decrease in cash reflects the outcome of business activity as if a *cash-basis* of accounting was in place.

The general framework of a cash flow statement and grouping of entries is shown in Table 2.12. We begin the construction of the statement of cash flows with items related to operations. This is logical since we are most interested in how the company has faired relative to the production and sale of goods and services. The accrual basis of the income statement can mask issues with conversion of revenue into cash, and the statement of cash flows gives insight on that aspect of operations.

The logic of Table 2.12 can be perplexing to the average engineer, commenting "Why not just adjust the entries on the income statement?" You can, of course, since there is no net effect on the net change in cash position for the period. Indeed, our simple start-up pro forma statements for the income statement Table 2.8 and statement of cash flows Table 2.11 illustrate the *direct methods* where some (but not all) adjusting entries are made on the income statement. This is ordinarily satisfactory when clarity of communication prevails, but we emphasize that such a situation is normally confined to

Cash Flow Statement
Five-Year Summary of Projected Cash Flows ($000) 31 December 20xx

CASH RECEIPTS	**Y1**	**Y2**	**Y3**	**Y4**	**Y5**
Cash from Product Sales	$259.5	$964.3	$2,303.7	$4,495.9	$6,593.1
Investments	$221.5	$1,076.0	$296.6	$0.0	$0.0
Total Cash Receipts	**$481.0**	**$2,040.3**	**$2,600.3**	**$4,495.9**	**$6,593.1**
CASH DISBURSEMENTS					
General and Administrative					
Salaries, wages, benefits, taxes	$0.0	$252.9	$265.6	$404.9	$425.2
Services (Audit, etc.)	$4.0	$4.0	$5.0	$8.0	$10.0
Office and warehouse leases	$30.9	$30.9	$38.6	$50.2	$60.3
Equipment rental	$1.6	$1.6	$1.7	$1.7	$1.8
Furniture and office equipt expense	$15.0	$10.0	$5.0	$5.0	$5.0
Insurance	$2.6	$9.6	$23.0	$45.0	$65.9
Utilities and telephone	$0.6	$0.6	$0.8	$1.0	$1.2
General travel and entertainment	$39.0	$45.0	$60.0	$60.0	$60.0
	$93.7	**$354.8**	**$399.7**	**$575.9**	**$629.4**
Sales and Marketing					
Salaries, wages, benefits, taxes	$0.0	$277.2	$291.1	$382.0	$401.1
Services, website	$5.0	$14.5	$34.6	$67.4	$98.9
	$5.0	**$291.7**	**$325.6**	**$449.5**	**$500.0**
Product Development					
Salaries, wages, benefits, taxes	$28.1	$117.8	$123.7	$129.9	$136.4
Materials, services, leases	$0.0	$0.0	$0.0	$0.0	$0.0
Capital Equipment Purchase	$25.0	$0.0	$100.0	$0.0	$0.0
	$53.1	**$117.8**	**$223.7**	**$129.9**	**$136.4**
Operations					
Salaries, wages, benefits, taxes	$29.7	$277.2	$334.7	$565.4	$593.7
Direct manufacturing costs					
Raw material and direct labor	$37.4	$231.3	$535.5	$1,049.1	$1,208.9
Less: Accounts Payable	$3.7	$23.1	$53.5	$104.9	$120.9
Equipment leases (2%)	$5.2	$19.3	$46.1	$89.9	$131.9
Royalties on units sold	$9.1	$33.8	$80.6	$157.4	$230.8
Transportation (1%)	$2.2	$9.6	$23.0	$45.0	$65.9
Prior year income taxes paid	$0.0	$10.8	$0.0	$67.4	$226.6
Prior year notes paid	$0.0	$17.0	$76.0	$12.0	$0.0
Miscellaneous (5%)	$13.0	$48.2	$115.2	$224.8	$329.7
	$96.6	**$647.2**	**$1,211.1**	**$2,211.0**	**$2,787.4**
	$248.3	**$1,411.5**	**$2,160.2**	**$3,366.2**	**$4,053.1**
NET CASH FLOW	**$232.6**	**$628.9**	**$440.2**	**$1,129.7**	**$2,539.9**
Opening Cash Balance	**$0.0**	**$232.6**	**$861.5**	**$1,301.7**	**$2,431.4**
Ending Cash Balance	**$232.6**	**$861.5**	**$1,301.7**	**$2,431.4**	**$4,971.3**

Table 2.11: Sample cash flow statement

Bethesda Imaging, Inc. Statement of cash flows for the period ending June 1, 20xx	
Cash flows from operations	
Net income	$100,000
Adjustments for non-cash items:	
Add: Depreciation expense	$5,500
Add: Current liabilities	$7,000
Subtract: Accounts receivable	($1,500)
Subtract: Inventory on-hand	($12,500)
Net cash used by operations	$98,500
Cash flow from investing activity	
Cash paid for capital equipment	($47,000)
Cash flow from financing activity	
Cash from sale of common stock	$40,000
Dividends	($1,000)
Net cash from financing	$39,000
Net change in cash position for the period	$90,500

Table 2.12: Simple example of a cash flow statement for a product firm.

moderately simple transactions. It is actually more customary for managers and analysts to use the *indirect method* of adjusting the cash flow statements.

The indirect method elucidates operational profitability on the income statement and reinforces several issues about cash flow that can be confusing. For instance, by looking only at the statement of cash flows, it might be thought that cash flow can be increased by increasing the amount of depreciation. Reflecting back on the income statement, adding depreciation would *decrease* net income featured on the cash flow statement and simply cancel out the increase in depreciation. Depreciation is not cash flow; it is simply a way to apportion a prior investment and match expenses to revenue created by assets. The logic of the indirect method becomes very clear when we want to take net income from the income statement and translate it into cash flow by adjusting for any line item that is not cash!

Cash flows from investing activities will normally be related to changes in assets reported on the balance sheet. Some of these investments might be in raw materials for production, and this raises the question of specifically what inventories should actually be reversed on the cash flow statement. Manufacturing engineers will understand the following sequence:

1. Production costs can be separated into (a) raw material purchases, (b) direct labor, and (c) overhead.

2. Raw materials inventory is drawn down as materials *work in process* (WIP) takes place to convert raw materials into finished goods. The value of WIP is an asset on the balance sheet.
3. When finished goods are produced, the goods become part of the finished goods inventory now available for sale, and the asset moves to a COGS expense item on the income statement.
4. The cash flow statement will clarify which of the costs or investments shown on the balance sheet or income statement were actually paid for in cash.

Cost accounting will enable distinctions to be made between manufacturing costs (materials, direct labor, and manufacturing overhead) and non-manufacturing costs (selling and administrative). Sometimes with retail enterprises it is easier to categorize *product cost* (versus manufacturing cost) and *period costs* (versus GSA).

Cash flow from financing activities may only involve a few items, but important items nonetheless! It may be essential for a special product line or start-up activity that the firm raise funds through the acquisition of long-term debt (bonds, to be discussed in Section 4.7) or through the sale of common or preferred stock. Such revenue items are not related to operations activity (but finance activities) and so would be incorporated in the balance sheet and not the income statement.

Gains and losses are another item generally unrelated to cash flow, and so the statement of cash flows would contain adjusting entries for these items.

2.9 Statement of owners' equity

Assets of the corporation are provided by non-owners and owners; the balance sheet discussion of Section 2.3.4 (p. 27) pointed out that the balance sheet is simply a listing of the resources (assets) and sources (liabilities and owners' equity) that are involved in the operations of the organization. The balance sheet is based on what is known as the *accounting equation*, and we categorized non-owner financing in terms of balance sheet liabilities and owner financing in terms of equity. In this section we discuss the *statement of owners' equity* since the hope is that is the value of owners' equity (OE) rises over time.

The two main components of OE are

1. Contributed capital, also known as *paid-in* capital, the amount of money invested by the owners of the company.
2. Retained earnings, the *cumulative* amount of net income ("earnings") that has been retained by the company over the life of the company.

In this case the owners of the company are stockholders and wealth is usually (but not always) allocated among owners by percentage stock holdings, but in the case of a limited-liability company (LLC), ownership would be established through the operating agreement of the LLC. Sometimes percentages are negotiated, as illustrated in Table 2.14. The maximum amount of stock a company is *authorized* to sell is set by the articles of incorporation the

Bethesda Imaging, Inc.
Statement of changes in OE
for the period ending June 1, 20xx

Contributed capital	
Beginning balance	$100,000
Common stock, $1 value, 10,000 shares authorized, 5,000 shares issued	$5,000
Additional paid-in capital in excess of par	$35,000
Contributed capital balance	$140,000
Retained earnings	
Beginning balance	$15,200
Net income	$41,850
Less: Dividends	($1,000)
Balance, end of period	$56,050
Total Owners' Equity	$196,050

Table 2.13: Simple example of changes in owners' equity (OE).

company files with the state in which they have chosen to incorporate. Shares authorized are different from shares *issued* during the course of raising funding for the company. Stock is often thought synonymous with *common stock*, the type of stock that would be offered to the general public at par value, or a stated value if stock is declared as no-par value. Par value is a legal "floor" to the value of the stock, but has been set to such a low value in many cases (between $0.01 and $1) that the idea of having stock sold at the equivalent of $10 per share as a mix of $1 par value and $9 per share paid-in capital is a bit out of date (but is shown in Table 2.13 for illustrative purposes).

It is important to remember that changes in OE are derived from the income statement, so it should not be assumed that the balance of retained earnings is *cash*, as we would have to look to the statement of cash flows to understand specific availability of cash. It is thus entirely possible for retained earnings to be *negative* and possibly require the owners to continue to contribute *paid-in* capital to ensure OE is non-zero. It may be necessary for a firm to issue *preferred* stock that had certain priority claims on assets in the case of bankruptcy, and the terms and condition of the preferred position are negotiated on a case-by-case basis. In the case of Table 2.14 there are often *planned increases* in the price of stock offered to investors, depending on the timing of their investment relative to the growth of the company.

Stockholders' equity has two main parts then, as illustrated in Table 2.13. The first major component is *contributed capital* of three basic types:

1. **Common stock**, that stock offered to the general public at par value or stated value.

2. **Preferred stock**, a special class of stock that has modified ownership rights articulated by agreement.
3. **Treasury stock**, another special stock representing what the *company purchased back* from stockholders.

The second component is *earned capital*, composed of

1. **Retained earnings** representing the *accumulated* net profit *not including* any dividends paid out to stockholders.
2. **Gains and losses** reflecting changes in equity that would not normally be found on the income statement (but could affect asset value changes through the balance sheet or statement of cash flows).

The statement of OE and the income statement are the two statements that link balance sheets between the beginning of the year and year-end statements. Ultimately, the actions of managers pass through the various financial statements, and the statement of OE reflects the wealth generation for the owners of the company.

2.10 Summary

Engineering management development often involves the acquisition of new skills that enable a person to be a functional part of the management team. Financial statements are the scorecard of a company, and an understanding of what the statements communicate to stakeholders is important. In this accelerated course of study it is essential that the manager understand the basic mechanics and procedures behind the statements, since without that understanding, decision-making is "crippled" since the manager has no insight or intuitive sense of cause-and-effect. It is the same as a hydraulic engineer who memorizes equations without knowing the meaning of terms; without that understanding there is no way to prioritize components that spark relevant questions. Engineers who do not understand the Navier-Stokes equation for flow simply cannot fathom what convection-dominated flow means. Managers who fail to understand the sequence of developing financial statements will watch revenue rise through the income statement without ever appreciating they are on the brink of insolvency as chartered by the cash flow statement. They may have completely missed the significance of ratios and thus lost track of what peers in the industry are accomplishing.

Shareholders
Projected Equity Ownership — **31 December 20xx**

	Initial Incorporation			**Round I**			**Round II**		
	Shares	Stake	Value	Shares	Stake	Value	Shares	Stake	Value
Stock Value Profile									
Corporate Valuation			$4,500			$1,004,500			$1,289,108
Shares Authorized	10,000			10,000			10,000		
Shares Outstanding	4,500			6,000			7,700		
Shares Remaining	5,500			4,000			2,300		
Stock Price			$1.00			$167.42			$167.42
Initial Assignments									
Executive - A	1,250	27.8%	$1,250	1,250	21%	$209,271	1,250	16.2%	$209,271
Executive - B	1,250	27.8%	$1,250	1,250	21%	$209,271	1,250	16.2%	$209,271
Executive - C	1,000	22.2%	$1,000	1,000	17%	$167,417	1,000	13.0%	$167,417
Executive - D	500	11.1%	$500	500	8%	$83,708	500	6.5%	$83,708
Executive - E	500	11.1%	$500	500	8%	$83,708	500	6.5%	$83,708
Follow-on Funding									
Partner - X	50%			750	13%	$125,563	750	9.7%	$125,563
Partner - Y	30%			450	8%	$75,338	450	5.8%	$75,338
Partner - Z	20%			300	5%	$50,225	300	3.9%	$50,225
	4,500	100.0%	**$4,500**	6,000	100%	$1,004,500	6,000	77.9%	$1,004,500
	Round I Dilution								
	Partnership Investment			1,500		$1,000,000			
	Ownership percentage			25%					
				Round II Dilution					
				Early Stage Fund I			700	9.1%	$117,192
				Early Stage Fund II			1,000	13.0%	$167,417
							1,700	22.1%	**$284,609**
							7,700	100%	$1,289,109

Table 2.14: Example of an equity deal flow to raise funds for a start-up.

2.11 Problems to explore

Problem 2.1
Discussion questions:

1. Describe "intangible" assets.
2. Is it possible for working capital to be negative? Retained earnings? Why?
3. The liquidity of assets is important to creditors. Does the liquidity of assets affect OE? Why?
4. What might be of consideration in an exchange of intellectual property (IP) for common stock?
5. Does preferred stock have an advantage in bankruptcy proceedings?

Problem 2.2
It seems that there are many ways for fraud and abuse to take place in the field of accounting. How can this occur when there are so many regulatory bodies and professional accounting standards in place?

Problem 2.3
In several accounting situations it seems like there is more than one acceptable way to account for a transaction. Is this a significant issue? Are there situations where it can lead to ethical problems?

Problem 2.4
Assume you work for an accounting firm and you are part of a team that does work for two clients that are competitors. Describe some ways in which ethical considerations might be an important part of day-to-day work.

Problem 2.5
Managers seem to be concerned about EBITDA quite a bit. What is it that EBITDA really communicates to managers?

Problem 2.6
Adjusting entries are often required in accounting. Please describe the various types of adjusting entries and give at least three examples.

Problem 2.7
Sometimes financial statements will show "unearned revenue." How does this arise? What type of transactions might be typical examples?

Problem 2.8
Two key performance indicators for a company are profitability and solvency. Why are these so important? Is the value of accrual accounting more evident when centered on profitability or solvency?

Problem 2.9
What does it mean to recommend that an expense or revenue should be "accrued?" Would this be considered "normal?"

Problem 2.10
What is the implication of the statement "the books are not in balance"? What corrective action should be taken?

Problem 2.11
Items in the table below need to be categorized (if relevant) as part of an income statement. Indicate whether each of the accounts falls under the category of operating expense, cost of goods sold, net sales, "other" or is not found on the income statement.

a.	Unearned revenue
b.	Cost of sales
c.	Owners' capital
d.	Interest earned
e.	Interest expense
f.	Depreciation expense
g.	Leased equipment

Problem 2.12
Data extracted from a year-end balance sheet are shown below. Compute the working capital for this firm. What would the current ratio be, and what is the significance for this firm?

Accounts payable	$6,250
Accounts receivable	$15,000
Cash	$4,500
Securities	$10,000
Inventory	$8,760
Long-term bonds	$5,000
Plant equipment	$68,000
Owners' capital	$40,000

Problem 2.13
Describe how trend analysis can provide insight on whether a firm's financial position is improving or deteriorating.

Problem 2.14
Would the combination of a high current ratio and a low inventory turnover ratio suggest a firm is maintaining too high an inventory level? Why or why not?

Problem 2.15
A senior VP has proposed that "we pay too much tax, so we should issue new debt and use the proceeds to buy back common stock." What would be the logic behind this statement?

Problem 2.16
Jameson Polymers, Inc. has common stock that is currently selling at $100 per share. If Jameson has 100 shares of common stock outstanding, what is its return on total assets (ROA)? Supplementary financial data are given below.

Price/Earning Ratio	11.5%
ROE	21%
Debt Ratio	55%

Problem 2.17
Blair Company has $5 million in total assets. The companys assets are financed with $1 million of debt and $4 million of common equity. The companys income statement is summarized below:

EBIT	$1,000,000
Interest Expense	$100,000
EBT	$900,000
Taxes (40%)	$360,000
Net Income	$540,000

The company wants to increase its assets by $1 million, and it plans to finance this increase by issuing $1 million in new debt. This action will double the companys interest expense, but its operating income will remain at 20 percent of its total assets, and its average tax rate will remain at 40 percent. What is the net effect if the company takes this action?

Problem 2.18

A healthcare home-care firm has a profit margin of 15 percent on visiting nurse sales of $20,000,000. If the firm has debt of $7,500,000, total assets of $22,500,000, and an after-tax interest cost on total debt of 5 percent, what is the firm's ROA?

Problem 2.19

Independent Nursing Consultants (INC) has the following data:

Sales	$10,000
Net income	$240
Total assets	$6,000
Debt ratio	75%
BEP ratio	13.33%
TIE ration	2.0
Current ratio	1.2

If INC could streamline operations, cut operating costs, and raise net income to $300, without affecting sales or the balance sheet (the additional profits will be paid out as dividends), by how much would its ROE increase?

Problem 2.20

Dogway Candy Company sells candy on consignment to "big-box" stores and currently has $1,000,000 in accounts receivable. Its days sales outstanding (DSO) is 50 days (based on a 365-day year). The company wants to reduce its DSO to the industry average of 32 days by pressuring more of its customers to pay their bills on time. The company's CFO estimates that if this policy is adopted, the company's average sales will fall by 10 percent. Assuming that the company adopts this change and succeeds in reducing its DSO to 32 days and does lose 10 percent of its sales, what will be the level of accounts receivable following the change?

Problem 2.21

Custom Baseball Hats sells all its merchandise on credit. It has a profit margin of 4 percent, days sales outstanding equal to 60 days (based on a 365-day year), receivables of $147,945.20, total assets of $3 million, and a debt ratio of 0.64. What is the firm's ROE?

Problem 2.22

ABC, Inc. had the following partial balance sheet and partial annual income statement:

Partial balance sheet:

Cash	$20
A/R	$1,000
Inventories	$2,000
Total current assets	$3,020
Net fixed assets	$2,980
Total assets	$6,000

Partial income statement:

Sales	$10,000
COGS	$9,200
EBIT	$800
Interest (10%)	$400
EBT	$400
Taxes (40%)	$160
Net income	$240

The industry average DSO is 30 (based on a 365-day year). ABC plans to change its credit policy so as to cause its DSO to equal the industry average. If the cash generated from reducing receivables is used to retire debt (which was outstanding all last year and which has a 10 percent interest rate), what will ABC's debt ratio (Total debt / Total assets) be after the change in DSO is reflected in the balance sheet?

Problem 2.23

Raster Imaging Company has a new management team that has developed an operating plan to improve upon last year's ROE. The new plan would place the debt ratio at 55 percent, which will result in interest charges of $7,000 per year. EBIT is projected to be $25,000 on sales of $270,000, and it expects to have a total assets turnover ratio of 3.0. The average tax rate will be 40 percent. What does Raster Imaging expect return on equity to be following the changes?

Problem 2.24

Circuit Board Systems (CBS) has the following data: Assets: $100,000; Profit margin: 6.0%; Tax rate: 4.0%; Debt ratio: 40.0%; Interest rate: 8.0%: Total assets turnover: 3.0. What is CBS's EBIT?

Problem 2.25

For the balance sheet for Bethesda Imaging, shown in Table 2.6, p. 53, identify three financial ratios that might be of interest to an investor in the company. What do the 5-year trends for the ratios suggest?

Problem 2.26

An advanced engineering computer was purchased for $22,750 and its expected life is 4 years. It is estimated the company could sell the computer for $500 after 5 years. Compute and compare the book value of the computer using straight-line and MACRS depreciation for a 5-year life. If the computer sold for $10,000 in year 2, what would be the gain or loss?

Problem 2.27

Chase Publishing Company purchased the copyright to a financial textbook for $25,000. Although the author thought the book would serve academia for 10 years, the reality is that the typical life of a textbook is 4 years. Still, the copyright has a length of 75 years. Compute the annual amortization of the copyright.

Problem 2.28

Chase Publishing was disappointed with the financial textbook and decided to purchase the trademark from a well-known record store for $250,000. How can the cost of the trademark be accounted for if management thinks the trademark has indefinite life?

References

[1] Daniel Kahneman and Dan Lovallo. Timid choices and bold forecasts: A cognitive perspective on risk-taking. *Management Science*, 39(6), January 1993.

[2] Michael J. Mauboussin. *Think twice: Harnessing the power of counterintuition.* Harvard University Press, 2009.

[3] Mohamed Hegazy and Rasha Kassem. Fraudulent financial reporting: Do red flags really help? *Journal of Economics and Engineering*, (4):69–79, January 2010.

[4] Nicky A. Dauber, Joel Siegel, and Jae K. Shim. *The vest-pocket CPA.* Prentice Hall, 1996.

[5] Joel G. Siegel and Jae K. Shim. *Accounting handbook.* Barrons, 4th edition, 2006.

Chapter 3

Time Value of Money

3.1 Learning objectives

Most engineers understand the significance of consistent systems of units in engineering analysis; problem parameter unit-of-measure conversions are often made in preparatory work to ensure equivalent units of measure in subsequent calculations. In accounting, finance, and engineering economics the same basic principle applies. Calculations and comparisons must be made for situations where *equivalent* units of measure are in place, but the situation is slightly different because the basic units of measure are time, risk, and the *time value of money*. Given that typical investment choices are being made over long periods of time, we have to account for the fact that risk, inflation, and the *cost* of money (interest rate) affect estimated values. In this chapter, we describe the tools and techniques for converting time series of cash or cash flows into a form that permits reasonable assessments of value and decisions to ensue. There is a standard set of formulas for which a complete understanding of the homework problems should leave the student comfortable with this subject and ready to move on to more complex applications.

After reading and discussion, the student should be able to:

1. Describe the concept of "equivalence"
2. Explain the concept of the "time value of money"
3. Know the difference between simple and compound interest calculations
4. Be able to draw cash-flow diagrams
5. Analyze a series of transactions to establish present value and future value
6. Know the difference between nominal and effective interest rates

3.2 Equivalence is key

The concept of equivalence arises in engineering in many different ways. Consider an I-beam flexure problem with bending M and torsion T. There arises an equivalent bending moment that is the sum of M and that portion of T expressible in terms of an effective bending moment. This problem requires understanding and categorizing beam loading components so that a proper mathematical characterization is established. There could also be a bending moment M expressed in terms of *ft-lb* and a torque T in terms of newton meter *Nm*. Engineers learn to spot immediately such an inconsistency in units (you did notice, right?) and convert data prior to calculations. Problem context plays a role too. For instance, we might have a different safety factor for static loading of a pedestrian bridge over a creek compared to the safety factor for dynamic loading of a suspended pedestrian bridge in a convention center. Often there are many different ways to accomplish tasks, and establishing the best alternative can be a challenge. Tangible and intangible engineering factors play a large role. You can begin to see why an engineer (rather than an accountant) might possibly learn a little about finance to help solve a technology-intense business problem. It is generally easier for the engineer to learn the relevant financial calculations required to address a business problem than for the accountant to learn and understand the full context of technical issues. Indeed, it is prudent that economic concerns about a project be factored in as early as possible in a project; equivalence of options in financial decision-making terms is central to project activity.

We have often heard the expression "time is money," and for us this is just another way to describe the *time value of money* (TVM). There are three basic components to TVM calculations:
Asset: The sum of money we have available to earn, hold, or lend to others.
Period: The period of time over which events takes place.
Interest: The cost of using (or derived by loaning) the money over time.
Interest is a critical concept to understand. Let's say you had a project in mind whereby you could "double your money" within a year. Your colleague Ajit is interested in this project and indicates it is worth \$15 for him to let you have the use of \$100 for a year to fund your idea (Ajit is not interested in becoming a partner, just a creditor!). The "interest" for such a lender could be expressed as

$$\begin{aligned} \textit{Amount owed} &= \textit{Principal} + \textit{Interest} \\ \$115 &= \$100 + \$15 \end{aligned}$$

In this quite trivial situation, a project anticipated to last a year involves a loan with a \$100 original investment (principal) and an *interest rate* of

$$\frac{\$15}{\$100} = 0.15 = 15\%$$

Is this situation – the use of cash assets – any different for Ajit if, as an alternative to your deal, the local bank has a "special" advertised in which new deposits will earn 15%? Ignoring the possible differences in *investment risk*, then really it makes no difference if we invest in our colleagues' venture or put the money in the bank. However, if the bank rate were only 10%, then indeed the $100 investment is no longer equivalent in both situations. There may be good reasons for the difference (the bank may claim their investment is a "safer" option and less risk of loss exists), but we understand that different options can produce different outcomes for the same $100.

Generally we say that $1 is worth more today than tomorrow, and this derives from the opportunity that exists to make your money work for you. For instance, with the bank interest rate of 10% it does not matter if I am given a gift of $100 today or $110 a year from now since the situations are equivalent. But if I am given a $100 gift and I simply hold the cash and do nothing with it for a year, then I have lost the opportunity to earn $10; the *opportunity cost* of $10 is what erodes the power of funds in my possession. This is why given the choice, people would rather have $100 today instead of tomorrow. Again, all other things remaining equal, then at an interest rate of 12%, the situation of $100 today, $112 a year from now, or $89.29 a year ago are all equivalent situations. Re-read the previous sentence and ask yourself if you *really do* understand the underlying meaning. Can you explain it to a friend? Do you need to look anything up, or could you do a sample calculation on the chalkboard (whiteboard) if asked?

Equivalence is really no more complicated than quantifying the economic value of money (an asset) for various durations of time (the period) at a specific interest rate (cost). You have available many, many different accounting or engineering economy books (and millions of Internet websites) describing TVM from many different perspectives and providing a wide variety of sample calculations. It can be overwhelming to some students but does not have to be. We'll work through more concepts and examples later in this chapter to clarify and supplement what you might read elsewhere. Quite frankly, once you understand the basics then all the website descriptions seem to be the same.

3.3 A few symbols: The "fab five"

One of the wonderful aspects of engineering economics and many finance and accounting books is that there seems to be fairly consistent definition and usage of the key terms used for time value of money (TVM) calculations. The "fabulous five" parameters are:

- P, a sum of money at the *present*, expressed as dollars, euros, rupees, etc.
- F, a sum of money in the *future*, expressed as dollars, euros, rupees, etc.
- A, a series of payments or *annuity*; consecutive, equal, end-of-period amounts
- i, interest rate for a stated period (days, months, years)
- N, number of consecutive periods

Typical TVM calculations involve four out of the "fab five" parameters; further, in any given problem, the *choice of a singular* TVM equation requires that three out of the four parameters be known. We have to remember the challenge is less than that of finding the *unknown* value as much as establishing an equivalent value framework and understanding interest rate options! Let's take a moment to address simple versus compound interest rate options before moving deeper into more problem-solving and decision-making tips.

3.4 Compound interest is the norm

In Section 3.2 we expressed "interest" as the difference between the loan principal and the total owed. Further, we identified the interest rate as a fractional amount and simply stated that interest rate is simply the fee we pay for the right to use someone else's money. We assume our use of funds involves such a clever or unique idea that we'll come out ahead since our product or service will produce more money than we pay in interest. There are two basic ways to compute interest and we need to differentiate between the two since the differences can be significant:

- When *only* the original principal amount owed is used in the interest calculation, then this is called a simple interest calculation.
- When *both* the original principal amount and the amount of interest accumulated for prior periods, then this is called a compound interest calculation.

The difference is subtle, but the example below provides a mathematical interpretation of the same statement. Not clear what the statements *really* mean? Well, Google the phrase "simple interest versus compound interest" if you would like other perspectives; there were over 450,000 hits when I searched that phrase and after 17 minutes you'll get the message. More seriously, though, a conceptual challenge is a difference in the way we "think" and "calculate." Most business activity uses compound interest as a *calculation basis*, but most people *think* in terms of simple interest. Thus an unintended consequence is that people will normally underestimate the value of saving as much as they will underestimate the cost of borrowing. Literally, the word "compounding" means to "form by combining parts" of both interest and principal in subsequent calculations. It is helpful that the use of simple interest calculations is actually quite rare in business, so by default everything moving forward uses compound interest calculations. Still, let's work through an example to illustrate the differences.

Example:

Suppose you borrow $10,000 with an interest rate of 12%; how much do you owe at the end of 3 years if a simple interest calculation is used?

$$\begin{aligned} X &= \$10,000 + (\$10,000)(0.12)_{Y1} + (\$10,000)(0.12)_{Y2} + (\$10,000)(0.12)_{Y3} \\ &= \$10,000 + \$1,200 + \$1,200 + \$1,200 \\ &= \$13,600 \end{aligned}$$

If we include interest from the prior period in the calculation of subsequent prior interest, then we have compounded principal with interest and

$$\begin{aligned}X &= \$10,000(1+0.12)_{Y1}(1+0.12)_{Y2}(1+0.12)_{Y3} \\ &= \$10,000(1+0.12)^3 \\ &= \$14,049.28\end{aligned}$$

The difference of $449.28 might seem moderately small, but the time value of money is powerful and the difference would be an additional $62,463 in interest paid over the life of a 20-year loan! Derivations of the compound interest formula are readily available in several textbooks; below, we illustrate the compound interest calculation in tabular form for a $1,000 loan at a 10% interest rate to make the point from another perspective.

Year	Principal Start of Year	Interest Earned	Principal End of Year
1	$1,000	$100	$1,100
2	$1,100	$110	$1,210
3	$1,210	$121	$1,331

It only takes a few moments to work out the numbers in the table above. If at this point the difference between compound and simple interest escapes you, then re-read this section (3.4) one more time. Then carefully dissect the initial definitions and *work through* (yes, calculate on a separate piece of paper) the related calculations for the $10,000 loan examples.

Compounding so significantly enhances the time value of money and works so much in favor of the person in control of the assets that one rarely encounters simple interest problems in practice. Unless otherwise stated, always assume an interest calculation is based on the compound interest principle. We will return to the specifics of compound interest calculations later in this chapter when we discuss TVM computational factors.

3.5 Cash flow diagrams

Cash flow diagrams simplify the solution of many TVM problems. Very "wordy" problems can be slightly confusing and graphic representation helps the analyst identify what is known and unknown in the problem. Given that a majority of TVM problems end up being "one equation with one unknown," the cash-flow diagram helps organize problem data. Figure 3.1 is a classic cash-flow diagram depicting a loan of amount P to be repaid with a lump-sum cash payment of F after 5 years.

There are several things to note about the cash flows and other items shown in the diagram:

- Cash can flow into or out of a project, and the direction of the arrows for cash flow is based on the analysis *frame of reference.* Adopting the perspective of the lender, then a loan of P is a disbursement and thus

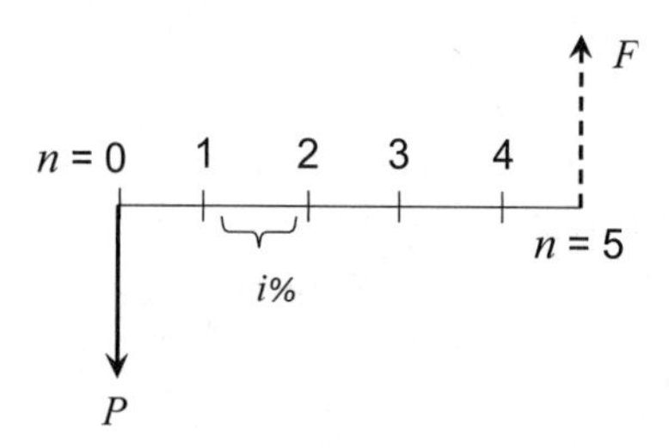

Figure 3.1: Cash-flow for diagram for TVM problems

the vector points downward in Figure 3.1. When the load is repaid by the amount F at the end of year 5, the funds are received by the lender and thus the vector points upward.

- Payments and activities are most likely to occur throughout an analysis period, but, for convenience, we adopt an *end of period* convention to correspond with the end of period interest computation procedure. If the granularity of a problem dictates, say, a weekly analysis, then it is just a matter of defining periods of performance and interest calculations on a weekly basis.
- The periods of performance are uniform and the interest rate stated must correspond with the period of performance. Thus, in Figure 3.1, if the overall project period is 5 years and calculations are to be performed on an annual basis, then the interest rate shown should be an annual interest rate. There are situations where modifiers to the nature of the interest rate are important (say, 8% nominal interest compounded quarterly), and that should be noted on the diagram in such a case. It helps to clarify situations in which the interest rate might be a corporate investment rate, not a bank rate. One should not be bashful at making the diagram as descriptive as possible, a self-contained story.
- Drawing a cash diagram is not just for novices. Actually, the cash flow diagram is a wonderful tool for communicating in a concise fashion, even to the extent where assumptions about the problem are visible and easy to question or change if a parametric study is warranted. The use of this diagram in financial analysis is as useful as the "free body diagram" quite natural to use in the study of mechanics of materials. It might seem silly to draw the cash flow diagrams for the simple problems at the end of this chapter, but the experience will be helpful in Chapter 4.
- We generally start the diagram time-line with N=0 to denote the beginning of the first period. N=1 is the beginning of the second period, and so on. Thus the end of period 1 (at N=1) also marks the beginning of period 2. Cash flows are placed at the end of each month.

There is the practical matter that some TVM problems involve a degree of complexity that warrants the use of a computer spreadsheet solution. Still, situations involving pre-payments, partial payments, gradients, deferred payments, and tax scenarios will be useful to sketch out on a cash flow diagram for clarity in visualizing the problem, even if computationally it is more convenient to perform detailed calculations with a spreadsheet.

3.6 Engineering economy computational factors

In the "old days," *engineering economy computational factors* provided simplicity in calculations; tabulated values of these factors expedited problem solutions. With the popularity of spreadsheets it seems questionable that time should be spent on the concept. While it is true that today's "real" problems involve a degree of complexity best solved with a computer, there is no question about the value derived from insight into the problem at hand when the framing of the problem and its initial layout are based on fundamentals. Noting that "the purpose of computing is insight, not numbers," the analyst will find it useful to bound the upper and lower limits of a problem quite quickly using computational factors even if the ultimate solution demands granularity best provided with a spreadsheet. Let's proceed and demonstrate by example.

Earlier, we linked the future lump-sum payment F to the present value P through the formula

$$\begin{aligned} F &= P(1+i)_{Y1}(1+i)_{Y2}(1+i)_{Y3} \\ &= P(1+i)^3 \end{aligned}$$

which, for the more general case of N periods, is quite simply

$$F = P(1+i)^N \tag{3.1}$$

in which the quantity

$$(1+i)^N \tag{3.2}$$

is known as the *single payment compound amount factor* (SPCAF), enabling one to readily calculate F given the parameters P, i, and N. Values of the SPCAF can be found in the appendix of many, many textbooks and handbooks. Table 3.1 provides just a small sample of tabulated values for various values of i and N. Functional symbol standardized convention is given in the form

$$(F/P, i\%, N) = (1+i)^N \tag{3.3}$$

If we wanted to know now what the value P of a future payment F is to us, then we simply invert Equation 3.1 to read

$$P = \frac{F}{(1+i)^N} \tag{3.4}$$

from which the *single payment present worth factor* has the functional symbol standardized convention given by

$$(P/F, i\%, N) = \frac{1}{(1+i)^N} \tag{3.5}$$

n	F/P	P/F	A/F	A/P
Factor table for $i = 6\%$				
1	1.060	0.9434	1.0000	1.0600
2	1.124	0.8900	0.4854	0.5454
3	1.191	0.8396	0.3141	0.3741
4	1.262	0.7921	0.2286	0.2886
5	1.338	0.7473	0.1774	0.2374
6	1.418	0.7050	0.1434	0.2034
7	1.504	0.6651	0.1191	0.1791
8	1.594	0.6274	0.1010	0.1610
9	1.689	0.5919	0.0870	0.1470
10	1.791	0.5584	0.0759	0.1359
20	3.207	0.3118	0.0272	0.0872
30	5.743	0.1741	0.0126	0.0726
40	10.286	0.0972	0.0065	0.0665
50	18.420	0.0543	0.0034	0.0634
60	32.988	0.0303	0.0019	0.0619
100	339.302	0.0029	0.0002	0.0602

Table 3.1: Sample computational factors for $i = 6\%$.

Equations 3.1 and 3.4 involve simple calculations and are the key single payment formulas applicable to a wide variety of situations. Although it is likely that, in practice, engineering managers might possibly calculate the factors from the basic equations (probably buried in a spread sheet), we note that many textbooks (as well as the National Council of Examiners for Engineering that administers the FE exam) provide tables like the one shown in Figure 3.1. Such tables become much more useful when annuity calculations are involved, as we discuss shortly.

Notice the general form of the factor

$$\left(\frac{X}{Y}, i\%, N\right) \tag{3.6}$$

Whereby the letter X represents what you want to find and the letter Y represents what is known. This can be read as "Find X given Y at the nominal interest rate of $i\%$ and a period N." The correct form of the factor is easy if we visualize the algebraic concept

$$X = \not{Y}\left(\frac{X}{\not{Y}}, i\%, N\right) \tag{3.7}$$

Let's apply this to the situation where $X = A$. Although P and F are quite useful, one of the most popular "fab-five" parameters in cash-flow diagrams is A, an annuity or series of consecutive, equal, end-of-period amounts; two reasons for such popularity are (a) the preference to pay for things "on terms" and (b) some moderately complex decision-making analyses are simplified when an equivalent uniform annual cost is computed (more later). For now, the cash-flow diagram is as shown in Figure 3.2.

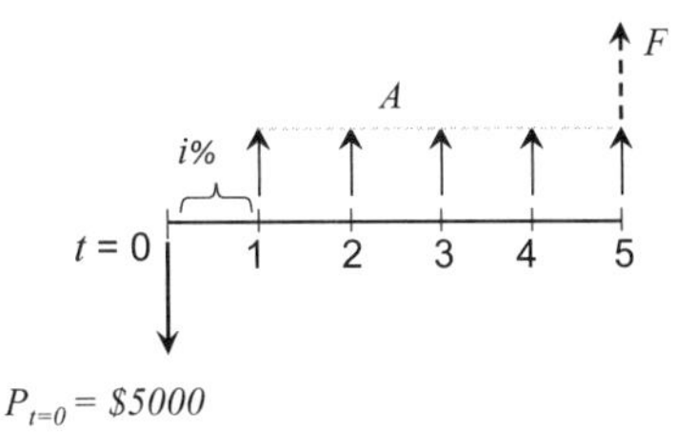

Figure 3.2: Cash-flow diagram for TVM problems with annuity factors.

If we consider the application of Equations 3.4 to Table 3.2 with A replacing F, then

$$P = \frac{A}{(1+i)^1} + \frac{A}{(1+i)^2} + \frac{A}{(1+i)^3} + \frac{A}{(1+i)^4} + \frac{A}{(1+i)^5} + \ldots \tag{3.8}$$

which reduces to

$$P = A\left\{\frac{(1+i)^N - 1}{i(1+i)^N}\right\}, i \neq 0 \tag{3.9}$$

and for which the uniform series, present-worth factor is now expressed by

$$P = A\left(\frac{P}{A}, i\%, N\right) \tag{3.10}$$

From Equation 3.9 the equivalent uniform annual cost for an investment P is simply

$$A = P\left\{\frac{i(1+i)^N}{(1+i)^N - 1}\right\} \tag{3.11}$$

Table 3.2 summarizes several combinations of P, F, and A, along with the common factor description and common notation. A more comprehensive set of factors and descriptions is available from the NCEES website `www.ncees.org`; many students will have already registered with the site as part of the process of becoming a Professional Engineer. In case you've skipped or don't intend to take the NCEES/FE, then it is recommended you register anyway to avail yourself of the numerous sample worked problems and test problems on TVM. Working NCEES problems is a useful way to build competency.

Section 3.6.1 and Section 3.7 are *very short but very important*; most cases where students have a failed understanding of advanced concepts in Chapter 4 and Chapter 5 can be traced back to weak competence in TVM basics. In particular, the difference between nominal and effective interest rates will confound understanding bond problems (jump ahead to Section 4.7.4, p. 126).

An excellent way to develop a good working understanding of TVM problems is to do *every* problem using *both* the computational factor method and by spreadsheet. First, a spreadsheet can help check the problem you worked by hand, or conversely, the hand calculation will help you figure out the idiosyncrasies of some of the built-in spreadsheet functions.

Second, as you build an understanding of how spreadsheet templates can be constructed, you may want to develop your own "toolbox" of spreadsheets within a workbook. Explore the spreadsheets that accompany this book to get an idea of how you may want to frame your own library.

Third, it will initially take longer to solve problems if you do them using computational factors and with a spreadsheet, but soon you'll start to develop an intuitive sense of when a spreadsheet is too much hassle or a computational factor may be too limiting (gradient calculations always come to mind). Build proficiency with easy problems and you will find that the more (seemingly) complicated problems are actually quite manageable.

Factor Name	Converts	Notation	Formula
Single Payment Compound Amount	to F given P	$(F/P, i\%, n)$	Eqn 3.12
Single Payment Present Worth	to P given F	$(P/F, i\%, n)$	Eqn 3.13
Uniform Series Sinking Fund	to A given F	$(A/F, i\%, n)$	Eqn 3.14
Capital Recovery	to A given P	$(A/P, i\%, n)$	Eqn 3.15
Uniform Series Compound Amount	to F given A	$(F/A, i\%, n)$	Eqn 3.16
Uniform Series Present Worth	to P given A	$(P/A, i\%, n)$	Eqn 3.17

Table 3.2: Notations and shorthand notations for commonly used discrete compounding interest factors.

Equations for the discrete compounding interest factors are

$$F = P\left(\frac{F}{P}, i\%, n\right) = P \ (1+i)^n \tag{3.12}$$

$$P = F\left(\frac{P}{F}, i\%, n\right) = F \ (1+i)^{-n} \tag{3.13}$$

$$A = F\left(\frac{A}{F}, i\%, n\right) = F \ \frac{i}{(1+i)^n - 1} \tag{3.14}$$

$$A = P\left(\frac{A}{P}, i\%, n\right) = P \ \frac{i(1+i)^n}{(1+i)^n - 1} \tag{3.15}$$

$$F = A\left(\frac{F}{A}, i\%, n\right) = A \ \frac{(1+i)^n - 1}{i} \tag{3.16}$$

$$P = A\left(\frac{P}{A}, i\%, n\right) = A \ \frac{(1+i)^n - 1}{i(1+i)^n} \tag{3.17}$$

3.6.1 Simple worked problems

There are three classic, quite simple types of TVM problems, gaining a full understanding of the simple problems is essential as a step to building on the understanding of more complicated problems. In this situation, there is no better way to get practice than to get a copy of the NCEES Study Materials (`www.ncees.org`) and work through problem after problem until the patterns of each question are clear and your activities become routine.

Sample Problem. Simple present worth (PW) problem, find P given F, N, i.

Joyce would like to buy a new car that is expected to cost $14,999 when she graduates from college in 5 years. Today, she is holding $9,000 in gifts from high school graduation. Will she have enough money if she deposits the money in a savings account that pays 12% compounded annually? What if the savings account only had an interest rate of 4%?

Let's begin with the simple cash flow diagram:

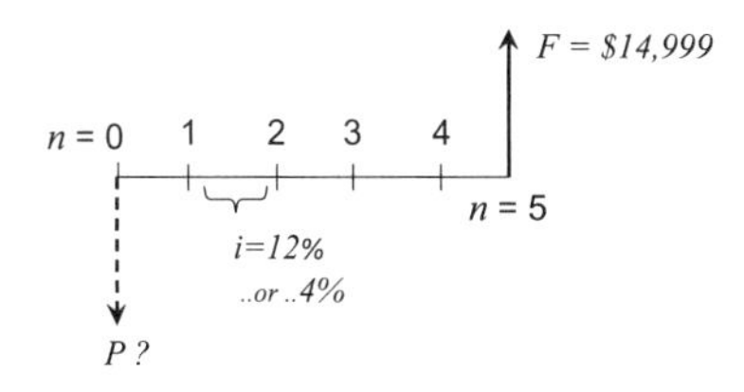

from which we select the single payment present worth factor, yielding

$$P = F(P/F, i\%, N) = \$14,999(P/F, 12\%, 5) = \$14,999(0.5674) = \$8,510.43$$

$$P = F(P/F, i\%, N) = \$14,999(P/F, 4\%, 5) = \$14,999(0.8219) = \$12,328.08$$

Clearly, Joyce will have enough to deposit if the prevailing interest rate $i = 12\%$, but not if $i = 4\%$.

Sample Problem. Simple future worth problem, find F given P, N, i.

Joyce's uncle Louis suggests, "It is ridiculous to save to buy a new car that will depreciate 30% once you drive it off the lot. You are better putting the $9,000 in a retirement account. Even at 10.0% I estimate you'll be a millionaire if you keep the money in the bank for 25 years!" Is Uncle Louis correct?

$$F = P(F/P, i\%, N) = \$9,000(F/P, 10\%, 25) = \$9,000(10.835) = \$97,515.00$$

Well, so maybe not a *million* dollars, but still, having nearly $100K is not a bad element to a retirement portfolio.

Sample Problem. Uniform series compound worth, find F given A, N, i.

Despite the momentary excitement about the possibility of using the $9,000 to be a millionaire, Joyce's father explains, "Actually, Louis had part of the story right, but more typically you *can* become a millionaire if you deposit just a little in the bank every year. For instance, suppose rather than going away on a trip every spring, you took the $1,000 and deposited that into the bank. I think that in that case you would be worth a million!" So now,

$$\begin{aligned} F &= A(F/A, i\%, N) = \$1,000(F/A, 10\%, 25) \\ &= \$1,000(98.347) = \$98,347.00 \end{aligned}$$

Well, once again, Joyce is a bit below the mark! Still the idea of putting aside $1,000 a year does not seem "too big a deal," and Joyce is curious what savings it *would* take to become a millionaire. She decides to let the time value of money work for her, and at 18 years of age she's willing to wait till she retires at 63 to "collect." In this case we simply are looking for the equivalence of a series of uniform payments and future lump sum:

$$\begin{aligned} A &= F(A/F, i\%, N) = \$1,000,000(A/F, 10\%, 45) \\ &= \$1,000,000(0.00139) = \$1,390 \end{aligned}$$

What a difference time makes! By placing "spring break" money every year into an account paying 10% compounded annually, Joyce *will* become a millionaire by the time she retires!

3.7 Nominal and effective interest rates

Nominal and effective interest rates are often the source of conceptual confusion and also the root of many TVM calculation errors. A distinction between the two is essential and must be fully understood, not just memorized.

To begin, let's reflect on the adjective used to describe the interest rate. Most engineers (hoping to see how quickly they can compute the wrong answer) simply *lock in* on the word "interest" and immediately look for any interest rate number to "plug in" and compute. As a result, they may be oblivious to the *implied* meaning, *specific* use of terms, and *context* that define the complete meaning of the phrase - I mean, we are engineers used to much more complicated calculations (right?) and, after all, this *is only* finance, involving the simplest of equations, so *how hard can it be*? Let's examine closely the term *nominal*. As seems often the case, a dictionary[1] definition can help us, and provides insight about what "nominal" means:

nominal ... 2. Existing in name only, not real or factual

Hopefully your attention is drawn immediately to the part that says *not real or factual*! Think of the *nominal* interest rate as a representative interest rate that *may or may not* correspond with the actual interest you pay or earn. So why even have the term around to confuse us? We have to understand that the nominal interest rate is based on *simple interest* calculations (*not* compounded), and so if we have been quoted a 12% nominal yearly interest rate, this is the same as a 1% nominal monthly interest rate. It should be clear: simply remember that nominal interest rate is synonymous with simple interest calculation. And there is some convenience, too, if we note that to change from an annual to any other period, we *simply divide* the annual interest rate by the number of periods; that is, monthly is 1/12 annual, quarterly is 1/4 annual. Not complicated, but we've traded off simplicity for accuracy. We also have to remember that whether it is *explicitly stated or not*, the basis period is *annual.*

What can be perplexing and might add a little confusion is that many states and countries have consumer protection laws that mandate a lender disclose an *annual percentage rate, (APR)*. And while the protection laws may have standardized nomenclature, this does not mean finance was made any clearer, as an APR annualized interest rate is not based on *simple* interest (all payments made are based on *compound* rates!). You would have thought that the consumer protection laws helped reveal the actual rates and fees, and in a very indirect way they do, as long as you understand the meaning of the definitions!

As an example, let's assume that the bank offers a credit card deal in which you have a monthly interest of 2.4% that has $APR = 2.4\% \times 12 = 28.8\%$ (obviously your credit score is below 690). Anyway, you might be able to justify this until you realize the 2.4% is compounded monthly, and that if you have a beginning credit card balance of \$4,500, then at the end of the year you will owe

$$F = P(F/P, i\%, N) = P(1+i)^N = \$4,500(1+0.024)^{12} = \$5,981.52 \quad (3.18)$$

Which, of course means that you accrued \$1,481.52 in interest owed over the course of the year. It is a simple calculation to realize that the ratio of interest to principal over the course of the year is *not* the nominal 28.8% but instead 32.9%:

$$i_{effective} = \frac{\$1,481.52}{\$4,500.00} = 0.3292 = 32.9\% \quad (3.19)$$

This example illustrates that there is a difference between the *nominal* interest rate that is quoted and the *actual* or *effective* interest rate that you pay. It is not that one is wrong and the other is right, but that there is a *difference in terminology* for which it is important to be able to convert between so you know what your actual payments are. The good news is that convention across textbooks and disciplines is quite standardized and the following can be expected:

- **Nominal interest rate**, denoted by the symbol r, based by default on an *annual* period, which does *not* consider the impact of compounding.
- **Annual percentage rate**, denoted by the symbol APR, originating with consumer protection laws and based by default on an *annual* period (but could be for any period, say, monthly, as may be the case with consumer credit), which also does *not* consider the impact of compounding.
- **Effective interest rate**, denoted by the symbol i, or sometimes i_{eff} based on a *stated* period, which *does* consider the impact of compounding and is needed to compute the actual sum of interest one pays.

Many textbooks and handbooks derive a mathematical relation between *nominal interest rate* and *effective annual interest rate* i_{eff} that looks similar to the following:

$$i_{eff} = \left\{1 + \frac{r}{m}\right\}^m - 1 \tag{3.20}$$

where

- r – nominal interest rate per year
- m – number of compounding sub-periods per year

As an example, our earlier problem Equation 3.19 can be expressed in terms of Equation 3.20 as

$$i_{eff} = \left\{1 + \frac{0.288}{12}\right\}^{12} - 1 = 0.3292 = 32.92\% \tag{3.21}$$

Because the effective interest rate shows you the *actual* return from a given interest rate and compounding period and is the *true* cost of financing, the *effective annual interest rate* i_{eff} is always the interest rate that should be used in calculations. In hindsight, we see that the very simplistic calculations in Section 3.6.1 did not make much of a fuss describing what i was in the calculations; we just used the "interest rate" given and proceeded with calculations. Now, of course, it is clear that all calculations were for periods of 1 year, so in that case $i_{eff} = i$.

There are several basic ways that interest rates can be expressed, and problem statements might be quite vague, but in the absence of other guidance it may be useful to recognize descriptive patterns. We provide two general rules of thumb in the event that interest and compounding attributes are not explicit for a problem statement.

Type 1 If no compounding period is specifically stated, then the compounding period is the same as the stated interest period and the interest rate is assumed to be an *effective* interest rate.

- When stated "$i = 12\%$ per year"
  then assume "$i = 12\%$ *effective* per year, compounded yearly"
- When stated "$i = 4\%$ per quarter"
  then assume "$i = 4\%$ *effective* per quarter, compounded quarterly"

Type 2 A compounding period *is* specifically stated, so the compounding period is the same as the stated period. If the interest rate has no adjective to specify type, then always assume a *nominal* interest rate prevails.

- When stated "$i = 6\%$ per quarter, compounded monthly"
  then assume "$i = 6\%$ *nominal* per quarter, compounded monthly"
- When stated "$i = 24\%$ per year, compounded semi-annually"
  then assume "$i = 24\%$ *nominal* per year, compounded semi-annually"
- When stated "$i = 12\%$ per year, compounded monthly"
  then assume "$i = 12\%$ *nominal* per year, compounded monthly"

Table 3.3 underscores the power of compounding on effective interest rates.

Nominal rate, r	Semi-annually $m = 2$	Quarterly $m = 4$	Monthly $m = 12$
1	1.003	1.004	1.005
6	6.090	6.136	6.168
12	12.360	12.551	12.683
24	25.440	26.248	26.824

Table 3.3: Impact of compounding on effective interest rates.

3.8 Summary

The ability to fluently speak about the "fab five" and *quickly* determine which of the compound factors applies to a problem is essential prior to moving on to the next chapter. It is very difficult to grasp the subtleties of the application of TVM to realistic situations if the current chapter presents difficulties. Clever narratives are no substitute for practice and your first-hand experience. The problems on the following pages are just a sample of the general type of problems you should be able to solve; it is recommended that you examine the National Council of Examiners for Engineering website (that administers the EIT exam) and download the sample exams and problems they have to offer. You know you have mastered the basics when

- All the problems begin to have the same "look."
- The drawing of the cash-flow diagram takes less than 12 seconds after reading the problem.
- The choice of nominal or effective interest rates does not make you pause.
- You find that a "good table" actually is twice as fast as using your cell phone for calculations.

3.9 Problems to work

Problem 3.1
You just received an unexpected bonus at work of $10,000. While you plan on using a bit of it to celebrate by purchasing all the accounting books you ever dreamed of, how much should you deposit in an account earning 6% per year if you'd like to have $10,000 in the account in 10 years?

Problem 3.2
Your colleague is excited about your good fortune (Problem 3.1) at work, but she only got the promise of a watch or $300 cash. You convince her that she will be better in the long run by just taking the money and trying to find an account earning 12% per year to invest her money. How much will she have accumulated in this account after 25 years? What if she waits 40 years?

Problem 3.3
What is the future equivalent of $1,000 invested at 8% simple interest for 3 years?

Problem 3.4
What lump-sum amount of interest will be paid on a $10,000 loan that was made on 1 June 2011 and repaid on 1 September 2015, with ordinary simple interest at 10% per year?

Problem 3.5
You borrow $495 from your brother-in-law to buy an iPad and agree to pay it back when you get your tax refund (6 months away). Because you are family, but not trusted, you are being charged simple interest at the rate of 6% per month. How much will you owe after 6 months?

Problem 3.6
You invest $17,000 in a mutual fund recommended by a fellow Dallas Mavericks sports fan. The fund is known to have "highs and lows," but your friend virtually guarantees you will beat the market by 2% and earn 12% per year in the "long run." How much should your investment be worth in 20 years?

Problem 3.7
Aaron loans Victoria $10,000 with interest compounded at a rate of 8% annually. How much will Victoria owe Aaron if she repays the entire loan at the end of five years?

Problem 3.8
Approximately how long does it take to quadruple an investment of $1,000 when the interest rate is 15% per year?

Problem 3.9
Suppose you make 15 equal annual deposits of $1,000 each into a bank account paying 5% interest per year. The first deposit will be made one year from today. How much money can be withdrawn from this bank account immediately after the 15th deposit?

Problem 3.10
Jim makes a deposit of $12,000 in a bank account. The deposit is to earn interest annually at the rate of 9% percent for seven years. (a) How much will Jim have on deposit at the end of seven years? (b) Assuming the deposit earned a 9% rate of

interest compounded quarterly, how much would he have at the end of seven years? (c) In comparing parts (a) and (b), what are the respective effective annual yields? Which alternative is better?

Problem 3.11
John is considering the purchase of a lot. He can buy the lot today and expects the price to rise to $15,000 at the end of 10 years. He believes that he should earn an investment yield of 10% annually on this investment. The asking price for the lot is $7,000. Should he buy it? What is the annual yield (internal rate of return) of the investment if John purchases the property for $7,000 and is able to sell it 10 years later for $15,000?

Problem 3.12
An investor can make an investment in a real estate development and receive an expected cash return of $45,000 after six years. Based on a careful study of other investment alternatives, she believes that an 18% annual return compounded quarterly is a reasonable return to earn on this investment. How much should she pay for it today?

Problem 3.13
A loan of $50,000 is due 10 years from today. The borrower wants to make annual payments at the end of each year into a sinking fund that will earn interest at an annual rate of 10 percent. What will the annual payments have to be?

Problem 3.14
The Dallas Development Corporation is considering the purchase of an apartment project for $100,000. They estimate that they will receive $15,000 at the end of each year for the next 10 years. At the end of the 10th year, the apartment project will be worth nothing. If Dallas purchases the project, what will be its internal rate of return? If the company insists on a 9% return compounded annually on its investment, is this a good investment?

Problem 3.15
Consider an investment that will pay $680 per month for the next 15 years and will be worth $28,000 at the end of that time. How much is this investment worth to you today at a 5.25% discount rate?

Problem 3.16
You currently owe $18,000 on a car loan at 9.5% interest. If you make monthly payments of $576.59 per month, how long will it take you to fully repay the loan?

Problem 3.17
You have just borrowed $10,000 and will be required to make monthly payments of $227.53 for the next five years in order to fully repay the loan. What is the implicit interest rate on this loan?

Problem 3.18
Your uncle has given you a bond that will pay $500 at the end of each year forever into the future. If the market yield on this bond is 8.25%, how much is it worth today?

Problem 3.19
What monthly payment would a college senior make to pay off a used car loan of $2,000 at 12% by the end of the year?

Problem 3.20
You are considering financing a new car which cost \$51,300 with an amortized loan. The nominal rate is 2.9% per annum, the term of the loan is 6 years, and you will make monthly payments. How much will each payment be?

Problem 3.21
You want to retire in 30 years. You are starting to invest in a growth income fund that promises an ambitious rate of 15%. You can put in \$200 per month. How much will you have in 30 years?

Problem 3.22
Suppose that you make an investment that will cost \$1,000 and will pay you interest of \$100 per year for the next 20 years. Then at the end of the 20 years, the investment will pay \$1,500. If you purchase this investment, what is your compound average annual rate of return?

Problem 3.23
A biologist decided that she wanted to have \$54,267.89 in her account in 10 years, and she found a bank which compounded monthly at 6%. What are her monthly payments to achieve her goal of \$54,267.89?

Problem 3.24
How many years will it take for your savings account to accumulate \$1,000,000 if it pays 4% interest per annum compounded semiannually and you deposit \$10,000 every 6-months at the end of the 6 month period?

References

[1] William Morris. *The American Heritage Dictionary.* Houghton Mifflin, 1978.

Chapter 4

Applications of the Time Value of Money

4.1 Learning objectives

Previous work developing a basic understanding of the time value of money (TVM) is now applied to decision-making about a single project (choosing among *multiple* project options is addressed in Chapter 5). The current chapter frames the basic aspects of most contemporary methods for determining project desirability. To fully appreciate the basic TVM application strategies presented here, the reader should already be fluent in the use of the concepts and tables (discussed in Chapter 3) that provide normalized TVM factors as a function of N and i, for single, uniform, and non-uniform payments. Bond analysis sometimes seems like a "dated" topic, but we will see how central a role bonds play in project financing - there is so much more in the way of policy that has bearing on investment decisions. Studying this chapter will result in improved analysis skills, measured by your ability to perform the following tasks:

1. Compare and contrast the primary TVM tools:
   (a) Net present value (NPV) calculations
   (b) Equivalent uniform annual worth (EUAW) calculations
   (c) Rate of return calculations
   (d) Payback Period calculations
2. Describe considerations for setting a Minimum Attractive Rate of Return (MARR).
3. Distinguish between internal rate of return (IRR), MARR, and other rate-of-return methods.
4. Perform simple bond analysis and describe how and why bond "prices" fluctuate in the market.

4.1.1 Supplementary reading

Much has been written about the applications of TVM, frequently under the categories of capital budgeting, decision-making, or business investment decisions. It helps broaden your perspective to review a few documents (even if you just skim through them) in parallel with the current work and to look at a few articles from the press to better appreciate the multiple facets and potential complexity of TVM applications. Over time, it can be surprising how relatively few concepts are really needed to illuminate and guide "everyday" decisions. That pragmatic perspective prevails in this chapter.

Take a moment to explore the following links and print something of interest to read off-line, even if just to understand the relevance of the current chapter to real-world situations.

1. FORBES (search newsfeed for IRR): `www.forbes.com`
2. Investors Business Daily editorials: `www.investors.com/editorial/`
3. And how could we not include: `www.wsj.com`

Passive approaches to studying applications of time value of money can be dull (i.e., simply reading a book) unless you deliberately work to link and compare concepts (like MARR) to the newspaper, internet, or other media. We understand that a key component of critical thinking skills is "viewing a problem from many perspectives," and this applies to TVM applications, too. After reading this chapter, try to gain a slightly different perspective (and test your competencies) by reviewing, for instance, the work of Newnan [1], Chapter 16, "Business Investment Decisions." An "a-ha" moment for study is when you can decipher a common pattern of analysis and decision-making paths from a *variety* of sources of information. Documents cited here are just a sample! Chapter 9 of this book will discuss capital budgeting in more detail.

We would be living under a rock if it was not noted that structural changes in education have resulted in "course-ware" sites with freely available information. Content dissemination through the internet is not without some controversy, with purchased content (like this book!) an increasing challenge for content creators. Still, for topics like TVM that have been around for *decades*, content variety can facilitate learning.[1]

Once you get a grasp of TVM application concepts, it helps to reach out to peer-reviewed articles that can stretch your thinking a bit further. Two I recommend are: Murphy et al. [2], "Enhancing Commercial outcomes from R&D" and Miller and Ireland [3], "Intuition in Strategic Decision-Making: Friend or Foe in the Fast-Paced 21st Century?" There are, of course, many, other articles in the literature you could choose from.

[1]For instance, tutorials written by Michael Roberts, Wharton School of Management, touch on key TVM application concepts; see: `http://fnce.wharton.upenn.edu/people/faculty.cfm?id=973`. His teaching notes are posted under his "Teaching" links; this is just an example of what can be found on many university sites and an internet search would reveal dozens more.

4.1.2 Questions to consider

A technology leader has to explore tactical and strategic moves for product (or service) investment within the constraints of the resources allocated by the organization. More often than not, this simply means the leader has to (draw on Shakespeare and), ask, "to invest or not to invest?" Decisions have to be made that trigger subsequent project activities whose outcome (success) occurs in the future. Although the manager might feel that a post-investment decision is subsequently a game of "wait and see," experience shows that preparatory work and due diligence in validating data build the confidence that the "right" choice was made with the best available data at the time action was required. Many times a decision might seem unique to the company, but it is not necessarily unique to the industry itself (recall the "outside view" of Section 2.3). In such a case, risk is mitigated by precedent (other people and cases where a similar decision had to be made), and it helps to know the answer to a few simple questions:

1. How can I characterize the benefit, if any, of an investment?
2. How much is the benefit, and does it meet the expectations of my organization?
3. What are the common metrics for investment expectations?
4. Does it make sense to "do nothing" as an option?
5. What tools should I use to make predictions about the future?
6. How reliable are my predictions?

These are insightful questions that a technology leader should (a) have the critical thinking skills to probe individually, (b) draw on networking skills to seek industry-specific insight, and (c) have the emotional intelligence (see Goleman et al. [4]) to discuss with senior leadership.

4.1.3 Five chapter concepts

Our chapter learning objectives (recall page 93) are slightly broad in scope, so in this section we provide an "elevator pitch" for each of the five major chapter concepts.

Net present value

NPV characterizes the "net sum" of the PV of a series of cash inflows and outflows over a period of time. NPV is used to estimate whether a series of net cash flows is positive (favorable, $NPV > 0$) or negative (unfavorable $NPV < 0$) for a project. Usually $NPV > 0$ is preferred as this is a "good" investment.

Equivalent uniform annual worth

EUAW serves a slightly different purpose than NPV. Whereas NPV provides the cost *in today's dollars* of all expenses incurred over the life of a project, EUAW provides insight on the *cash flow requirements* of a project. NPV provides the value we would pay if we paid all expenses *up front*, but we know as a practical matter that many expenses are paid monthly or annually. Cash flow may be more critical to a company than total investment, in the same way you can be profitable, but not solvent. EUAW is delightfully simple (in comparison to NPV) when comparing projects with different life expectancies.

Rate of return

Return on investment is a common measure of economic desirability, and the interplay between the *minimally acceptable rate of return* (MARR) for management and the corresponding value of the NPV often takes center stage. We examine the *internal rate of return* (IRR) for the special case NPV = 0 and discuss the implications of computing an IRR that is significantly higher than generally accepted investment returns.

Payback period

Potentially - but not always - one of the easily computed project investment parameters, *payback period* is a method of expressing the desirability of a project, often expressed as the length of time that it takes for an investment in a project to be recovered. Essentially, you are answering the question "How long will it take for this investment to pay for itself?" There are many variations on this basic theme, and in this chapter we illustrate where outcomes can be misleading.

Bond analysis

Bond analysis (however old-fashioned it may seem) is also discussed in this chapter to assist in remarking on the difference between investing and financing a project. Bond calculations are a classic use of TVM for public project financing, offering a non-intuitive trade-off between PV and interest rate that everyone should have the chance to explore. Bond investments set a financial "floor" against which the opportunity cost of pursuing project options can be measured.

4.2 Net present value

Section 3.2 outlined the critical need for *equivalence* of information for project decision-making. This is often a matter of converting all data relevant to an analysis in equivalent form so that project analysis outcomes are meaningful. As mentioned earlier, engineers routinely exercise similar concepts for computing equivalent physical units for technical design problems; the current section intends to illustrate TVM methods to harmonize time-dependent data in application to net present value (NPV) problems. We will see shortly that we are simply finding the "net sum" of the PV of a series of cash inflows and outflows over a project period. NPV estimates whether a series of net cash flows is favorable ($NPV > 0$) or unfavorable ($NPV < 0$) for a given investment situation.

Three types of information are involved in the NPV calculation:

1. **Cash flow** - Sums of money that represent either a positive cash flow (benefit, revenue, or income as a result of an investment) or a negative cash flow (usually the initial cash investment to acquire the asset or get the project in place).

2. **Project duration** - Cash flow events that occur at different points along a time-line spanning the *project period* or *project duration* are often expressed as $t = N$ years.
3. **Cost of capital** - An opportunity cost exists for the use of funds today (expense) over another project or the receipt of funds (income) in the future, expressed here in terms of an interest rate $i\%$.

Present worth or *present value* analysis is fundamentally an exercise in conversion; a sum of money to be spent or earned in the future ($t = N$) is converted into an equivalent value at the present time ($t = 0$). Typical calculations convey that money to be received far in the future has an *equivalent* value *today* less in value than the future value. This can be a little confusing, but the promise of money in the future is eroded by the risk that the returns will be as predicted; there is a cost to waiting for future gains compared to having that same sum of money to use today. Promises can be abandoned and the purchasing power of money erodes over time, so the present value is a discounted future value. The idea of *discounting* was presented through a few simple problems in Section 3.6.1, the present worth method of future cash flows being synonymous with the *discounted cash flow* methodology. Project contributions are positive or negative on a cash-flow diagram and are more easily understood as either a benefit or a cost. Many projects are very similar in form, typically involving less than a dozen parameters. Table 4.1 illustrates.

Some of the items in Table 4.1 require subjective decisions. For instance, the

Transaction	Value	Timing
Initial project investment, P	$3,000	t = 0
Expected duration of use, N	8 years	
Discount rate per period, i	10%	t = 1,2,...N
Annual revenue from use, A	$750	t = 1,2,...N
Salvage value of system, SV	$300	t = N

Table 4.1: Typical project investment scenario.

project investment discount rate, $i\%$, will have a significant impact on NPV. We often assume the interest rate to be the highest value we might obtain by investing activity (i.e., stocks) with the understanding that the higher the rate, the lower the computed PV; that is, the higher the interest rate, the less expensive the present value of the project. If interest is estimated too high, the PV is unnecessarily low; conversely, too low an interest rate results in an artificially attractive PV. Salvage value is also subjective and subject to change. Technological obsolescence (versus just "wear and tear") might result in scrapping a piece of high-tech electronics after 3 years that might once have had a perceived value equal to 10% of the purchase price at $N = 8$ years. If you can distill all project information in the form suggested by Table 4.1 and Figure 4.1, then the calculation procedure is quite straight-forward.

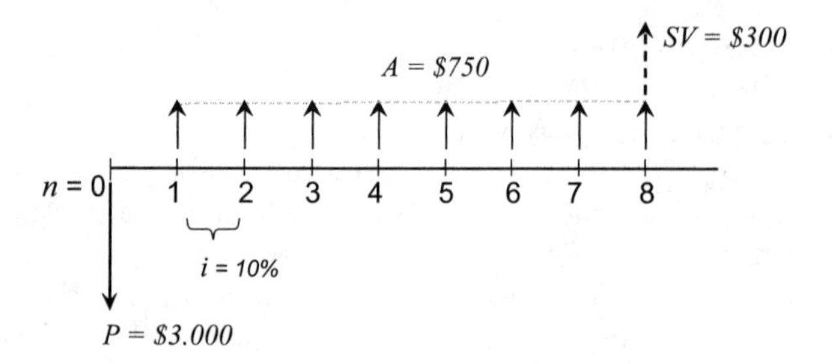

Figure 4.1: Cash-flow diagram for a simple project NPV set-up.

$$NPV = \sum_{n=1}^{N} PV_n = \left(\sum_{n=1}^{N} PV_n\right)_{\text{Revenue}} - \left(\sum_{n=1}^{N} PV_n\right)_{\text{Expenses}} \tag{4.1}$$

In practice the series of hand calculations implied by Equation 4.1 are often replaced with spreadsheets, especially when a large number of different parameters are involved. In the present work we outline the calculations for hand calculation to just illustrate the procedure. Approaches vary, but it makes sense to follow the same simple three-step PV pattern to be sure every component is included:

- Account for the initial investment, paying attention to the sign (expense vs. revenue).
- Translate all annuities A to the present.
- Move the one-time expenses and revenues to the present.

For the example above this procedure results in the following:

$$\begin{aligned} NPV = &- \$3,000 && \text{Present value of initial investment, } P \\ &+ \$750(P/A, i\%, N) && \text{Present value of the annuity, } A \\ &+ \$300(P/F, i\%, N) && \text{Present value of the salvage value, } SV \end{aligned}$$

$$\begin{aligned} NPV = &- \$3,000 \\ &+ \$750(P/A, 10\%, 8) \\ &+ \$300(P/F, 10\%, 8) \end{aligned}$$

And this leads to

$$\begin{aligned} NPV &= - \$3,000 + \$750(5.3349) + \$300(0.4665) \\ &= - \$3,000 + \$4,001.17 + \$139.95 = \$1,141.12 \end{aligned}$$

Observe that

$$\begin{aligned} \$750(5.3349) &= \$4,001.17 && << && 8 \text{ years x } \$750/\text{year} = \$6,000 \\ \$300(0.4665) &= \$139.95 && << && \$300 \end{aligned}$$

This result highlights that when a sum of money at a future time is converted into an equivalent value at the present time, the net result is a *discounting* of future values.

A common variation of the NPV calculation involves recurring expenses associated with a project that might offset revenue.

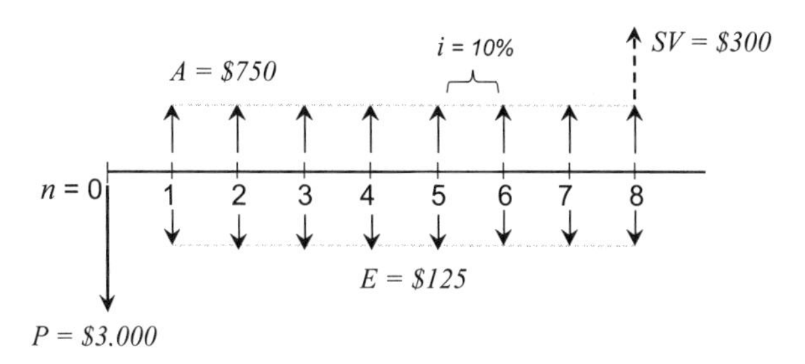

Figure 4.2: Cash-flow diagram for a simple project including annual expense related to the investment.

The inclusion is easy, and let's assume we need to add an expense line item in Table 4.1.

Transaction	Value	Timing
Annual expense to maintain, E	$125	t = 1,2,...N

Table 4.2: Project investment scenario: added expense.

This leads to a modest change in calculation procedure:

$$\begin{aligned} NPV = &-\$3,000 && \text{Present value of initial investment, } P \\ &+\$750(P/A, i\%, N) && \text{Present value of the annuity, } A \\ &-\$125(P/A, i\%, N) && \text{Present value of the maintenance expense, } E \\ &+\$300(P/F, i\%, N) && \text{Present value of the salvage value, } SV \end{aligned}$$

$$\begin{aligned} NPV = &-\$3,000 \\ &+\$750(P/A, 10\%, 8) \\ &-\$125(P/A, 10\%, 8) \\ &+\$300(P/F, 10\%, 8) \end{aligned}$$

$$\begin{aligned} NPV &= -\$3,000 + \$750(5.3349) - \$125(5.4439) + \$300(0.4665) \\ &= -\$3,000 + \$4,001.17 - \$681.12 + \$139.95 = \$459.00 \end{aligned}$$

We see the impact of maintenance, sometimes overlooked. Think beyond classic machinery maintenance, understanding that IT projects must include long-term annual license fees. Figure 4.3 illustrates project annuity delays or the situation where installation of an asset could take several years to occur. This is reasonable in an enterprise resource planning (ERP) project where the conversion of legacy data and installation of new software and databases could require several years of work prior to "go-live."

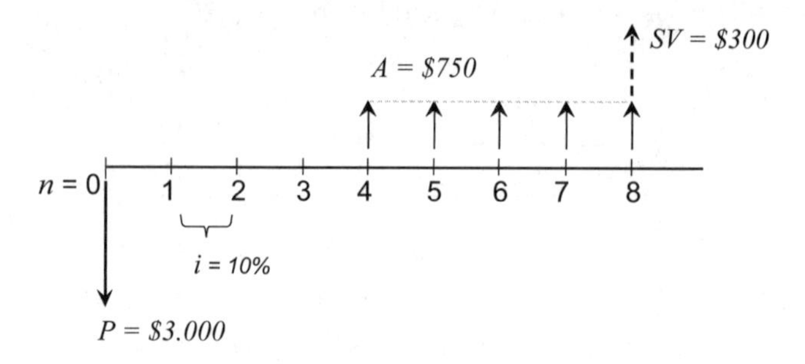

Figure 4.3: NPV for slight variation in scenario parameters.

There are two ways to adjust the NPV calculation for the delayed annuity, A, and we'll illustrate the easier of the two methods as it simply involves a two-step procedure to convert a future annuity to an equivalent present worth value. First, take a look at Figure 4.4.

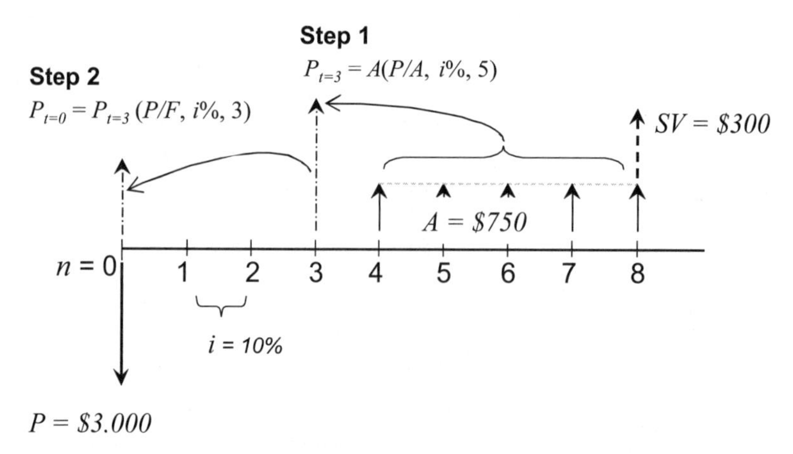

Figure 4.4: Two-step procedure for converting a future annuity to the present value.

In step 1, convert the annuity A into a single lump-sum value at year $t = 3$ using the present worth factor:

$$P_{t=3} = A(P/A, i\%, N) = \$750(P/A, 10\%, 5) = \$750(3.7908) = \$2,843.10$$

In step 2, we convert the lump-sum $P_{t=3}$ to its equivalent $P_{t=0}$ value by applying the present worth factor again:

$$\begin{aligned} P_{t=0} =& \ F(P/F, i\%, N) = \$2,843.10(P/F, 10\%, 3) \\ =& \ \$2,843.10(0.7513) = \$2,136.02 \end{aligned}$$

Now, the remainder of the calculation looks much the same as before:

$$\begin{aligned} NPV =& - \$3000 + \$2,136.02 + \$300(0.4665) \\ =& - \$3000 + \$2,136.02 + \$139.95 = -\$724.03 \end{aligned}$$

The negative consequences of the delay are vividly clear ($NPV < 0$) since the present value of the annuity has dropped by about 50%. This example is a simple case of non-uniform revenue streams, in which project revenue is either $A = \$0$ or $A = \$750$.

In the more complicated case where *every* value is different, the series of calculations is tedious and clumsy, not hard. For the most part the use of spreadsheets dominates practice (even when the calculations are easy) since the "book-keeping" of all the parameters becomes easier and less error-prone. The spreadsheet layout for a typical problem is illustrated in Table 4.3. Such spreadsheets are really not that difficult to create, whether you use Microsoft Excel or Open Office.

Interestingly, the *Excel* function NPV computes the PV of a cash flow, not the NPV. This is exaggerated somewhat for illustrative purposes in Table 4.3. Cell `C25` computes the PV of the cash flows in cells `C12:C19` with the interest rate from cell `C7`. The initial capital investment occurs at $t = 0$, so this cannot be included in the *Excel* NPV function, as shown in cell `C27`. By whatever method, it is clear that the present value $NPV = \$1,141.15 > 0$ of a one-time project provides wealth to the investors while the implementation delay of 3 years yields $NPV = -\$723.99 < 0$ and does not reach the "break even" point over the project life and is thus unfavorable. It is left to the reader to investigate adjusting the expected life of the project to a dozen years and computing that $NPV = \$340.72 > 0$.

Parametric analysis

Earlier, we pointed out that some of the items in Table 4.3 involved subjective decisions. In such we may want to explore parameter variances and "bracket" high and low values of baseline outcomes. Spreadsheets are handy for computing the best-case and worst-case scenarios. Consider Table 4.4.

Several "what-if" questions are easily explored with the format illustrated in Table 4.4. Three parameters are varied slightly to produce the values shown:

- Interest rate, with the worst-case being a slightly higher cost of capital than expected.
- Obsolescence of an asset (pessimistic) or the ability to extend useful life (optimistic).
- Variance in the benefit or revenue from the investment.

When several parameters are changed at once, the situation can be made quite favorable or bleak!

2	A	B	C	D
3				
4			**On-Time**	**Delayed**
5			**Go-Live**	**Go-Live**
6		Capital investment	$3,000	$3,000
7		Annual revenue	$750	$750
8		Asset life	8	8
9		Cost of capital	10.0%	10.0%
10		Salvage value	$300	$300
11				
12		Period	Cash Flow	Cash Flow
13		0	-$3,000.00	-$3,000.00
14		1	$750.00	$0.00
15		2	$750.00	$0.00
16		3	$750.00	$0.00
17		4	$750.00	$750.00
18		5	$750.00	$750.00
19		6	$750.00	$750.00
20		7	$750.00	$750.00
21		8	$750.00	$750.00
22		10		
23		Residual	$300.00	$300.00
24				
25		PV		
26		Principal	($3,000.00)	($3,000.00)
27		Revenue	$4,001.19	$2,136.06
28		Residual	$139.95	$139.95
29		**NPV**	**$1,141.15**	**-$723.99**

Table 4.3: Spreadsheet for equivalent present value of non-uniform project revenue.

	Pessimistic	Baseline	Optimistic
Capital investment	$3,000	$3,000	$3,000
Annual revenue	$700	$750	$750
Asset life	8	8	9
Cost of capital	11.0%	10.0%	9.0%
Salvage value	$300	$300	$300
Period	Cash Flow	Cash Flow	Cash Flow
0	-$3,000.00	-$3,000.00	-$3,000.00
1	$700.00	$750.00	$750.00
2	$700.00	$750.00	$750.00
3	$700.00	$750.00	$750.00
4	$700.00	$750.00	$750.00
5	$700.00	$750.00	$750.00
6	$700.00	$750.00	$750.00
7	$700.00	$750.00	$750.00
8	$700.00	$750.00	$750.00
9			$750.00
Residual	$300.00	$300.00	$300.00
PV			
Principal	($3,000.00)	($3,000.00)	($3,000.00)
Revenue	$3,602.29	$4,001.19	$4,496.44
Residual	$130.18	$139.95	$138.13
NPV	**$732.46**	**$1,141.15**	**$1,634.56**

Table 4.4: NPV for slight variation in scenario parameters.

4.3 Equivalent uniform annual worth

Equivalent uniform annual worth (EUAW) is based on the concept of annual worth (AW) (discussed in Section 3.6, p. 79). EUAW provides a slightly different perspective on cash flow streams than NPV. EUAW offers a popular and convenient procedure for the comparison of project alternatives, discussed more extensively in Chapter 5 for multiple project options. In the current section our goal is to introduce the EUAW calculation methodology for a *single project* study, serving as a stepping-stone to the more complex project comparison situations of Chapter 5.

Two features of EUAW are attractive for both single project analysis and multiple project comparisons:

- Whereas NPV provides the cost *in today's dollars* of all expenses *incurred over the life of a project*, EUAW provides insight on the annual periodic *cash flow* requirements of a project, simulating slightly more realistic project cash flow outlays.
- EUAW offers some computational simplicity (in comparison to NPV) when comparing projects with *different life expectancies.* Conceptually, the idea that unequal asset life can be "fixed" by the "repeating asset use" assumption may not always be a practical option for NPV calculations, and in that situation the EUAW is a preferred alternative.

Situations arise where project cash flow may be more critical in managerial decision-making than total investment. For instance, your enterprise might be profitable, but weakly (or not) solvent, and it is on this basis that some individuals rationalize leasing rather than buying a vehicle, again, reflecting possible sensitivity to cash flow. In principle the NPV provides a single value we would pay *today* if we paid all expenses *up front*, but we know as a practical matter that many expenses are apportioned over time and paid on a periodic monthly or annual basis. From a day-to-day operations perspective we may be interested in a better understanding of the convenient (though approximate) *equivalent periodic costs* of a project and this is where EUAW can play a role in facilitating decision-making.

EUAW is often part of calculations for project alternative comparisons (again, especially when the service life of the alternatives is different), but we should note it can be handy for single-project analysis, too. Simple capital investment projects assume machines will have an assigned *useful operational life* in terms of years of service. More complicated investments involve *systems* in which the components of the system may have a distinct probability of unequal useful service life. For instance, consider an IT network investment involving building wiring, network switching boxes, routers, backup power supply modules, and server hardware. It might be reasonable for the wiring and backup power supply infrastructure to have a well-vetted assumption of service life that equals the duration of the project life, but it is also reasonable that some parts of the system (say, the server and router hardware) require periodic maintenance, upgrades, or replacement.

EUAW can help in the development of internal pro forma financials for decision-making. We know a frequently occurring error in financial statements is the treatment and inclusion of operational expenses that might be more appropriately capitalized. In this case, the proper technique is to place such an asset on the balance sheet, then apportion the asset investment over time into the form of operational expenses through depreciation schedules (assuming, of course, the asset is used in business operations in some way to produce revenue). Now, for some internal project assessment, we may need to estimate a burdened "overhead rate" that reflects what each strategic business unit (SBU) can expect *on the average* to be charged for the life of the project. That is, we would like a simple *a priori* estimate of what the depreciation expense would be over the life of the project (for which a simple straight-line depreciation overlooks the time value of money). Items like expected overhead become a tricky part of large company infrastructure projects to estimate but are important in understanding whether it makes sense to internalize an activity or to outsource. For instance, a decade or so ago it would have seemed odd that a company might lease office furniture and carpeting for a new office, but more recently the "outsourcing" of such an asset utilization accompanied by, say, early contract termination clauses might make this a convenient way to minimize financial exposure on new SBU ventures.

Let's use a typical cash-flow diagram to clarify the steps involved in converting cash flows into EUAW. Table 4.5 illustrates some expenses and revenue considered as part of a company expansion plan.

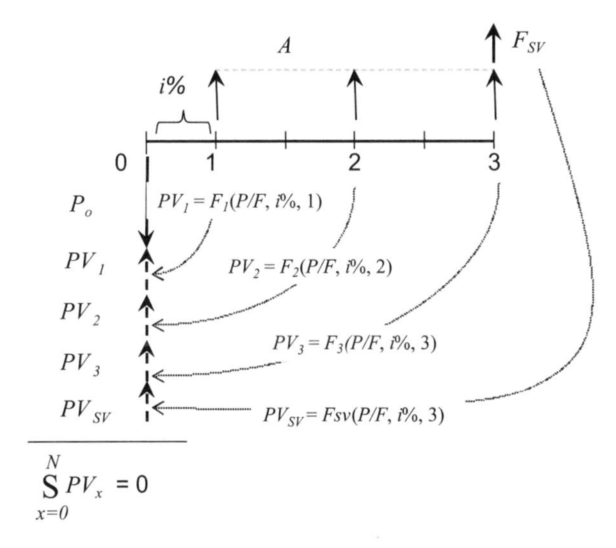

Figure 4.5: Cash-flow diagram for a building IT expansion plan.

EUAW can provide insight on how to reflect aggregate changes in cash flow in a simple way which elucidates the impact of variances on an SBU annual cash flow pro forma. Table 4.6 illustrates the conversion of project cash flows into EUAW form.

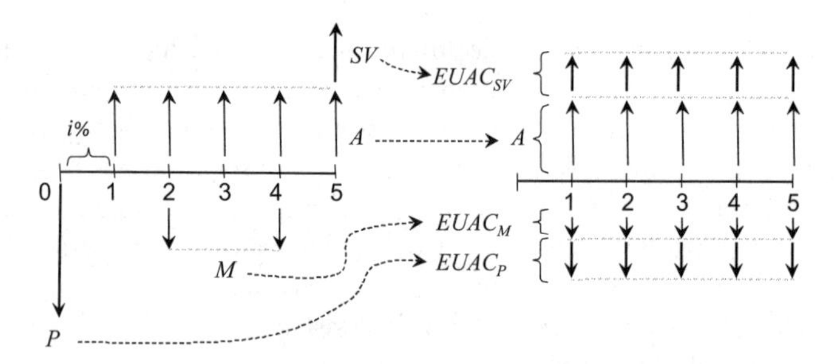

Figure 4.6: Conversion of cash-flow diagram into EUAW.

Let's take a closer look at the calculation steps. For each component of the cash-flow diagram in Table 4.6 we need to compute the annualized worth of each cash item. Typically this means converting the cash item with a capital recovery or sinking fund factor. Recall Equation 3.11, for capital recovery:

$$A = P\left\{\frac{i(1+i)^N}{(1+i)^N - 1}\right\}$$

Clearly, this provides the conversion of a present value P in terms of an annuity, A. Expressing Equation 3.11 in terms of the usual computational factor shorthand,

$$A = P\left(\frac{A}{P}, i\%, N\right)$$

from which the EUAW for a capital investment P_X is established by

$$\texttt{EUAW}_X = P_X\left(\frac{A_X}{P_X}, i\%, N\right) \tag{4.2}$$

In a similar way, recall from Chapter 3 the sinking fund factor

$$A = F\left(\frac{A}{F}, i\%, N\right)$$

from which the EUAW for a lump sum F_Y is established by

$$\texttt{EUAW}_Y = F_Y\left(\frac{A_Y}{F_Y}, i\%, N\right) \tag{4.3}$$

Most common TVM applications for EUAW involve some combination of Equations 4.2 and 4.3. To illustrate further, consider the IT project shown in Table 4.5.

Now, in terms of the cash-flow diagram shown in Figure 4.6:

$$\begin{aligned}
\texttt{EUAW}_P &= -\$17,000(A/P, i\%, N) && \text{Annualized project capital investment,} P\\
\texttt{EUAW}_S &= -\$6,000(P/A, i\%, N) && \text{Annualized salvage value,} SV\\
\texttt{EUAW}_A &= +\$4,950 && \text{Overhead recovery revenue from SBU,} A
\end{aligned}$$

Transaction	Value	Timing
Initial project investment, P	$17,000	t = 0
Expected duration of use, N	5 years	
Discount rate per period, i	8%	t = 1,2,...N
Annual SBU overhead recovery, A	$4,950	t = 1,2,...N
Periodic server upgrades, 4	$1,250	t = 2,4
Salvage value of system, SV	$6,000	t = N

Table 4.5: IT project investment scenario.

The conversion of the maintenance costs takes a bit of care since a two-step process is involved. The value of M at each year must *first* be converted to a present value, *after which* the capital recovery factor is applied. Many times EUAW calculations are in error since the first step is skipped and only the capital recovery factor applied. The two-step process is simple, however, and is concisely captured as follows:

$$\begin{aligned}\texttt{EUAW}_M &= -\$1,250(P/F, i\%, N)_{t=2} \\ &\quad - \$1,250(P/F, i\%, N)_{t=4}(A/P, i\%, N) \\ &= -\$1,250(P/F, 8\%, 2)(A/P, 8\%, 5) \\ &\quad - \$1,250(P/F, 8\%, 4)(A/P, 8\%, 5)\end{aligned} \tag{4.4}$$

So the composite value of EUAW is computed as

$$\begin{aligned}\texttt{EUAW} =& -\$17,000(A/P, 8\%, 5) + \$6,000(A/F, 8\%, 5) + \$4,950 \\ & - \$1,250(P/F, 8\%, 2)(A/P, 8\%, 5) - \$1,250(P/F, 8\%, 4)(A/P, 8\%, 5)\end{aligned}$$

And with the conversion factors substituted,

$$\begin{aligned}\texttt{EUAW} =& -\$17,000(0.2505) + \$6,000(0.1705) + \$4,950 \\ & - \$1,250(0.8573)(0.2505) - \$1,250(0.7350)(0.2505) \\ =& -\$4,258.50 + \$1,023.00 + \$4,950 - \$268.44 - \$230.14 \\ =& +\$1,215.92\end{aligned}$$

EUAW and NPV assessments provide equivalent results. For instance, the NPV of the $1,215.92 EUAW just computed is simply

$$\texttt{NPV} = +\$1,215.92(P/A, 8\%, 5) = +\$1,215.92(3.992) = +\$4,855.16$$

If we now compute NPV for each of the values in Table 4.3, we find that

$$\begin{aligned}\texttt{NPV} =& -\$17,000 - \$1,250(0.8573) - \$1,250(0.7350) + \$6,000(0.6806) + \$4,950(3.992) \\ =& -\$17,000 - \$1,071.67 - \$918.79 + \$4,083.50 + \$19,763.86 \\ =& +\$4,856.90\end{aligned}$$

In principle the two NPV values should be numerically identical, the differences here the result of round-off error in computational factors (differing significant digits) and the use of a hand calculator. The overall equivalence of results should be clear.

EUAW calculations are not difficult as much as they require good organization of information. This will become quite clear later in Section 4.6 (page 119) when we look at an EUAW refinement of the payback concept. As mentioned earlier, EUAW is not as popular as NPV for single project analysis, but we will explore single project analysis in this chapter for completeness as well as serving a preparatory role for the discussions on multiple alternative selection to follow in Chapter 5. Spreadsheets can simplify EUAW calculations, especially if we want to explore the impact of final values of EUAW for various assumed values of physical life N. Consider a case where a machine with an expected physical life of 6 years has an expected maintenance cost of \$2,000 at the end of the first year of service that increases by \$250 annually for the next 5 years. As the calculations on page 106 suggest, there can be many parts of an EUAW calculation to track, so a spreadsheet like that shown in Table 4.6 can be helpful.

	A	B	C	D	E	F
1	Expense		PW of Annual Expense		Equivalent Uniform Annual Cost	
2	Year	Expense	Value	Equation for "C"	Value	Equation for "E"
3	0					
4	1	\$ 2,000	\$ 1,818	= B4/(1+\$B\$10)Â4	\$ 2,000	= PMT(\$B\$10, A4, C4)
5	2	\$ 2,250	\$ 1,860	= B5/(1+\$B\$10)Â5	\$ 2,119	= PMT(\$B\$10, A5, C4:C5)
6	3	\$ 2,500	\$ 1,878	= B6/(1+\$B\$10)Â6	\$ 2,234	= PMT(\$B\$10, A6, C4:C6)
7	4	\$ 2,750	\$ 1,878	= B7/(1+\$B\$10)Â7	\$ 2,345	= PMT(\$B\$10, A7, C4:C7)
8	5	\$ 3,000	\$ 1,863	= B8/(1+\$B\$10)Â8	\$ 2,453	= PMT(\$B\$10, A8, C4:C8)
9	6	\$ 3,250	\$ 1,835	= B9/(1+\$B\$10)Â9	\$ 2,556	= PMT(\$B\$10, A9, C4:C9)
10	MARR = 10%					

Table 4.6: EUAW calculations for expected machine maintenance cost.

The calculation process illustrated in Table 4.6 is important and has several subtle components. We have included the Excel worksheet equations to underscore several points. First, column "B" provides the estimated maintenance costs over the 6-year period. For computational simplicity we have assumed the expenses are paid at the end of each year. Next, column "C" computes the present worth (PW) of the value of the maintenance expense for each year. The calculation is the present worth factor for a single payment in that year. It is not the PW of cumulative maintenance expenses, again, just the value for the year N converted into an equivalent value for $t = 0$.

The third step can be just a bit confusing, so let's examine the values in Table 4.6 carefully, noting the slight differences for the formula in column "F." In column "F" we wish to compute the EUAW for those maintenance expenses associated with all years, up to and including the present year. For instance, to compute EUAW for $N = 3$ we use the PMT function in Excel for the sum of the PW values for $N = 1, 2, 3$ as shown by

```
EUAW = PMT($B$10, A6, C4:C6)
```

which is the same as summing the computational factor result `A/P` for each year (where P is the maintenance cost brought forward to the present):

$$\begin{aligned} \texttt{EUAW}_{N=3} &= +\$1,818(A/P, 10\%, 1) \\ &\quad +\$1,860(A/P, 10\%, 2) \\ &\quad +\$1,878(A/P, 10\%, 3) = \$2,234 \end{aligned}$$

In this situation we find that the EUAW for various life spans continues to rise over the years, but notice the rise in EUAW is not as steep as the maintenance cost increase. This reflects the benefit of lower maintenance costs early in the life of the machine relative to the higher maintenance costs as the machine ages. To make this point quite clear, suppose the maintenance costs were *uniform* over the life of the machine. As shown in Table 4.7, this results in the trivial result that EUAW is $2,000 in each year. Why? Examine the formula and calculation sequence and convince yourself the PW value in column "C" varies but the "cumulative" EUAC of column "E" is identically equal to the maintenance expense from column "B" in any given year.

	A	B	C	D	E	F
1	Expense		PW of Annual Expense		Equivalent Uniform Annual Cost	
2	Year	Expense	Value	Equation for "C"	Value	Equation for "E"
3	0					
4	1	$ 2,000	$ 1,818	= B4/(1+$B$10)Â4	$ 2,000	= PMT($B$10, A4, C4)
5	2	$ 2,000	$ 1,653	= B5/(1+$B$10)Â5	$ 2,000	= PMT($B$10, A5, C4:C5)
6	3	$ 2,000	$ 1,503	= B6/(1+$B$10)Â6	$ 2,000	= PMT($B$10, A6, C4:C6)
7	4	$ 2,000	$ 1,366	= B7/(1+$B$10)Â7	$ 2,000	= PMT($B$10, A7, C4:C7)
8	5	$ 2,000	$ 1,242	= B8/(1+$B$10)Â8	$ 2,000	= PMT($B$10, A8, C4:C8)
9	6	$ 2,000	$ 1,129	= B9/(1+$B$10)Â9	$ 2,000	= PMT($B$10, A9, C4:C9)
10	MARR = 10%					

Table 4.7: EUAW calculations for uniform machine maintenance cost.

Essentially, the two-part calculation sequence can be represented by:

$$\texttt{EUAW} = AW_n \left(\sum_{n=1}^{N} PV_n \right)_{\text{maintenance}} \tag{4.5}$$

4.4 Rate of return

Financial decisions involve contexts that can vary widely, and as a result, a variety of indices, parameters, and calculations are drawn upon during the decision-making process. For any specific situation, some measures are more dominant than others. We have already noted as a measure of success the popular NPV > 0 analysis of Section 4.2, described as the sum of the equivalent present values of a series of cash flows:

$$NPV = \sum_{n=1}^{N} \frac{F_1}{(1+i)^n} \tag{4.6}$$

$$NPV = \sum_{n=1}^{N} PV_n = \left(\sum_{n=1}^{N} PV_n\right)_{\text{Revenue}} - \left(\sum_{n=1}^{N} PV_n\right)_{\text{Expenses}} \tag{4.7}$$

NPV calculations are based on a *given* interest rate of return, $i\%$, assumed to be linked to the cost of capital (discussed earlier in Section 3.2, page 73). Favorable project outcomes are marked by $NPV > 0$, but, we've examined several cases where slight changes in the interest rate of return (cost of capital) $i\%$ can dramatically impact outcomes (refer to Table 4.4). The purpose of this section is to turn things around a bit and invert the NPV problem to allow a focus on *rate of return* as an outcome. Interest rates are *relative* measures and offer a alternative view of project success in comparison to the *absolute* measures obtained through NPV calculations. We'll focus on two special values of $i\%$ that are frequently helpful in comparing and characterizing estimated project outcomes.

Minimally acceptable rate of return (MARR)

Also known as the *hurdle rate* and the minimally *attractive* rate of return, the MARR serves the same basic purpose: to set the lower limit of an acceptable rate of return for a project investment to be considered attractive. MARR establishes the risk threshold of opportunities available to management, reflecting an assumed business acumen in managing projects. Methods to set the MARR vary. Some companies set the MARR a few percentage points higher than the S&P 500, and some couple MARR to the magnitude of the capital investment. Historical corporate policy might also set MARR.

Internal rate of return (IRR)

While MARR tends to result from corporate policy, IRR is a *computed quantity*. The process is to reverse the conventional NPV calculation and identify the rate of return for which $NPV = 0$ for an assumed series of cash flows. With Equation 4.7 in mind, the IRR is that rate of return in which the present value of the estimated cash outflows is equal to the present value of the estimated inflows.

4.4.1 Minimally acceptable rate of return

While there is no universal method to arrive at a corporate MARR, the final value for a given project *must equal or exceed* the cost of capital for a firm for that project. The cost of capital as a "lower bound" criteria is logical when we look beyond the attractiveness of a project and begin to think about where we will get the funds to make the investment. A companys financial condition will influence specific debt-to-equity ratios when there is a need to raise money for investments, or even if they have the capacity to invest at all.

Setting the MARR is not an exact science. MARR can be expected to vary as a result of any combination of the following:

1. Project-to-project variance based on perceived risk
2. The presence of alternate investment opportunities
3. Shifts in company competitiveness and economic position
4. Economic health of the company and availability of capital
5. Prevailing tax incentives and deal level
6. Rates of return other companies are using for similar projects

Capital for investment projects may be raised from some combination of equity (retained earnings, stock) or debt (bonds, credit lines). Government policy can affect MARR through the cost of capital charged to banks (monetary policy).

If all the capital needed by a company were derived from stockholders, then the required rate of return on equity would dominate the MARR decision. However, large projects (say, $P >$ \$5 million) could be imagined as a combination of debt, equity, and economic development tax incentives. When project funding is a mix of debt and equity, the *weighted average cost of capital* (WACC) establishes the MARR floor. We will demonstrate later in Chapter 9 that the WACC is obtained by "weighting" the various components of capital. A quick calculation illustrates. Suppose we have the following capital structure:

W_e	$= 75\%$	fraction of project funding through equity
W_d	$= 25\%$	fraction of project funding through debt
T	$= 40\%$	combined federal and state tax rate
r_e	$= 13\%$	prevailing equity returns to stock investments
r_d	$= 8\%$	before-tax cost of debt

Then from Equation 9.1, p. 274,

$$\begin{aligned} WACC &= W_e r_e + W_d r_d (1 - T) \\ &= 0.75(.13) + 0.25(0.08)(1 - 0.4) \\ &= 0.1095 = 10.95\% \end{aligned}$$

Suppose that the product pipeline is weak and management views the need to be aggressive about expanding the business; in this case a MARR close to the "floor" might be considered acceptable for new projects and older projects might require a MARR twice as large. An investment committee might set the following:

- 11% MARR for investments in equipment enabling new products for new markets.
- 17% MARR for investments in IT projects to enhance website functionality.
- 23% MARR for investments to expand current production of older products.

We see that the MARR for each of the initiatives varies widely and is somewhere between the WACC (10.95%) and a highly aggressive upper limit (23%). In this case the low MARR for new products in new markets suggests a firm is trying to invest more in becoming competitive by encouraging market expansion. Opportunity cost plays a role, too. A company with obsolete equipment may want to discourage continued use of existing equipment and set a high MARR to encourage that opportunities be sought in new ways. Regardless of strategic intent, the MARR must cover the cost of capital for the investment under consideration. —indexWACC Little has been mentioned about the variation in capital cost for an international firm. Firms can only be competitive with investments involving a cost of capital favorable to competition in the areas they wish to operate. It can be expected that most modern corporations are already securing capital on a global basis, with costs modified more by local government policies than by the global market. Smaller companies often do not have a distinctive global advantage, and this dictates global activity for them is less likely to involve foreign direct investment.

In summary, the cost of capital involves factors the firm can and cannot control.

- Policies that *can* be controlled by the firm:
    1. Dividend policy - If dividends are "excessive" in a year that capital investments are required, then the costs of raising capital are higher and terms less favorable.
    2. Investment policy - Firms that depart from their historical investments might be exercising new management competencies and these are not without risk.
    3. Capital structure policy - Shifting between debt and equity will affect WACC, and if debt were to rise relative to equity, then leverage might diminish.
- Factors that *cannot* be controlled by the firm:
    1. Capital gains and tax rates - Taxes play a significant role in WACC; outside of hiring lobbyists there is little an individual firm can do to influence taxes.
    2. Perceived market risk - Overall economic conditions dominate investor perception of risk and probability of realizing returns for a given investment climate.
    3. Overall market interest rates - Many bank rates (LIBOR, U.S. bonds) seem to be triggers or catalysts for interest rate changes and opportunity cost for investors.

4.4.2 Internal rate of return

Assets are entrusted to project managers for investments that enhance operations. This was illustrated earlier in Figure 1.8 and previously discussed, beginning on page 11. Implied is that the benefits of an attractive project investment should exceed the total cost of the investment itself. How *much* aggregate benefits exceed the original principal investment can be measured by the *rate of return* that has been the focus throughout Section 4.4.

For those stakeholders *providing* assets, there is a minimum cost of capital given by the MARR, linked in the previous section to the WACC. By now it should be entirely clear that the *cost* of providing the assets (emphasized by discussing the WACC) is *not* the same as the benefits produced by decisions that managers make about asset *use* that presumably leverages the competencies of the organization in ways not possible for external investors or competitors. This underscores the premise that managerial knowledge, skills, and abilities can uniquely put to use assets in a way more productive and with *greater* value-add than simply "putting the money in the bank." *Internal rate of return* is a measure of project desirability and the ability of project managers to produce value above and beyond the market norm.

Up to this point we've examined investment situations where a sum of money to be invested has associated with it a *given* rate of return $i\%$, from which the TVM equations of Chapter 3 are applied to predict present or future value. IRR requires us to invert the usual NPV calculation procedure. In this case the investment and anticipated revenue streams are *given* and the rate of return to balance the investment and revenue stream is unknown. The *internal rate of return* (IRR) calculation is a special outcome of the PV equations when NPV = 0.

Consider the high-tech project investment shown in Table 4.8. Here, it is proposed that buying a wafer profilometer and then selling it after 3 years is less expensive than the existing service being subcontracted at $1,500/year. We want to answer the question "Is this a good idea?"

Transaction	Value	Timing
Initial project investment, P	$4,999	t = 0
Expected duration of use, N	3 years	
Annual revenue from use, F_t	$1,199	t = 1,2,3
Salvage value of system, SV	$3,000	t = 3

Table 4.8: Project investment scenario.

Ordinarily we'd say that the NPV of the project is simply

$$NPV = -P + \frac{F_1}{(1+i)^1} + \frac{F_2}{(1+i)^2} + \frac{F_3}{(1+i)^3} + \frac{F_{SV}}{(1+i)^3} \tag{4.8}$$

But, of course, we don't know the rate of return to use to compute the NPV, so we proceed to compute the IRR (associated with $NPV = 0$) that equates the present value of the benefits from the investment with the expense of the investment itself. Equation 4.8 is simply one equation with one unknown $i\%$.

$$0 = -P + \frac{F_1}{(1+i)^1} + \frac{F_2}{(1+i)^2} + \frac{F_3}{(1+i)^3} + \frac{F_{SV}}{(1+i)^3} \quad (4.9)$$

$$0 = -\$4,999 + \frac{\$1,199}{(1+i)^1} + \frac{\$1,199}{(1+i)^2} + \frac{\$1,199}{(1+i)^3} + \frac{\$3,000}{(1+i)^3} \quad (4.10)$$

It might help in understanding Equation 4.10 to visualize the cash-flow spreadsheet in Table 4.9. Equation 4.10 is a non-linear equation for which there is no simple algebraic solution; either a trial-and-error (iterative) calculation by hand or spreadsheet can be used to find a solution. A spreadsheet solution (shown in Table 4.10) has been created for this problem, and the Excel `NPV` function has been used to determine that $IRR = 15.4\%$. Table 4.9 is a cash-flow diagram for this calculation.

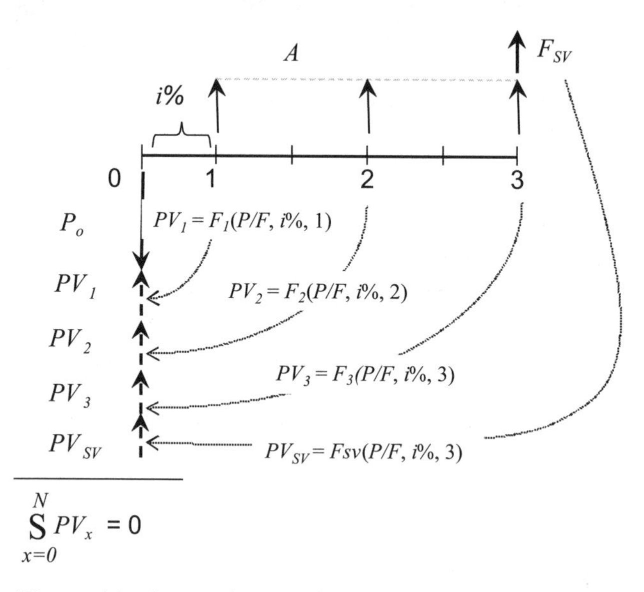

Figure 4.7: Internal rate of return cash flow diagram.

Let's assume Equation 4.8 (page 113) sets the $WACC = 10.95\%$ for this problem. We observe that $IRR > WACC$ and this leads to the conclusion "Yes, this is a favorable situation" and would suggest it is better to purchase the equipment and bring the asset in-house rather than continue leasing.

Consider an NPV perspective with $WACC = 10.95\%$. In this case

$$NPV = -\$5,000 + \frac{\$950}{(1+i)^1} + \frac{\$950}{(1+i)^2} + \frac{\$950}{(1+i)^3} + \frac{\$3,000}{(1+i)^3} = \$496.20 \quad (4.11)$$

While we've clearly met the criterion $NPV > 0$, the NPV is barely 10% of the capital investment, which is not much, really; this use of \$4,999 does not seem to have much of a "wow" factor. Actually, the situation is more impressive if the duration of the project is 7 years.

2	B	C	D	E	F
3	Capital Investment		$4,999		
4	Residual Market Value		$3,500	29.99%	Loss in value
5	Annual Benefit		$1,199		
6	Period		3	years	
7	Corporate Rate of Return		10.95%		
8					
9		Capital	Annual	Residual	Cash
10		Investment	Benefit	Value	Flow
11	0	-$4,999			-$4,999
12	1		$1,199		$1,199
13	2		$1,199		$1,199
14	3		$1,199	$3,500	$4,699
15					
16				**IRR**	**15.4%**
17					=IRR(F11:F14,+D7)
18	**Present Worth Method**				
19					
20	PW:	($4,999.00)	$2,932.56	$2,562.63	$496.20

Table 4.9: IRR calculation for a 3-year high-tech equipment purchase.

2	B	C	D	E	F
3	Capital Investment		$4,999		
4	Residual Market Value		$1,500	69.99%	Loss in value
5	Annual Benefit		$1,199		
6	Period		7	years	
7	Corporate Rate of Return		10.95%		
8					
9		Capital	Annual	Residual	Cash
10		Investment	Benefit	Value	Flow
11	0	-$4,999			-$4,999
12	1		$1,199		$1,199
13	2		$1,199		$1,199
14	3		$1,199		$1,199
15	4		$1,199		$1,199
16	5		$1,199		$1,199
17	6		$1,199		$1,199
18	7		$1,199	$1,500	$2,699
19					
20				**IRR**	**18.3%**
21					=IRR(F11:F14,+D7)

Table 4.10: IRR for the high-tech equipment purchase when the duration of the project is extended to 7 years.

4.5 Payback period

Practitioners often want to know a project *payback period* as an estimate of how long it takes a project investment to "pay for itself." Project investments are intended to produce benefits to an organization; the specific length of time for which the accrued revenue equals the project investment itself is known as the *payback period.* Usually the payback period is expressed in term of *years* since the project is ordinarily a significant capital expense whose utility spans several years and is thus a depreciable asset to the company. There are several different ways to compute payback period, the level of complexity depending on whether the time value of money, taxes, or interest is included in the calculation. In this chapter we explore just two variants: (a) the simple estimate and (b) the EUAW method. We have to keep in mind that payback period is generally not a preferred basis on which to *select* between investment alternatives, but it *can be* a *supplemental* piece of information that does go into the decision-making process.

Payback period for an asset is similar to, but often confused with, *break-even* for a product (or a company!). Break-even identifies the point at which revenues equal expenditures based on assumptions of fixed cost and variable cost. Break-even is usually attending to the profitability of an *entire* enterprise and is an expected calculation in business plans. References such as Dauber et al. [5] delineate break-even as a point of "no profit and no loss" (page 298) and payback as the length of time to recover an investment (page 340). Some people view the difference as academic; for instance, if you are computing payback on a services project investment that is the core of company operations it might be hard to tell the difference.

Let's take a closer look at payback period.

4.5.1 Simple estimate

As mentioned many times already, there are a variety of ratios and factors that can be incorporated in financial decision-making. Payback period is commonly used in practice but potentially limited in utility. Many business discussions warrant simple "first pass" estimates to vet economic value. If we are lucky, then project investment information is available that makes the calculation easy. In most other cases, data are ambiguous and the calculation is a little more tricky.

The payback period for an asset is the number of years the asset must be retained or held to recover the initial investment:

$$0 = -C + \sum_{n=1}^{N} R_t(P/F, i\%, n)$$

where

$$C = \text{initial capital investment}$$
$$R_t = \text{annual returns or annual benefit in year } t$$
$$N = \text{years (not necessarily an integer)}$$
$$i\% = \text{investment interest rate}$$

and the P/F factor is defined by Equation 3.4.

For some unknown historical reason it frequently seems acceptable to ignore the required rate of return (set $i\% = 0$), in which case you have the result (absent the P/F factor) that

$$0 = -C + \sum_{n=1}^{N} R_t$$

If we further assume the annual benefit R is *uniform*, then you have the resulting equation for the *simple estimate*:

$$\text{Payback}, n = \frac{C}{R} = \frac{\text{Capital Investment}}{\text{Uniform Annual Benefit}} \tag{4.12}$$

The simple estimate takes very little work to compute (maybe that's the reason for historical use). Simple can be good in the right context, but at the same time the result may have potentially little value in a rigorous economic study. No reason to ignore it, though. A typical business investment situation illustrates use.

Sample Problem. Electronic Health Records, Inc., would like to bid on a skilled nursing facility contract, but the uptime requirement demands much higher quality computer servers than EHR normally purchases. It is estimated that a server costing \$3,000 can bring in \$750/year in revenue. Since this is a high-paced technology business, management wants to know if the servers will pay for themselves in less time than the conventional 5-year depreciation.

$$\text{Payback} = \frac{\$3,000}{\$750} = 4 \text{ years}$$

If the benefit of the new servers is estimated too optimistically and the anticipated revenue drops to \$600, then the payback period is equal to the 5-year limit imposed by management. In this case the investment is marginally acceptable when viewed from this narrow perspective.

The simple payback period can be useful for framing the problem. For example, if the simple payback period were computed to be 10 years and the risk tolerance of management is 18 months, then clearly the project is a no-go. Reflecting on the situation above, it seems reasonable to include the time value of money ($i\% > 0$) in this case since the simple result $N = 4$ is within a reasonable range of the managerial metric $N = 5$, especially if there is any variance in the estimate of the revenue stream. If a company is facing major liquidity issues, then immediate payback might be of central concern (cash flow beyond the short term might be too uncertain to count on). A slightly more complicated situation stems from non-uniform revenue streams. To illustrate, consider the situation in Table 4.11 in which EHR heavily discounts their services for the first 3 years. Revenue is projected to build (very) slowly over time, finally reaching the target $750 annual revenue per server in year 5.

Year	Investment or Revenue	Cumulative Revenue	Net Cash Flow
0	-$3,000		-$3,000
1	$75	$75	-$2,925
2	$150	$225	-$2,775
3	$300	$525	-$2,475
4	$600	$1,125	-$1,875
5	$750	$1,875	-$1,125
6	$750	$2,625	-$375
7	$750	$3,375	$375
8	$750	$4,125	$1,125

Table 4.11: Example of non-uniform accrued revenue over an 8-year period from a $3,000 capital equipment investment.

Table 4.11 illustrates that it is not until the 6th year that we begin to realize a cumulative project revenue of $2,625 that is (somewhat) within range of the original $3,000 investment. The interpolation calculation to find the payback period estimate is simply

$$\text{Payback} = X + \frac{Y}{Z} \tag{4.13}$$

where

$$\begin{aligned} N &= \text{year (not necessarily an integer)} \\ X &= \text{last year N when net cash flow} < 0 \\ Y &= \text{absolute value of net cash flow at year X} \\ Z &= \text{expected revenue at year X+1} \end{aligned}$$

Using the data from our example we have

$$\text{Payback} = 6 + \left(\frac{375}{750}\right) = 6.5 \tag{4.14}$$

4.6 EUAW for payback period

Equivalence is a fundamental concept underpinning the evaluation of various revenue streams, expenses, and irregular project charges; it is essential to have all terms expressed on a common ("equivalent") basis. Recall that in the present worth (PW) method we sought to express all values in term of the present, that is, the *equivalent value today*, $t = 0$. As well, there is the future worth (FW) scenario where *all sums* are converted in terms of their equivalent values in the *future*, at some year $t = N$. Since PW and FW are equivalent mathematically, selection of one method over another is, in principle, immaterial to outcomes, but as a practical matter the choice of analysis method is often set by corporate precedent, project context, analyst habit, or simply personal preference. This section of the chapter expands on the ideas of Section 3.6 to produce a third option in which values are expressed for an *annual period*, not at a specific point in time. This method of analysis based on annualizing costs leads to the so-called *equivalent uniform annual worth*, (`EUAW`), method. When using the `AW` method and the focus of the project is *only on cost* then we have the *equivalent uniform annual cost* method, (`EUAC`). Often the terms `EUAW` and `EUAC` are used interchangeably since the calculation methods needed are similar; it is just that the intent of the analysis is a little different.

As conveyed in the name *equivalent uniform annual worth*, the `EUAW` method relies heavily on the concept of annual worth, (`AW`). Previously, `EUAW` was linked to `AW` in the discussion of Section 4.3, with the outcome featured in the form of Equation 4.2. The concept of `AW` should be familiar from Section 2.5 when we discussed special issues in long-term assets and from Section 3.6 when illustrating the equivalence of PW, FW, and AW calculations.

It is quite common to borrow a large sum of money to purchase a house or car, and then to compute the (uniform) annual payments that might be expected in order to see if we can afford the payment, based on the payment as a fraction of our disposable income. `EUAW` is a method to help us convert project calculation into that familiar mode of thinking.

The basic steps involved in computing `EUAW` were outlined earlier in Section 4.3, p. 104. For the IT project investment in Table 4.3, an expected project life of $N = 5$ years was specified, and `EUAW` $= \$1,215.92$ determined. What if the project life were different? Say, $N = 4$ years were of interest. Or, suppose that we wanted to see the parametric dependence of `EUAW` as a function of the full range of $N = 1, 2, 3, 4, 5$ much like the analysis of Table 4.6? For any desired value of N, the equations of Section 4.3 are all that is needed.

Let's reconsider the situation of Figure 4.5, but rather than assume the physical life of the machine is $N = 6$, we'll work through the calculations to establish the value for `EUAW` for each year $N = 1, 2, 3, 4, 5$. So, in this situation $\mathtt{EUAW}_3$ would be the computed value of `EUAW` for years 0-3, as discussed and outlined in Section 4.3, specifically, page 104. Table 4.12 provides the evolution of the value of `EUAW` as `N` increases, for which the values at $N = 5$ match the calculations results in Equation 4.4.

Year	Expense	Benefit	Maint	Value	PV	NPV	**EUAW**
0	$ (17,000)				$ (17,000)	$ (17,000)	
1		$ 4,950	$ -	$ 4,950	$ 4,583	$ (12,417)	$ (13,410)
2		$ 4,950	$ (1,250)	$ 3,700	$ 3,172	$ (9,245)	$ (5,184)
3		$ 4,950	$ -	$ 4,950	$ 3,929	$ (5,315)	$ (2,062)
4		$ 4,950	$ (1,250)	$ 3,700	$ 2,720	$ (2,595)	$ (784)
5	$ 6,000	$ 4,950	$ -	$ 10,950	$ 7,452	**$ 4,857**	**$ 1,216**
MARR =	**8%**						

Table 4.12: EUAW calculations by spreadsheet formula.

Notice in Table 4.12 that EUAW is only positive in year 5. This means that the asset life must be held by the owner *at least* 5 years for the project investment to make sense. To explain this further, consider Table 4.12, where the focus is just on *annual costs and benefits* and not on the initial investment or salvage value.

Year	Expense	Benefit	Maint	Value	PV	NPV	**EUAC**
0	$ -				$ -	$ -	
1		$ 4,950	$ -	$ 4,950	$ 4,583	$ 4,583	$ 4,950
2		$ 4,950	$ (1,250)	$ 3,700	$ 3,172	$ 7,755	$ 4,349
3		$ 4,950	$ -	$ 4,950	$ 3,929	$ 11,685	$ 4,534
4		$ 4,950	$ (1,250)	$ 3,700	$ 2,720	$ 14,405	$ 4,349
5	$ -	$ 4,950		$ 4,950	$ 3,369	**$ 17,773**	**$ 4,451**
MARR =	**8%**						

Table 4.13: EUAW calculations: Focus on annual cost.

Table 4.13 illustrates that the NPV of the system continues to increase as time passes, but observe how the EUAC reaches a minimum at N=2 and N=4. This situation reflects the impact of variations of expenses *and* the time value of money. It would be expected that over time, the market value of the equipment will drop (as it depreciates) and that to maintain the same level of productivity there need to be periodic investments in maintenance. When a point of minimum EUAC is reached, we say that we have reached the *economic life* of the investment.

The significance of Table 4.13 is that the *economic life* is not the same as the *physical life* of the equipment. And it is quite typical that the economic life of an asset is *not* the same as the physical life of the asset. There are two reasons this is very important:

1. Estimating economic life of an asset cues managers as to when it might be worthwhile to invest in *new* equipment.
2. Managers must plan to have sufficient funds available for replacement equipment once the useful life has been exhausted.

Also note there are situations where the "life" decision is made for you, as in the case where the IRS clearly spells out the recovery periods allowed for various asset classes (buildings, machinery, equipment)!

If the basic ideas behind the EUAW are understood, then a more meaningful exploration of finance to support decision-making can be conducted and many parameters that might seem challenging to estimate are not that difficult. Consider Table 4.12, for example. In this case we are looking at a more complex payback problem where values for EUAW and PV against MARR are of interest. In this situation a spreadsheet organizes information for convenient use and display.

Elapsed Period	Scenario X **TruSteel**	Scenario Y **QualityKing**	Scenario Z **FastTrac**			
	Income	Income	Income	Aggregate Investment	Present Value of Income	EUAW of X-value
Expense			A	B	C=NPV(A)	D=EUAW(C)
0	-$3,000.00	-$4,599.00	-$4,599.00			
Revenue						
1	$499.00	$699.00	$1,199.00	$3,400.00	$1,110.19	$1,199.00
2	$499.00	$699.00	$1,074.00	$2,326.00	$920.78	$1,138.90
3	$499.00	$699.00	$949.00	$1,377.00	$753.35	$1,080.41
4	$499.00	$699.00	$824.00	$553.00	$605.66	$1,023.51
5	$499.00	$699.00	$699.00	-$146.00	$475.73	$968.19
6	$499.00	$699.00	$574.00	-$720.00	$361.72	$914.46
7	$499.00	$699.00	$449.00	-$1,169.00	$261.99	$862.29
8	$499.00	$699.00	$324.00		$175.05	$811.68
9	$499.00	$699.00	$199.00		$99.55	$762.62
10	$499.00	$699.00	$74.00		$34.28	$715.09
Residual	$100.00	$750.00	$100.00		$46.32	$721.99
Simple Payback	**6.01**	**6.58**	**5.25**		EUAW Payback:	6.37
MARR	8.00%					
PW						
Principal	-$3,000.00	-$4,599.00	-$4,599.00			
Revenue	$3,348.33	$4,690.35	$4,798.28			
Residual	$46.32	$347.40	$46.32			
NPW	**$394.65**	**$438.74**	**$245.60**			

Table 4.14: Payback calculations for two product scenarios.

4.7 Bonds

Bonds have been a robust financial instrument for raising money for centuries. While the origin and common usage of bonds often makes them synonymous with public project activity, many other non-profits and for-profit entities take advantage of the bond "instrument" as a favorable way to issue debt with many institutional and individual investors. Earlier in Section 4.4.1, we noted that both stocks *and* bonds were used to raise money, but there are some fundamental differences relating to interest, guarantee of returns on the investment, risk of total loss, and legal standing:

1. Bonds issued are categorized as debt on the balance sheet. Bond performance is defined by terms and conditions that essentially represent a binding contract.
2. Stocks are categorized as equity on the balance sheet. Stocks provide the investor an ownership role in the company. Return on investment is implied, but not explicit.

It is easy to forget that financial instruments such as bonds are often just *paper* securities and may not be "backed" by specific assets. In such a case, no intrinsic value is possessed by the bond beyond our faith in the words and promises the security represents. This is quite unlike, say, a title to a piece of machinery that represents ownership of *real property* producing value for us *today* in the production of goods and services.

The true value of a paper security is created only by the perception, belief, and hope of a future interest payment revenue stream. This emphasizes the need for bond issues to be trustworthy and possess competence in the business they manage. Said another way, we want to be certain that organizations to whom we are lending money (buying a bond from) will not go bankrupt.

4.7.1 Bonds as a debt instrument to raise money

Bonds are a common way for governments to raise money for infrastructure projects. And it is the notion that governments do not normally go bankrupt that is the origin of the bond concept. After the 1795 Great Fire in Copenhagen there was a sudden need to rebuild homes in a large part of the city. The demand could not be met by local bankers (at reasonable rates, we might assume) so the government stepped in to generate a large pool of money for reconstruction needs (Ladekar[6]). In just a few years after the Copenhagen crisis the "Kreditkassen" mortgage banking framework was established, and some basics of the model are still evident in today's mortgage bond market. Most important is that government bonds are normally viewed as being very safe investments and the bond interest rate essentially is the "floor" of what it costs the government to borrow money.

Gjede[7](p.31) recites several features of the current Danish mortgage market that reach back to the 1700's:

1. Loans are fixed-interest long-term loans.
2. The loans are funded entirely though the issuance of bonds.
3. The bond investors have full knowledge of the security of the bonds, based in part on the mortgage on the real property, the [faith in] the legal framework, and the solidarity of the mortgage bank.

Bonds can play a unique role in the financial market, not only serving as a critical source of funds for development, but also as a secure investment, too. Sometimes the bond market is foundational for economic growth or rejuvenation in times of disaster. Notice also the role that *faith* plays in the market. We suggest that trust is a centerpiece of many components of the financial market, which is why ethics, transparency, and intent play such a significant role in the utility and success of the investment market.

Common to both stocks and bonds is the idea of the confidence one must place in the organization that future cash flows exist. For the purchaser of a bond the "market confidence" is the assurance the bond will provide interest payments to the bond owner (investor) in accordance with terms and conditions, and then eventually the principal of the bond is returned to the investor. Stocks are held with the hope of dividend payments and a rise in the market value of the shares held as the company grows and prospers. Normally the expectation is that the bond has a lower ROI than a stock because of the perception the stock is a riskier investment. If the corporation is dissolved, both a bondholder and a stockholder could lose their investment; however, if there are assets to be disbursed, a bondholder has a preferential position (as debt) and would be paid prior to a stock holder.

As with any financial instrument today, bonds come in a wide variety of forms. Sometimes bonds are called securities and the two descriptions are used interchangeably here.

Classification	Issuer	Remarks
Treasury Bonds	Federal government	Collateral is faith in the government.
Foreign Bonds	Foreign governments	Tend to have higher default risk
Municipal ("munis")	Local government	Collateral is anticipated tax revenue.
Debenture Bonds	Corporation	Collateral is faith in the company.
Convertible Bonds	Corporation	Swapped for stock at a later date.

Table 4.15: Typical bond classifications in the market today.

As an exercise, check the current business press to see what might affect confidence in bond purchases. At the time of this writing, the buzz in the press was that changes in US federal policy would affect investor preference for bonds as a safety net[8] or whether the European Union should create a eurobond market backed by the member countries. Although a benefit to some "muni" bonds is that interest earned may be tax-deductible, they are not without risk; some *do* default[9].

Some fairly universal terminology describes bond purchases and ownership:

1. The **face value (FV)** of the bond is the principal that is being borrowed. At one time when bonds were printed documents, the loan amount of $500, $1,000, $10,000, etc. was printed on the face of the bond document. The FV is also known as the par value of the bond.
2. Each bond has a **term**, also known as the *maturity*, which is simply the life of the bond; the *term* of a bond is very long, often greater than 10 years. There is the original term of the bond when first issued, but

we also use *term* to indicate the remaining life of the bond. A 10-year bond that has been held for 3 years has a remaining term of 7 years. Normally the face value of the bond is equal to the redemption value upon maturity.

3. The **coupon rate** is the interest paid by the bond at prescribed intervals over the life of the bond. Sometimes this is referred to as the *bond interest*, *bond dividend*, or *coupon payment*. The terminology *coupon* is reminiscent of an earlier era when the bond was like a payment booklet and the investor would cut off the "coupon" and redeem it for cash. This is the origin of the term "coupon clippers" to describe highly conservative investors. Payments are often set on a semi-annual or quarterly basis.
4. The **maturity** of a bond refers to the point in time at which the term of the bond has ended. At the *maturity date* the bond investor turns in the bond and receives back the original investment. A bond represents a special non-amortized type of loan since the only payment over the life of the loan is interest; no payment against principal is made during the life of the loan.
5. **Investor preferred yield rate** or simply *yield rate* for a bond relates to the return on investment that the bondholder hopes to achieve over the life of the bond. This is typically *not* equal to the coupon rate. As we know from Section 4.4, the *return* is quite simply what the investor receives from having made the investment. Each individual investor will have their own preferred yield (or MARR). PV calculations enter since the bond is non-amortized and it has to be that the combination of the PV of the face value of the bonds when combined with the PV of the coupon payments is favorable.

Clearly the variety of nomenclature used to describe a bond reflects the history of the investment security. In today's investment world the bond tends to be a contract with payments scheduled electronically to registered owners of the bond. Not very romantic or nostalgic, but that's the way it is.

4.7.2 Primary and secondary bond markets

The mechanics and complexity of the investment market are not quite within scope of this text, but three remarks can help you understand terminology and the use of bonds as an investment device (and the investment market in general).

1. We should think of the initial sale of the bond as occurring in the primary market, which would be a transaction that normally occurs between the initial purchaser of the bond and the issuer, that is, between the company and the person who purchases the bond. Because several thousand bonds might be sold to raise $12 million for a capital investment project, a *securities firm* might be used to manage the initial offering (on behalf of the company). Regardless, the *primary market* is characterized as the transaction that takes place in which the proceeds of the sale go directly to the company. Generally, the initial price paid for the bond is the face

value described above. Sometimes there is a bond auction that affects initial purchase price.

2. Activity surrounding the purchase and sale of bonds is an *investing* activity. Investing is simply the process of using current resources (cash) to purchase security instruments (stocks and bonds) with the promise of realizing future revenue whose equivalent worth today warrants forgoing the use of the funds for the period of the investment. As we have seen many times before, it is not the PV calculations that are difficult in themselves as much as the need to properly organize all related information.
3. Subsequent to the first sale, investors will sell bonds *to each other* in what we can think of as the *secondary market.* A majority of the "action" related to bonds concerns the trading of bonds in the secondary market. Here, the price of the bond is not likely to be the face value, but a price reflecting the market value of the remaining term of the bond. This difference between the market value and the face value is the source of much confusion in bond analysis since there is now a new *investor yield* that is entirely separate from the *coupon rate.*

Bonds often trade in a different "secondary market" than organized central exchanges created for stock transactions. Typically, large financial institutions, pension funds, and mutual funds will trade with each other in so-called "over the counter" (OTC) transactions using the telephone or an electronic network instead of what we might imagine is a trading floor as seen on the news. It is not often that the engineering manager becomes consumed with the nuances of bonds and the investment philosophy of the company. While it may be of most interest to assure the return on a project will exceed the cost of capital, the engineering manager must have *some* sense of the investment side of the company to better appreciate what might be a confounding factor in capital investments. For instance, investors may sell a bond at a discount (selling a $500 FV bond for $480) or at a premium (selling a $500 FV bond for $530), which has an indirect effect on the effective yield of the bond. Knowing the financial implications can help build credibility as a decision-maker.

For the majority of our bond calculations we assume trading in the secondary market. We emphasize that changes in the selling price relate to the investor's preferred yield or MARR, and has little to do with the coupon rate. We'll work through the logic and calculations for this in the following sections.

4.7.3 Coupon interest rate

Regardless of what happens to the market value (present worth) of a bond in the secondary market, we cannot forget that the *coupon rate* is set by what is essentially a contract the day the bond is initially issued. We know that the interest rate provided in most situations is ordinarily a *nominal* interest rate, and must be adjusted by the number of payment periods per year (see

Section 3.7, p.84). The periodic interest to be paid by the issuer is therefore:

$$I_b = FV \frac{b}{N} \tag{4.15}$$

where

I_b =periodic interest payment
FV =face value of the bond
b =coupon interest rate
N =frequency of bond payments each year

Sample calculation:

Scenario
Bethesda Imaging decides to fund a $1 million expansion project by issuing investment bonds for $10,000 with an 8% quarterly interest rate and a 10-year term. How much will an investor receive quarterly and at the time the bond reaches maturity?

To solve, substitute the bond data into Equation 4.15, readily producing the result:

$$I_b = \$10,000 \frac{0.08}{4} = \$200 \text{ per quarter}$$

This quarterly interest is paid by the issuer to the registered owner for the life of the bond. When the bond matures, the registered owner will redeem the bond for the face value of $10,000.

If we look in the paper it might seem odd that the corporate debenture rate of 8% quarterly seems quite high relative to, say, a US Treasury bond of 2.5%. This, of course, reflects the different bond categorizations shown earlier in Table 4.15.

4.7.4 What is a bond worth?

The process of determining the present worth of the bond (the *equivalent worth of all cash expressed in today's terms*) draws on the same PW calculations we outlined in Section 4.2. To emphasize coupon payment calculations let's change the problem by assuming coupon interest is paid semi-annually. First, the cash flow for this situation is shown in Figure 4.8.

The price of the bond follows the PV calculations that should be quite familiar by now:

$$PV = +I_b(P/A, i\%, N) + FV\ (P/F, i\%, N) \tag{4.16}$$

First, examine Figure 4.8 to be sure you are clear that the semi-annual coupon payment is $400 versus the quarterly payment of $200 computed in the previous section. Next, what is the preferred investor yield rate i? Let's assume

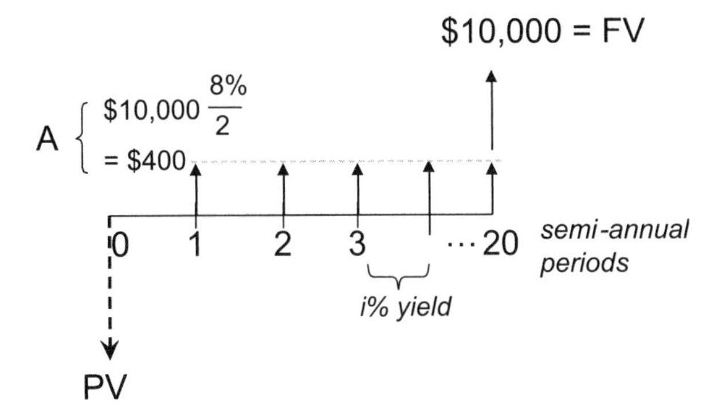

Figure 4.8: Typical bond problem: $10,000 bond with a nominal 8% coupon paid semi-annually and a 10-year maturity.

that at the time of the purchase, the company is offering a coupon rate competitive with prevailing investor yield rates:

$$\begin{aligned} PV &= +\$400(P/A, i\%, N) \quad && \text{PV of periodic coupon (interest) payments} \\ &\quad +\$10,000\ (P/F, i\%, N) \quad && \text{PV of the bond face value (FV)} \end{aligned}$$

Please note carefully the numerical substitutions from Figure 4.8 into the PV equations:

$$\begin{aligned} PV &= +\$400(P/A, 4\%, 20) \ + \ \$10,000(P/F, 4\%, 20) \\ &= +\$400(13.5903) \ + \ \$10,000(0.4564) \\ &= \$5,436.13 + \ \$4,563.87 \ = \ \$10,000 \end{aligned}$$

Does the result surprise you? Essentially the bond is worth today what is stated as the face value of the bond. In this situation, the semi-annual bond coupon payments equal the opportunity cost of delaying use of $10,000 for 10 years. It is thus a "safe" investment and also worked to earn you 8% per year! Indeed, the face value of this bond is worth paying.

It is highly *unlikely* that in the more general investment market the prevailing investor preferred yield rate would equal the coupon rate; the example above illustrates the PV of the bond in the idealistic case in which market rates do not fluctuate and for which a bond exactly satisfied investor risk levels. In reality, markets have shown themselves to be quite volatile and it is more likely the bond is being purchased to balance portfolio risk. Often the coupon rate is *far less* than prevailing investment preferences, commensurate with the investment risk involved. It can be confusing at first to be asked, "What would an investor pay for a bond?" when it seems obvious the initial price of the bond was quite clear and remains part of the bond security documentation. What is at play here is how bonds are traded by buyers and sellers in the secondary market *subsequent to the initial sale in the primary market*, and the role of intermediaries in setting prices in the OTC market.

Let's turn to a more realistic situation in which the prevailing interest rates fluctuate and thus have a large impact on the value of the bond to the investor. Let's look at a typical situation.

Scenario

Wing Yan wants to purchase a $5,000 corporate bond from a discount brokerage firm to balance her personal portfolio. She is not quite sure about the fees from the brokerage, but decides to check on that later. Due to some turbulence in the metals market, she thinks it might be a good idea to purchase some bonds issued from a mining company that used the bonds to acquire another mining company. The uncertainty in the market has other people who have been holding such bonds that want to sell. Bonds with a maturity of 20 years are available with 15 years remaining; coupon rate is 5% paid annually. Wing's MARR is 10%; she is not too aggressive but believes in "long-run" investing. What is the price she is willing to pay?

First, we compute the coupon payment:

$$I_b = \$5,000\,\frac{0.05}{1} = \$250 \text{ once a year}$$

And everything is on an *annual* basis, so from Equation 4.16 we have

$$\begin{aligned} PV &= \$250(P/A, 10\%, 15) + \$5,000(P/F, 10\%, 15) \\ &= \$250(7.6061) + \$5,000(0.2394) \\ &= \$1,901.53 + \$1,197.00 = \$3,098.53 \end{aligned}$$

This PV is less than the FV, illustrating the power of the TVM and what it means to defer the use of investment capital. Although Wing holds a 20-year bond, it can only produce revenue for the next 15 years so it makes sense she would pay less than FV.

Consider how the calculation procedure changes when the compounding periods are *unequal.* Suppose for Wing the coupon rate is nominally 5% but paid *quarterly* (effective rate of 1.25%) and Wing wants to earn a nominal 10% paid *quarterly* (effective 2.5%). Care must be taken to use the *effective* interest rates (if the difference is a bit fuzzy, consider reviewing Section 3.7, p. 84):

$$I_b = \$5,000\,\frac{0.05}{4} = \$62.50 \text{ four times a year}$$

Then, from Equation 4.16 we have

$$\begin{aligned} PV &= \$62.50(P/A, 1.25\%, 60) + \$5,000(P/F, 2.5\%, 60) \\ &= \$62.50(30.9087) + \$5,000(0.2273) \\ &= \$1,931.79 + \$1,136.42 = \$3,068.21 \end{aligned}$$

Not much of a change in PV, but a slight change nonetheless that would make a big difference in multimillion dollar currency transactions.

There are multiple permutations of unequal compounding frequency periods. The math is easy once you've chosen the right equations. Let's wrap up with a slightly tricky bond situation where dividend interest is compounded semi-annually and Wing wishes to have yields compounded quarterly. The first step is to compute Wing's bond coupon payment:

$$I_b = \$5,000 \, \frac{0.05}{2} = \$125 \text{ twice a year}$$

As will all PV problems, we have to be sure the same period and interest rate basis are used in the TVM calculations. We use the semi-annual (6-month) period as the TVM compounding frequency and need to ensure the interest rates are aligned, that is to say, that we are using the correct *effective* interest rates. In the PV calculation of the annuity (regular coupon payment) the set-up looks much as before, but before proceeding we must find the i_{eff} to use in Equation 4.16:

$$PV = + I_b(P/A, i_{eff}\%, N) + FV\ (P/F, i_{eff}\%, N)$$

For the desired investor yield, compounding is specified to occur on a quarterly basis, but the problem calculations are for a semi-annual basis. To accommodate this, a simple conversion is needed. Here the *nominal* 10% rate is 5% on a semi-annual basis but the stated compounding rate is *quarterly.* This means there are 2 compounding periods within the 6-month period over which the 5% nominal rate applies. We therefore have to compute the *effective* interest rate with Equation 3.20 on p. 86:

$$\begin{aligned} i_{eff} &= \left\{1 + \frac{r}{m}\right\}^m - 1 \\ &\left\{1 + \frac{0.05}{2}\right\}^2 - 1 = 5.063\% \end{aligned}$$

Clearly, the effective interest rate of 5.063% is just slightly higher than the nominal semi-annual rate of 5% due to the quarterly compounding providing accumulated interest over the first quarter (3-month period).

The important point is that the computation will take place an a semi-annual basis. Because the investor is using quarterly compounding for the yield and PV estimate, we have to use the *effective* interest rate of 5.063% in Equation 4.16 for the PV calculation. Summarizing data, then:

$$\begin{aligned} I_b &= \text{periodic interest payment}, \$125 \\ i_{eff} &= \text{effective coupon interest rate}, 5\%/2 = 2.5\% \\ N &= \text{investment term, in periods}, 10 \text{ years } x \ 2 = 20 \text{ periods} \\ FV &= \text{face value of the bond}, \$5,000 \end{aligned}$$

And so it follows that

$$\begin{aligned} PV &= +\$125.00\ (P/A, 5.063\%, 20)\ +\ \$5,000.00\ (P/F, 5.063\%, 20) \\ &\quad +\$125.00\ (12.3964)\ +\ \$5,000.00\ (0.3724)\ =\ \$3,411.71 \end{aligned}$$

By comparison with our previous scenario of semi-annual compounding PV = \$3,098.53 and by switching to quarterly compounding the non-linear nature of TVM yields a much higher PV = \$3,411.71.

Bond worth problems tend to follow a fairly standard calculation pattern, and a few key points are worth summarizing:

1. Clearly distinguish between coupon rate and investor yield rate in calculations. Often it is better to just use the coupon rate to compute the coupon annuity and then forget about the coupon rate in subsequent calculations.
2. Bond worth PV varies inversely with interest rate. As the investor yield rate increases, the bond PV will decrease.
3. Read the statement of interest rate very carefully so the correct effective interest rate is used. We see that it may be necessary to use Equation 3.20 if our required compounding period is *shorter* than the period for the calculation of PV. This most often takes the form of quarterly compounding with a semi-annual PV calculation.
4. Check to be sure the answer makes sense. If the prevailing investor MARR is higher than the bond coupon rate, the PV of the bond will be less than the FV of the bond.

4.8 Summary comments

This chapter is a critical "mile-marker" in gauging your understanding of the basics of the time value of money. Many engineers assume simple equations mean simple problems, and a goal of this chapter has been to emphasize that the *context* of a problem and *implied factors* can be just as important as the nominal data provided with a problem statement. An agility with the basic TVM tools outlined earlier in Chapter 3 should have been evident in the present chapter. Not being fluent in the use of PV, FV, and AW restricts the understanding of the derivative concepts discussed over the last 50 pages.

A couple of key points to remember are:

1. Rates of return are impacted by inflation, though often inflation is neglected when making a comparison between alternatives (since the impact on choice is a wash).
2. IRR is not without controversy, going back a very long time [10].
3. Simplistic measures have limitations that should be clear. For instance, the payback period formula does not look at the value of all returns. The formula for the net present value method may be used to close gaps in information to enable you to properly evaluate the best choice.
4. Although the net present value method may be preferable to determine long-term profitability, the payback period formula helps with cash-flow

analysis for short-term budgeting. Sometimes you really do have to concern yourself only with short-term cash flow in order to survive.

5. Payback ignores any benefits that occur after the payback period and, therefore, is not an adequate measure of profitability.
6. Despite the many, many limitations of IRR it remains a popular litmus test with many managers. This is unfortunate, but rather than fight tradition, the idea is that for any given problem or decision, the analyst must take the time to look at the problem from multiple perspectives and introduce a variety of financial measures. Don't immediately assume the board or team making the decision *only* wanted IRR; provide a richer response. It does not hurt to have an initial reputation that you're willing to do more than the minimum – you'll know soon enough what the culture encourages.

4.9 Problems to work

Problem 4.1

A project has a first cost of $16,999 with a life of 10 years and a salvage value of $2,500. If the MARR is 12% and provides benefits of about $2,750 annually, what is the PW of the project?

Problem 4.2

You are faced with making a decision on a large capital investment proposal. The capital investment amount is $640,000. Estimated annual revenue at the end of each year in the eight-year study period is $180,000. Estimated annual year-end expenses are $42,000, starting in year 1. These expenses begin *decreasing* by $4,000 per year at EOY 4 and continue decreasing through EOY 8. Assuming a $20,000 market value at EOY eight and a MARR = 12% per year, what is the present worth of this proposal? What is your conclusion about the acceptability of this proposal?

Problem 4.3

Evaluate machine XYZ on the basis of the PW method when the MARR is 12% per year. Pertinent cost data are as follows:

Investment cost	$13,000
Useful life	15 years
Market value	$3,000
Annual operating cost	$100
Overhaul cost - EOY 5	$200
Overhaul cost - EOY 10	$550

Problem 4.4

A testing laboratory requires a new DRIE machine with a cost of $499,000, and in this high-tech industry the machine will have a salvage value of $40,000 at the end of 10 years of service. Assuming a MARR of 18%, how much must be earned on an equivalent annual basis so the firm recovers its original investment *and* earns a return on the capital over the life of the DRIE machine?

Problem 4.5

A material processing oven has a first cost of $16,999 with a life of 10 years and a salvage value of $2,500. The corporate MARR is 12%. If the system provides about $2,750 annually, what is the PW of the project if the system is replaced by an identical system at the end of 10 years and then again at the end of 20 years?

Problem 4.6
A used conveyor dealer presents an investment package that requires you to pay $189.99 a month with $1,499.00 down and a nominal annual interest rate of 17%. What is the actual cost of the system? Assume monthly compounding.

Problem 4.7
A capital investment project involves the purchase of a 3-D soldering station for $15,000, producing an annual net benefit of $2,250. The system has an 8-year useful life with a salvage value of $6,200. What is the present worth of the project?

Problem 4.8
A tattoo shop has the following revenue stream over a 5-year period: $9,000, $17,000, $19,000, $11,000, and $22,000. What is the EUAW of this business (assume i=12%).

Problem 4.9
A dicing system has a first cost of $68,000 with a life of 8 years and a salvage value of probably not more than $9,000. The corporate MARR is 14%. If the system has a maintenance cost of about $10,000 annually, what is the EUAC of the system?

Problem 4.10
An observatory has a telescope that requires an initial investment of $25,000. A benefactor will provide a $7,500 gift every 5 years as long as the university makes the telescope available to the public. If the university MARR is 8%, what is the EUAW of this project?

Problem 4.11
What is the maximum price an investor will pay for a bond with a face value of $5,000 and a coupon rate of 12%, paid annually? Assume the investor seeks a yield of 10% and the bond will mature in 10 years.

Problem 4.12
An investor purchased $20,000 of U.S. Treasury bonds at face value. These bonds have a 25-year maturity period, and they pay 1.25% interest every three months. Unfortunately, interest rates for similar securities have since risen to a 6% APR because of an improving economy, and the Federal Reserve Board has acted to keep the economy from "overheating." What is the current value of the bonds after 1 year?

Problem 4.13
The current market interest rate on U-Store City Investment Building bonds is 10%, compounded semiannually. The original bonds are sold as $1,000, 12% interest rate notes. A batch of bonds is available that will mature in 8 years. What is today's market price for a $1,000 face value bond?

Problem 4.14
Ajax Savings Bank has certificate of deposit notes with a face value of $10,000 with a 10% interest rate compounded daily. What is the current market price of the CD if the term of the CD is 3 years?

Problem 4.15
A US Treasury bond has a maturity of 10 years and a face value of $10,000 with a 5% coupon rate. Trading of the bonds has been pretty active, and the current quoted price is $96,500. What yield can an investors expect if they buy these bonds and hold them full-term?

Problem 4.16
A new high-definition television costing $2,999 is on a rent-to-own plan and is to be paid in 18 end-of-month payments (starting one month from now). The monthly payments are determined as follows:

Loan principal = $2,999
Interest = 1.365% per month
Loan application fee = $149

What is the monthly payment for this system? What are the nominal and effective annual interest rates *actually* being paid by the owner?

Problem 4.17
It is said that "the higher the MARR, the higher the price that a company should be willing to pay for equipment that reduces annual operating expenses." Explain the reasoning behind this statement.

Problem 4.18
A homeowner found a preferred savings account book in the wall of a home he was remodelling. He went to the bank and discovered that the account was 25 years old and had $425,801 in the bank! The original investment was $20,000? What was the effective interest rate of the preferred account?

Problem 4.19
What is the rate of return for a $14,000, 20-year investment that provides a payout of $1,000 a year?

Problem 4.20
A neurosurgery resident wants to double her money in the time it takes her to complete a residency program (7 years) so she can buy a new car with cash upon graduation. What interest rate is needed to turn a $10,000 investment into $20,000.

Problem 4.21
What is the payback period for the project described below?

Machine investment	$24,999
MARR	10%
Annual benefit	$6,500
Annual maintenance	$2,000
Life	10 years
Salvage value	$2,000

Problem 4.22
What is the payback period for the project described in Problem 4.21 if the maintenance increases by $500/year (after the first year)?

Problem 4.23
What is the payback period for the project described below?

Machine investment	$24,999
MARR	18%
Annual benefit	$5,000
Annual maintenance	$2,500
Life	7 years
Salvage value	$1,000

References

[1] Donald G. Newnan, Jerome P. Lavelle, and Ted G. Eschenback. *Engineering economic analysis.* Oxford, 2009.

[2] L.M. Murphy, P. Jerde, L. Rutherford, and R. Barone. *Enhancing commercial outcomes from R&D.* National Renewable Energy Laboratory, 2007.

[3] C.C. Miller and R.D. Ireland. Intuition in strategic decision making: Friend or foe in the fastpaced 21st century. *Academy of Management Executive*, 16:19–30, 2005.

[4] Daniel Goleman, Richard E. Boyatzis, and Annie McKee. *Primal leadership: Learning to lead with emotional intelligence.* Harvard Business Review Press, 2004.

[5] Nicky A. Dauber, Joel Siegel, and Jae K. Shim. *The vest-pocket CPA.* Prentice Hall, 1996.

[6] Jeppe Ladekar. Safeguarding investment in Danish mortgage bonds. *Journal of Financial Regulation and Compliance*, 6(1):59–69, 1998.

[7] Torben Gjede. *Mortgage finance in Denmark.* Housing Finance International, 1999.

[8] Joe Light. *Where to invest in an era of risk.* Wall Street Journal, June 29, 2013, p. B7.

[9] Sanders Shanks Jr. Municipal bond defaults. *National Municipal Review*, 26(4):296–298, 2007.

[10] Tung Au. Overall rate of return as a profit measure for capital projects. *Engineering Costs and Production Economics*, (7):235–241, 1983.

Chapter 5

Comparing Alternatives

5.1 Learning objectives

A practical skill in finance and engineering economics is the comparison of alternatives. Usually we are comparing alternatives to (a) maximize the expansion of wealth, or (b) improve profitability through cost-reduction decisions. The foundation for this discussion has been set by the basic concepts in the time value of money (TVM), enriched with additional concepts involving depreciation, taxes, and probability. If you are not already fluent in the TVM equations and equivalence calculations then this chapter will be a challenge. The application of decision trees also plays a role for comparing alternatives, and can be a handy tool for moderately complex problems. The classic "lease versus buy" decision will be explored in some depth as we anticipate that FASB guideline changes will soon be in place that will improve transparency of off-balance-sheet assets. Our treatment of depreciation and taxes will be moderately brief - any number of textbooks describe these concepts well - to focus more on the context of problems that foster critical thinking skills. After reading and discussion sessions, the student should be able to:

1. Understand the basic concepts of selecting alternatives, including:
    (a) Mutual exclusivity
    (b) Cash-flow diagrams
    (c) Ensuring a comparable basis between alternatives
2. Understand how to handle simple after-tax-cash-flow (ATCF) problems.
3. Argue the "pros and cons" of contemporary lease-versus-buy decisions.
4. Illustrate how parametric studies can influence the outcome of a selection between alternatives.

5.2 Framework for comparing alternatives

Engineering managers must become accustomed to a steady diet of identifying, analyzing, and choosing between alternatives, even if a decision is made to just "keep doing what we're doing." Engineers new to the management role might feel uncomfortable making specific choices or declining to act on certain project options, especially if the decisions they must make involve ambiguous information. Normally this reflects discomfort with the decision-making process, exacerbated if analytic rigor is absent and the manager cannot defend a decision relative to specific organizational success metrics. Such tasks are clearly easier when the decision is framed with some form of objective criterion and the manager is fluent with the quantitative aspects of the decision-making – essentially decisions are informed and relevant to mission, not perceived as just personal opinion.

The framework for analysis and context of the outcome tend to derive from four basic decision-making scenarios:

1. **Scenario A - Single project decisions**. This is the classic "go" or "no-go" decision for a single project. The decision involved is conducted somewhat in isolation (relative to other project options), with principal project merit measured with respect to corporate benchmarks such as IRR, MARR, NPW $\geq$ X, etc.
2. **Scenario B - Choice between two options.** Here, the manager must choose between two alternatives in which the non-trivial "do nothing" is not an explicit option. The majority of our attention in this chapter will be spent on this scenario. We will examine in detail the *incremental rate of return* since this is the basis by which choices in this and other scenarios are best made.
3. **Scenario C - Multiple solutions from multiple options**. Situations involving multiple alternatives, where more than one option might satisfy a baseline criterion (say, MARR). It may not seem immediately obvious why this scenario is of interest; some projects have aesthetic or another non-financial basis which constrains the problem.
4. **Scenario D - Best solution from multiple options**. Situations where a *single best* option is to be flushed out of multiple alternatives. In this scenario spreadsheets are essential to keep the "bookkeeping" straight, though we'll notice the technique of Scenario C is essential in working through options to a final solution.

In each case, many of the tools we have developed in prior chapters (ROI, NPW, EUAW) will now be applied in a more complete managerial context that will underscore their value in more complicated analyses. Some analysis methods mix and match tools or algorithms and this may lead to confusion by the reader if the concepts, nuances, and limitations of IRR, PW, or EUAW are still unclear.

Scenario A: Single project decisions

Single project decisions have been discussed in Chapter 4, particularly MARR in Section 4.4.1 (p.111) and Section 4.4.2 (p.113) in the discussion of IRR. It was clear that these ratios alone might lead to erroneous decisions, and that the analyst must also consider, say, NPW when examining scenario data.

We will not explore single project scenarios analysis much further in this chapter beyond what has already been presented in the last chapter. Previous exercises (see sample problems 4.2-4) make clear that, say, MARR and NPW, set pre-defined institutional criteria from which the project can be determined as acceptable" or not. The single project decision is fairly unique in most large capital equipment decisions since such projects ordinarily have *many* different solutions (options) available, and (as we shall see) simply comparing the PW or IRR or each option individually is inadequate. Still, it has become popular in recent years that corporate finance may require the manager to write a justification form to support a purchase requisition, and that the justification must include an acceptable ROI or MARR for even moderately small equipment purchases. For instance, a purchase requisition many require an estimate of the total cost of ownership (TCO) for a printer offered at the introductory price of \$59 since – you guessed it – the annual cost of ink cartridges are an order of magnitude greater than the capital cost! Or more often, a piece of equipment in the lab may be "desired" but no specific thought given to how the equipment will pay for itself. It is perfectly fine to want lab upgrades to ensure competitiveness, but in budget-constrained environments there is nothing wrong with asking lab personnel to think about how such a purchase ensures value is relevant. If any aspect of the single project decision from Chapter 4 is unclear, a quick review is recommended prior to reading further.

Scenario B: Choice between two options

The bulk of this chapter will focus on methods to make a choice between two viable options. Textbooks vary somewhat on this topic, with some using only the so-called "incremental" analysis, others preferring graphical methods, and yet others using a combination of incremental, graphical, and differential methods. In the current work we will primarily use the *incremental rate-of-return method.* While this may not be viewed by some as the most comprehensive approach (with fewer graphs to entertain us), the method is nonetheless quite simple, and once understood the key concepts involved make understanding the other approaches that much easier. We bring into play the reasoning behind specifying *mutual exclusivity* of options, as well as outline why the incremental analysis technique helps avoid errors of declaring a "winner" when comparing against a predetermined corporate standard versus comparing the options against *each other.* We will distinguish between a choice between two options and the special case of *replacement analysis* that involves a special basis of analysis (and is a chapter in itself, Chapter 9).

Scenario C: Multiple solutions from multiple options

Project circumstances can steer us to two basic scenarios where more than one qualified solution meets a corporate need. In one situation, a problem is presented where the decision-making process does not yield a single best option, but rather multiple acceptable (qualified) courses of action emerge. In the narrative to follow later in this chapter we will see how to reduce such a cohort to a *single, optimal* option; for now let's simply acknowledge that the character of certain problems is that several solution pathways can be justified, especially if based on pre-set financial criteria. This is reasonable if we consider that some projects involve *more* than financial criteria during decision-making. For instance, it may be that aesthetics, customer preference, or regulatory affairs open the problem to several admissible financial solutions and we need to embark on a secondary selection process to converge on a single best solution.

R&D portfolio investment strategy is another situation where multiple solutions are (actually) preferred. Imagine that during a corporate annual meeting the board of directors suggests that lagging sales are attributable to a weak product pipeline. After considering industry comparables, the board might allocate $1 million to be invested in high-risk, high-return projects that have a minimum ROI of 25%. In this case, a request for proposals to researchers may elicit 10 project proposals meeting the ROI criteria and the R&D manager must convene a team to inaugurate a down-select process. Indeed, any project exceeding the cost of capital might be acceptable, and if too few projects are funded, then there is an underutilization of the funds set aside to meet corporate objectives. This has the look and feel of a capital budgeting discussion, to be discussed further in Chapter 9. We will discuss later how these projects can be characterized as *independent*, not mutually exclusive. Independent alternatives benchmarked against a financial standard are not compared against each other.

Scenario D: Best solution from multiple options

As a project's capital investment budget rises, so will the number of options people may offer as a potential solution. This is common sense, actually, and desirable, too. We seek to find the most productive use of capital that satisfies the needs of the organization. This requires that we scan the horizon for any relevant solutions and incorporate them in the decision-making process. We want to be sure that we've viewed our problem from enough different perspectives that we've harvested the best thinking and potential solutions to the problem. Once presented, each candidate solution must undergo a structured process to narrow the field, eventually leading to a single best option. Ultimately the options must be mutually exclusive; the selection of the "best" is the one that precludes all the rest.

In the case of multiple alternatives to a problem solution, the tendency is to first rank solutions in order of capital investment, and then go through a series of analyses where combination of pairs of options are studied. Incremental investment decisions are made until there emerges a single best "challenger" to each "defender." Our discussion here will be very helpful when we get to Chapter 9 and discuss *replacement analysis*. We will find out this is another example of how a simple set of tools can be re-purposed to assist in other problem domains.

5.3 Mutual exclusivity and independence

We have highlighted that for a given investment activity ("task"), there are many different possible proposals, methods, and activities (including "do nothing") that are simply a set of alternatives that can accomplish the task. Further we have alluded to three basic types of alternatives:

1. Mutually exclusive: The selection of one alternative precludes another.
2. Independent: The selection of one alternative has no bearing on the acceptability of another.
3. Dependent: The selection of one alternative requires the selection of another.

Mutually exclusive projects have an impact on NPV calculations in the comparison and ranking of alternatives. For a given financial objective, the selection of one option might provide a more favorable IRR and less favorable NPW relative to another (precluded) option. That the opportunity has not been *exhausted* means that the available capital is underutilized. The individual project may therefore meet some, but not all of the investment objectives. In this way, mutually exclusive projects affect cash flow while independent projects do not. The more complete capital budgeting discussion of Chapter 9 explores this further.

Sometimes the difference between *independent* and *mutually exclusive* projects is hard to grasp. In some textbooks the use of the term "disjoint" is synonymous with mutual exclusivity to underscore that mutually exclusive options have nothing in common. Consider that two projects X and Y are *independent*; from a probability perspective,

$$Pr[X \cap Y] = Pr[X]Pr[Y] \tag{5.1}$$

which tells us that for independent projects the probability that both projects X and Y can intersect (occur) is equal to the product of the probability X will occur and the probability Y will occur. In the case of mutually exclusive projects A and B the joint probability is null:

$$Pr[A \cap B] = 0 \tag{5.2}$$

This underscores the explanation that the selection of A precludes the selection of B.

5.4 Evaluation of alternatives

There are four basic steps involved in comparing alternatives with the evaluation method. After outlining these steps it helps to work through several examples. Sometimes it seems easier to convey key ideas and calculation sequences with specific examples than to try to architect a general theory.

Traditionally, textbook calculations for the assessment of alternatives provide separate sections for each of the graphical, incremental, and rate-of-return computational techniques. Engineering practice seems to have drifted away from graphical solutions and other methods that at one time provided calculation convenience. With the prevalence of computer "netbooks" and a variety of tablet computers in common use at the executive level, computing power is such that analysis techniques that might once have expedited interpretation of data might now actually be clumsy. Fitting with the "field guide" theme of this book, we therefore choose to draw on worksheets to illustrate the outcomes of calculations whose procedures have been explained in prior chapters; this enables us to focus instead on the essentials of decision making. The four-step procedure in evaluation of alternatives follows.

Step 1: List alternatives in order of increasing capital cost

Reflecting back on Section 1.3, the fiduciary responsibility of an engineering manager will impart a conservative slant to capital project investments and therefore the manager will want to arrange options in *increasing order of initial capital outlay.* Even if the decision-making process eventually points to a higher-capital-cost option as having an overall lower *system* cost, it is conventional to list in increasing order, as illustrated in Table 5.1. The reason for this is that we always select that option that *minimizes* the capital investment unless *the incremental investment for a higher capital cost alternative can be justified.* Option A will represent the *starting* option that will "defend" its preferred position against a "challenger."

In Table 5.1, computations begin with Option A as the defender and Option B as the challenger. Sometimes the defender is referred to as the *base* or *baseline* alternative.

Project Parameter (8% MARR)	Option A	Option B	Option C
Initial project investment, P	$10,000	$20,000	$30,000
Expected duration of use, N	7 years	7 years	9 years
Estimated annual net revenue	$2,500	$5,150	$1,200

Table 5.1: Capital project investment options

Even with the use of a spreadsheet you will want to organize information in this way so that tracking of options and outcomes is simplified.

Step 2: Harmonize data: Ensure expected-duration-of-use is equal for all options.

Equitable comparisons require the useful life of each option to be for equivalent periods of analysis. In Table 5.1 we see that data for Options A and B have the same 7-year expected life, but Option C has a 9-year life. Mutually exclusive options must be based on equivalent study periods. If the study periods are unequal, it is necessary to either *truncate* data when the study period is shorter than the expected useful life, or use the *repeatability assumption* if the study period is longer than the expected useful life. Adjust all cash flows to correspond with the study period. The way to do this is discussed further in Section 5.6. As a practical matter, note that this issue is simplified in the case of AW versus PV calculations.

Step 3: Identify the current defender and challenger

Decision-making begins by identifying the two alternatives for subsequent analysis. As mentioned above, by defualt, Option A will represent the *starting* option that will "‘defend" its preferred position against a "‘challenger." In Table 5.1, computations begin with Option A as the defender and Option B as the challenger. Sometimes the defender is referred to as the *base* or *baseline* alternative.

Step 4: Compute the PW equivalent worth and IRR values for each option

Internal rate of return concepts and calculations were discussed in Section 4.4.2. As well, PW equivalent worth calculations and NPV concepts were illustrated in Section 4.2, and these techniques are easily applied when we have alternatives with equal useful physical life. As indicated earlier in Section 4.3, many unequal-life challenger/defender calculations might more handily be performed with EUAW and EUAC techniques instead of PV methods of analysis.

The four steps are actually very easy to implement, and the pattern of analysis becomes obvious just by working through several problems. Three comments before discussing a worked problem:

1. Several worksheets (including for the sample problems discussed here) accompany this chapter. The reader is encouraged to download and explore these worksheets to assist with understanding detailed calculations.
2. A very conventional and typical "textbook" comparison is between *only* two options "A" and "B" and the difference between the PW for options, as we will see in an example below (Table 5.2) The idea is to introduce the so-called "delta" method to show that the PW for Option A is equal to the difference between the PW for Option B and the Delta(A-B).
3. IRR differences are consistent with the difference in magnitude of the PW values. A limitation of this type of calculation is the implicit assumption in IRR calculations that the *reinvestment* at the IRR is *actually possible.* Not a problem for our test case, but we stretch the bound-

aries of reason when, say, $IRR > 35\%$ are computed. This general topic is presented in all textbooks and has been an issue with analysts for decades [1], but IRR is simple and remains popular with managers.

Example 1 Two years into a 10-year project, Consolidated Shipping Associates is considering an upgrade to a package conveyor line. A local supplier has two systems on sale as shown in Table 5.2; the QuickLine series is a close-out special (discounted 30% from list of $15,999), and the RFID Red Rider is built with RFID tracking as an option, though "as-is" the floor model does not have the RFID electronics ($10,000 up-charge) included. Because both systems are close-outs, there is no expectation the conveyor lines will be of much value at the end of the 8-year project period. Which system should be chosen?

Project Parameter (15% MARR)	QuickLine	RFID Red Rider	Difference
Initial project investment, P	$11,199	$19,995	$8,796
Expected duration of use, N	8 years	8 years	None
Estimated annual net revenue	$2,500	$5,150	$2,650

Table 5.2: Capital project investment: Quickline vs. Red Rider.

Using methods from Chapter 4, the NPV of each option is easy to find:

$$\begin{aligned} NPV_{QuickLine} &= -\ \$11,199 + \$2,500(P/A, 15\%, 8) \\ &= -\ \$11,199 + \$2,500(4.487) = \$18.50 \end{aligned}$$

$$\begin{aligned} NPV_{RedRider} &= -\ \$19,995 + \$5,150(P/A, 15\%, 8) \\ &= -\ \$19,995 + \$5,150(4.487) = \$3,114.75 \end{aligned}$$

Results are summarized below, along with a computation of the IRR (calculation not shown for brevity) in each case. The obvious answer might be that

	Quick Line "Q" Conveyor	Red Rider "R" Conveyor
Net present worth	$19	$3,115
Annual net return	22.3%	25.6%
IRR	15.1%	19.6%

Table 5.3: Conveyor line upgrade options

System R has the greatest NPV and highest IRR, so that should be the system chosen. But does the answer make sense? Some questions jump to mind that need further exploration.

First, what is the source of the difference in the annual net revenue of Conveyor R being *twice* that of Conveyor Q? Possibly the basis was that

Conveyor R is RFID ready and would improve the productivity of the system. If that is the case, then we would need to adjust the initial capital investment by $10,000 for the cost of the software; in such a case you can substitute $29,995 for the $19,995 listed and what results is a NPW = -$6,885 (< 0!) and an IRR of 7.6%, which would fail the IRR > MARR test.

Second, why is the life of the more expensive system the same as for the less expensive conveyor system? In fact, why is the life of the conveyor system so short to begin with? Good conveyor systems that are maintained properly can have a life of 30,000 to 40,000 hours (more than 12 years). We need to inquire further about salvage value. Even though we expect to have the production line up for another 8 years, suppose Conveyor Q has a residual value of $3,000; this will boost the PV to $1,000 and the IRR will rise to 17.3%.

If we *double* the life of Conveyor R, then a 16-year life repositions *both* systems to have much higher NPV; for Conveyor Q the NPV is about $3,600, and for Conveyor R it is over $10,000. What was once a marginal investment scenario for Conveyor R is now more attractive.

Third, suppose that after all other downsides are considered, management still likes the idea of the more robust conveyor line ("robust rollers reap reliability") and they simply like the idea of having an RFID-capable system even if it is not used. In that case, the annual net benefit has to drop to $4,460 to have the same NPV (of $19) as Conveyor Q. And there might be very good reasons for this. Again, we consider that there may be uncertainty in the future and management wants to reserve the option to switch or expand the system in the future. It may be that there are market, industry, or regulatory issues on the horizon that warrant a hardware upgrade in place today. A good example might be lot control in food processing, whereby the residence time of frozen foods needs to be tracked for safety. Such a regulation may not be in place today, but possibly the senior VP of sales has been networking and the word on the street is that product tracking may be mandatory to be a supplier for certain distribution outlets. Later in Section 8.5.3 we expand on the use of decision trees to help us manage the risk of entering new markets.

This discussion encourages you to do the following for any real problem of practical interest.

1. Poke and prod at different assumptions or parameters in a problem. Don't lose sight of the fact that most are simply set to bound a problem, but that does not mean in practice that the "as given" parameters reflect reality.
2. Even if the parameters are assumed to be fixed and correct, performing a parametric analysis will enable you to answer *how far off* do the baseline parameters have to be for a good decision to turn into a bad one. This is not all that burdensome: modifying a few parameters up or down 10% will give you an idea of the scenario's robustness.
3. The ease with which calculations can be made leaves no reason why you wouldn't know IRR, PW, AW, etc. These parameters are simple to include on spreadsheets and assist in reminding the audience to examine the problem from many perspectives.

Avoid reading this chapter passively. Bring out the spreadsheets that accompany this volume and test the effect of changing numbers. Seeing trends helps underscore the understanding of the (very simple) PV and IRR calculation, and you may even be motivated to modify the worksheet to suit problems more difficult than presented in this book.

Example 2 At the end of FY20xx MRI Imaging has determined they need to increase their imaging capabilities to compete in the growing sports health management arena. Their closest competitor has three times the imaging capital equipment and they run at 82% capacity. MRI's board of directors estimates they could safely add $500,000 in capacity and the regional market would still not be saturated. The funds for the project are to be obtained by allocating $350,000 in retained earnings from the prior year and useing $150,000 from their line of credit, if needed. The company was expecting to use the same imaging vendor they've always used and the $350,000 investment in System K would produce a net rate of return of 20%; the return is considered fairly good in the down service economy where the company MARR has dropped to 8%. A new imaging vendor hears about the potential sale and offers a more advanced System Z that costs $500,000. The vendor claims the higher net rate of return of 22% is "worth it in the long run." The argument is that the higher cost enables faster imaging and greater patient throughput. "You'll make up the difference in the capital cost in three years," exclaims the salesman. MRI is a conservative company, and the board is inclined to spend as little on capital equipment as possible. Can it be argued that the more expensive system is worth it?

Using methods from Chapter 4, we find the NPV of each option:

$$\begin{aligned} NPV_K &= -\ \$350,000 + \$70,000(P/A, 8\%, 7) \\ &= -\ \$350,000 + \$70,000(5.2064) = \$14,448 \\ NPV_Z &= -\ \$500,000 + \$115,000(P/A, 8\%, 7) = \$72,700 \end{aligned}$$

Now, NPV calculations and Table 5.6 are used to construct Table 5.7. From a "classic" NPV perspective, Option K not only has the lowest cost, but the NPV > 0, so it meets the basic criteria of being an acceptable investment. But is there the possibility that *relative* to Option Z, the default is the *best* alternative? The annual net return of both options is comparable and reasonable, far exceeding the somewhat low 8% MARR set by the company. Note that were we to increase MARR by just 1.2% the NPV ≈ 0 and thus we really have a situation with a low hurdle rate – a pause for concern. The hurdle rate is below stock market returns for Q1CY13 and the 15% YTD S&P 500. Might the money be better put to use in the market? Maybe, and maybe the expected returns are *intentionally* conservative since the company is adding capacity and it may take time for marketing and patient referrals to build to utilize capacity.

Table 5.7 demonstrates that Option Z meets the PW and the IRR "sanity check," but there still is quite a management decision to make. Suppose that

QuickLine

Capital Investment	$11,199
Residual Market Value	$0
Annual Benefit	$2,500
Period	8 years
Corporate Rate of Return	15%

	Capital Investment	Annual Benefit	Residual Value	Net Cash Flow
0	-$11,199			-$11,199
1		$2,500		$2,500
2		$2,500		$2,500
3		$2,500		$2,500
4		$2,500		$2,500
5		$2,500		$2,500
6		$2,500		$2,500
7		$2,500		$2,500
8		$2,500	$0	$2,500
			IRR	**15.1%**
Present Worth Method				
PW:	($11,199.00)	$11,218.30	$0.00	**$19**

RFID Rider

Capital Investment	$19,995
Residual Market Value	$0
Annual Benefit	$5,150
Period	8 years
Corporate Rate of Return	15%

	Capital Investment	Annual Benefit	Residual Value	Net Cash Flow		**Delta (Y-X)**
0	-$19,995			-$19,995		-$8,796
1		$5,150		$5,150		$2,650
2		$5,150		$5,150		$2,650
3		$5,150		$5,150		$2,650
4		$5,150		$5,150		$2,650
5		$5,150		$5,150		$2,650
6		$5,150		$5,150		$2,650
7		$5,150		$5,150		$2,650
8		$5,150	$0	$5,150		$2,650
			IRR	**19.6%**	**PW:**	**$3,095**
Present Worth Method						
PW:	($19,995.00)	$23,109.71	$0.00	**$3,115**		

Table 5.4: Example 1 equivalence and IRR calculations.

Project Parameter (8% MARR)	System K	System Z	Difference
Initial project investment, P	$350,000	$500,000	$150,000
Expected duration of use, N	7 years	7 years	None
Anticipated annual net return	20%	22%	2%
Annual net revenue	$70,000	$110,000	$40,000

Table 5.5: MRI Imaging upgrade options.

Project Parameter (8% MARR)	System K	System Z
Net present worth	$14,448	$72,700
Annual net return	20%	22%
IRR	9.2%	12.1%

Table 5.6: NPV of MRI Imaging Upgrade Options

each $10,000 in net revenue requires 5 new patients. So to achieve $110,000 in new net revenue there has to be an increase of 55 patients annually, just a little over 4 patients a month. Given that the nearest competitor runs consistently at 82% capacity, it does not seem to be much of a "reach" to expect the business could expand to meet or exceed the minimum usage level.

Still, there is a significant difference in capital equipment cost, and the absolute cost of the machine may cause issues with finance. But in this case Bethesda is using up to $350,000 in *retained earnings* to finance the project, and using a line of credit to fill the 30% gap in capital budgeting. The commitment to reinvestment in the business with retained earnings is consistent with the conservative nature of the company. All around, then, an upgrade may help Bethesda cross the "critical" mass of new capacity and productivity.

If MRI purchases the equipment with $350K in cash and $150K in a loan, then the balance sheet becomes:

The debt ratio is

$$\texttt{Debit Ratio} = \frac{\texttt{Current Liabilities} + \texttt{Long-term Liabilities}}{\texttt{Total Assets}}$$

And from the two balance sheets:

$$\text{DR}_{before} = \frac{\$75,000 + \$65,000}{\$950,000} = 14.7\%$$

$$\text{DR}_{after} = \frac{\$75,000 + \$215,000}{\$1,100,000} = 26.4\%$$

Although the percentage has risen almost by a factor of two, the DR after the acquisition is still less than 30% and is not unrealistic for a capital-intensive company.

Back to the question: Can it be argued that the more expensive system is worth it?

1	A	B	C	D	E
2					
3	Capital Investment		$350,000		
4	Residual Market Value		$0		
5	Annual Benefit		$70,000		
6	Period		7	years	
7	Corporate Rate of Return		8%		
8					
9		Capital	Annual	Residual	Net Cash
10		Investment	Benefit	Value	Flow
11	0	-$350,000			-$350,000
12	1		$70,000		$70,000
13	2		$70,000		$70,000
14	3		$70,000		$70,000
15	4		$70,000		$70,000
16	5		$70,000		$70,000
17	6		$70,000		$70,000
18	7		$70,000	$0	$70,000
19					
20			**IRR**		**9.2%**
21					=IRR(E11:F17,+C7)
22	**Present Worth Method**				
23					
24	PW:	($350,000.00)	$364,445.90	$0.00	$14,445.90
25			=PV(C7,C6,-C12)	=PV(C7,C6,-D18)	

Table 5.7: IRR for Selection K of MRI Imaging.

1. For either system the IRR criteria is met, though it seems the more expensive system will provide a more robust return on investment.
2. The NPV of the lower cost system is essentially $0, and this is of concern since the hurdle rate MARR=8% is not aggressive at all. The MARR is quite close to the cost of capital and this may need a closer look.
3. Both systems provide a reasonable annual net return to the business (> 20%) and this seems not to be a issue.
4. The $500K system can be estimated to only require just a little over 4 new patients a month in the medical practice, which seems reasonable given the high level of capacity utilization of other firms.
5. Under either scenario the debt ratio is at a conservative 30%; long-term debt-to-equity hovers around 26%, too, and this does not seem to place the company in any liquidity risk. The company has chosen to use a line of credit to conserve cash on hand.

Stepping back and considering how the market is growing and the belief the company has that they can appropriately use retained earnings to fund 70% of the $500K capital acquisition, the recommendation to pursue the more expensive system seems reasonable at this time.

It is not expected that the reader needs to dig in as deeply as we have here for a typical PV or IRR calculation, and the remarks and balance sheet discussions are simply to show a close link in other concerns in decision-making. At a minimum we place what is otherwise a fairly "sterile" calculation in a

Bethesda Imaging, Inc.
Condensed Balance Statement

Current Assets		Liabilities and OE	
Cash	$425,000	Current Liabilities ..	$75,000
Accounts Receivable	$50,000	Long-term Liabilities	$65,000
Inventory	$25,000	Owners' Equity	$810,000
Equipment	$450,000		
	$950,000		$950,000

Table 5.8: MRI Imaging, Inc. condensed balance sheet *before* equipment purchase.

Bethesda Imaging, Inc.
Condensed Balance Statement

Current Assets		Liabilities and OE	
Cash	$75,000	Current Liabilities ..	$75,000
Accounts Receivable	$50,000	Long-term Liabilities	$215,000
Inventory	$25,000	Owners' Equity	$810,000
Equipment	$950,000		
	$1,100,000		$1,100,000

Table 5.9: MRI Imaging, Inc. condensed balance sheet *after* equipment purchase.

richer context. As you work in teams on chapter problems, it is wonderful review to ask how each of the book's chapters pertain to the dilemma or decision at hand.

The next section rounds out this chapter by exploring the impact of taxes on decision-making.

5.5 Before-tax and after-tax issues

Most people are quite familiar with payment of sales and personal income taxes. To the extent that a corporation has a legal identity, it has to pay or deal with sales and income taxes, too. We briefly explored taxes in the context of financial statements in Section 2.3.4 and Section 2.3.5. There are four main types of taxes we are generally concerned with:

1. **Income taxes** are a tax on earnings and can make up a significant fraction of the cost of doing business. Income taxes can be *progressive, proportional*, or *regressive* and a prudent engineering manager understands the need for corporate tax planning. Certain aspects of "lease versus buy" or replacement analysis have tax implications that should be considered when making investment decisions.
2. **Sales taxes** are normally levied on goods and services provided to the end-user. If the product is purchased by one company from another

company (so-called business-to-business transactions) leading with the intent to produce a product for re-sale, then a sales tax many not apply. Some sales taxes are not collected by the seller, but the end-user pays the tax directly to the government ("use-tax"). Sales taxes are not normally in the mix of concerns an engineering manager has when weighing capital equipment alternatives.

3. **Property taxes** are paid by owners of real property, whether the owner is an individual or a corporation. Rules and rates vary widely with jurisdiction, and taxes levied on inventory have the look and feel of a property ownership tax. Taxes of this kind are often viewed as a barrier to doing business in certain states, and become the subject of special incentives by legislatures, intended to spur business investment. The same property may be subject to taxes by several different agencies of the local, state, and federal government.
4. **Excise taxes** are a tax on a very specific product *within* a country, in comparison with special taxes on *imported* goods known as *customs duties*. Excise taxes often add to the normal sales tax levied on a product. Many times the excise tax has a very high rate that targets a specific product, such as fuel, tobacco, or alcohol. These type of taxes generally have little bearing on equipment investment choices.

Reflecting on Section 1.2.2 we indicated the treatment of taxes would be light in this field guide as the topic has many detailed aspects a typical company must deal with on an on-going basis. The works of Lasher[2], Meigs and Meigs [3], Powers [4], and others dive into this important topic. When thinking of replacement problems or leasing decisions, some tax topics are very distant to decision-making issues. For instance, in replacement analysis we assume the outputs are comparable, and thus the replacement of one item with another would not impact sales; sales tax discussions are ordinarily not that relevant for such a decision.

Generally, the idea of an on-going concern (recall the balance sheet discussion, Section 2.3.4, p. 27) means there is a history of output already having certain taxes levied. Replacement of one asset with another equivalent asset does not have a profound impact on all possible taxes. In some cases, the linkage between new capital equipment and new sales is speculative, so the conservative analyst will leave this out. Similar thinking prevails for excise taxes. Slightly different and much more confusing are the array of possible property taxes. There are quite a large variety of local and state rules and special credits that can lead to confusion about the accounting of capital equipment that qualifies (and it not always does) as real property. This could be of concern if a new piece of equipment has a significantly different valuation than its predecessor. Impact will vary widely and *a priori* guidance that is absent knowledge of the locality is difficult. As well, the level of concern will vary depending on the value of the equipment relative to the existing production system investment. If the equipment is less than 10% of the value of the overall operation, then property tax issues are most likely incremental. Our work here centers primarily on the impact a decision will have on *income taxes*.

It is entertaining to think that at the time of the passage of the Sixteenth Amendment to the US Constitution in 1913, personal income was taxed at 1%! It has risen to over 50% at various times. All of this suggests that tax considerations are a worthwhile part of a complete financial analysis. What becomes evident is that taxes play two roles:

1. Obtain revenue for the operation of the government.
2. Influencing behavior of companies and individuals.

Notwithstanding the on-going legislative changes that affect tax schedules, in engineering economics we generally assume a corporate income tax rate of between 35% and 50%, with clear significance relative to the scope of decisions of interest in the present work. An issue to mention here is how the organizational form of the corporation plays into the effective tax rate. A sole proprietorship is likely to simply use a Schedule C form to report business income as part of the filing of personal income taxes. This pass-through means that taxes are levied once at the time the net income for the person is determined.

In comparison, a corporation has an identity and pays income taxes as mentioned earlier. But this is just the tax on the corporation. Owners of the company will receive earnings from the company *after* the corporate income has been taxed. Thus, we have *double taxation* of revenue to the owner, once at the corporate level and then again at the personal level. Thus, the specific organizational form discussed in Section A.3 should be given some thought early on in the start-up stage. Once the firm is under way, corporate form is dictated by current corporate standing and is generally outside tactical financial activities; this is interesting to think about but is somewhat academic since there is little to act on in practice.

Of the many possible pathways to accounting for the impact of income taxes, our comparison of alternatives will draw on just three aspects of taxes most closely coupled to capital equipment decision-making:

1. Depreciation impact on income statement.
2. Capital gain or loss on residual sale.
3. Tax savings for allowable expenses.

Depreciation tax credit. While many problems are solved on a before-tax-cash-flow (BTCF) basis a very simple example illustrates the impact of depreciation on after-tax cash flow (ATCF). Consider an enterprise with $50,000 in revenue and $20,000 cost of goods sold (COGS) and a nominal 40% tax on net income. Ignoring many of the items that one might include in an income statement (refer back to Table 2.10 on p. 61), the BTCF and ATCF would be:

There is thus a 23% decrease ($2,800) in tax liability on net income, equal to 40% of the allowable depreciation (0.40% x $7,000). This can have significant bearing on comparison of outcomes and decision-making, so often that a depreciation tax credit is included in comparison problems since accounting for depreciation has the effect of reducing the net expense of owning new equipment. **Capital gain or loss on asset sale**. Depreciation schedules provide

	BTCF	ATCF
Net Sales Revenue	$50,000	$50,000
Expenses		
COGS	$20,000	$20,000
Depreciation		$7,000
Net Income	$30,000	$23,000
40% tax	$12,000	$9,200
Net Income after Taxes	$18,000	$13,800

Table 5.10: Impact of depreciation on net income.

acceptable or traditional expensing of assets, as discussed in Section 2.5.1, p. 43, in accordance with proper expense realization principles. We do not expect this forecasting of the future value of an asset to accurately reflect *current* market conditions. Normally, then, there is a difference between the residual value of an asset and the price it will command if sold on the market today. Asset depreciation is an expense that lowers the effective taxes for a firm. At the time when one asset is replaced with another, there will be a capital gain or loss on the sale.

Consider the $350,000 investment for System K being contemplated by Bethesda Imaging back in Table 5.5, p. 146. Presume that the company used straight-line depreciation over the 7-year life of the system that has a salvage value of $70,000. This suggests a total depreciation of $280,000 and thus a $40,000 annual depreciation allowance. If the company wished to replace the system in year 5, the book value of the system would be

$$BV = \$350,000 - 5(\$40,000) = \$150,000$$

If market conditions or technology has changed rapidly since the time of the purchase (and may be why it is being replaced), suppose the asset is only worth 75% of its book value in year 5. When sold, this would be recorded as a loss:

Current market value	$112,500	$112,500
Book value (Y5)	$150,000	
Loss	($37,500)	
40% tax	($15,000)	$15,000
Net sale w/tax credit		$127,500

It can easily be shown that if the market value of the system were 6% higher than the book value, the *gain* on the sale of the system would be $9,000 and the *additional* tax of $3,600 would erode net cash proceeds to

$$\texttt{Net cash} = (\$159,000) - 0.4(\$9,000) = \$159,000 - \$3,600 = \$155,400$$

Because the change in the depreciation tax shield will vary widely, it is wise to even consider a rough estimate in preliminary calculations to bracket the possible issues.

Tax savings for allowable expenses. While it is entirely reasonable in considering alternatives to have the expectation that revenue streams will be similar for comparable service to production processes, many times the operation and maintenance costs will be quite different. There may have been a conscious decision to invest in a more expensive piece of equipment for the benefit of lower recurring operational costs. When expenses decrease, the tax liability will increase, and when expenses increase, there are higher tax deductions allowed. Labor, maintenance, and downtime can add up, as shown below:

	System K	System Z
Expense savings		
Labor	$20,000	$10,000
Maintenance	$12,000	$4,000
Downtime	$ 6,000	$2,000
Total savings	$38,000	$16,000
Tax increase	$15,200	$6,400
Net expense	$22,800	$9,600

Despite the fact that the allowable tax deduction for operational expenses has diminished and *more* taxes are paid, over the course of a 7-year period overall operational savings provide a benefit to the company. Later, in Section 9.5.3, p. 313, we work through an after-tax replacement problem to illustrate another calculation comparison in detail.

5.6 Comparable basis issues

A basic assumption made in earlier discussion and analysis is that the mutually exclusive equipment options under consideration are "comparable." In Section 5.4, the idea that alternatives should be listed in order of increasing capital cost implies the *technical* performance (output) was comparable and that it is the *financial* aspect that differentiates choices. Alternatives may not always be presented on a comparable basis for a variety of (good) reasons. Expectations or predictions of system life, performance, output, and efficiencies are bound to vary (even a small amount) and must be considered during the decision-making process as vendors compete for business supplying equipment for company projects and other capital equipment replacement needs.

Comparability must be examined at several levels:

1. Physical, operational, and financial life.
2. Equity of quality when making repeatability assumptions.
3. Useful life in unequal study analysis periods.

5.6.1 Technical comparability

If proposed systems *are* identically equal, then the technical comparison is a non-issue and the decision strictly a financial one. For decades, advances in technology or process improvements have been at the heart of incremental and disruptive innovations that enable competitive offerings in the marketplace[5]. Competitiveness in the capital goods industry often emerges from some combination of knowledge, competencies, and innovative offering; consider the following factors:

1. Customer knowledge, reflecting a vendor's intimate understanding of your process or product needs and a willingness to "collaborate to compete."
2. The ability to solve tough customer problems. If the problem is "easy," then the choice is "among commodities" and the evaluation of options must move beyond the trivial.
3. Innovation. To be one step ahead is to be at the point where you can offer "better, faster, and cheaper" by the time competitors are at your doorstep.

The equivalence of technical performance should be verified. Later, in Table A.1 we will consider the topic of decision-making and ensure that for every option, step 2 (collect data), step 3 (analyze data), and step 4 (determine plan of action) have been performed.

It is not out of the realm of possibilities that one vendor is offering a product that is superior to its competitors, but the vendor is undercutting price to get the job. There are strategic reasons a supplier might do this. More than a century ago, George Westinghouse (who purchased Nicola Tesla's AC patents) and Thomas Edison (DC current backed by J.P. Morgan) engaged in the "War of Currents" culminating in the Niagara Falls equipment contract to Westinghouse[6], so symbolic to the credibility of Westinghouse in this emerging industry he lost money on the contract just to "build brand."

5.6.2 Period comparability: Repeatability assumption

Comparison of assets with unequal physical lives requires several assumptions to be invoked to ensure calculations are for options of comparable periods. Consider the situation shown in the table below. Given that Option M has a

Project Parameter (8% MARR)	Option K	Option M
Initial project investment, P	\$12,500	\$19,995
Expected duration of use, N	5 years	10 years
Estimated annual net expense	\$2,500	\$2,999

life twice as long as Option K, we use the *repeatability* assumption to construct a problem where the analysis is for two options with equal lives. In this case, if we repeat the use of Option K for a second cycle, then we have two 10-year periods of data to use in analysis. Figure 5.1 illustrates this situation.

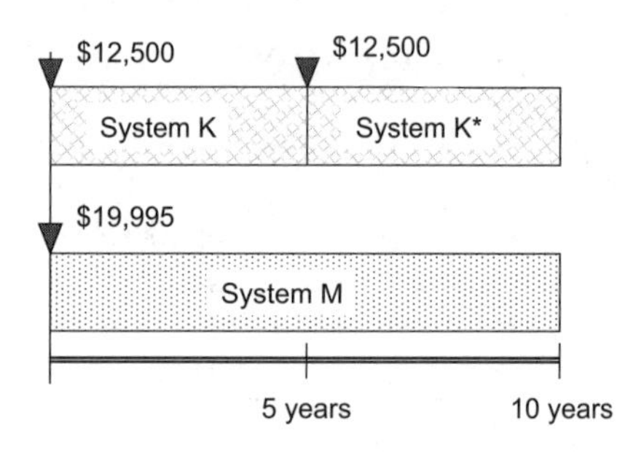

Figure 5.1: Repeatability assumption to simulate equal lives (K* is option K repeated).

From this, the PW calculation is simply

$$\begin{aligned} PV_K &= -\ \$12,500 - \$12,500(P/F, 8\%, 5) - \$2,500(P/A, 8\%, 10) \\ &= -\ \$21,007.29 \ - \ \$16,775.20 \ = \ -\$37,782.49 \end{aligned}$$

$$\begin{aligned} PV_M &= -\ \$19,995 - \$2,999(P/A, 8\%, 10) \\ &= -\ \$19,995.00 \ - \ \$20,123.53 \ = \ -\$40,118.53 \end{aligned}$$

Despite the fact that the initial capital cost of Option M might be lower, the overall expense of Option K is less and this would be the preferred alternative. The difference in payment would be small, actually (with a 4.5% incentive loan, the Option K payment is $233 a month versus the Option M $207 monthly payment).

Note that if we used the AW method there would not need to be an additional calculation to account for Option K repeated as K*. IN this case, we only have to operate under the *assurance* that all conditions could be repeated and data for the second cycle would be comparable to data from the first cycle.

In cases where there is not a simple doubling of one period to equate the other, the common multiple is used. If System K had a life of 3 years and System M a life of 5 years, then the common multiple of 15 years would be the basis of the PV calculation.

5.6.3 Repeatability of useful life

The validity of "repeatability of useful life" is sometimes called into question. The focus of many arguments is that for an extended period of performance, it would be unlikely that an identical piece of equipment could be brought in at the same price and performance. Further, the simplistic calculation in the previous section does not account for fluctuations in expenses. Over a 10-year period the PW comparisons only differed by less than $2,500! As a practical matter a more realistic analysis would account for a number of other important factors that would change over the course of the life of the project. Said another way, the repeatability assumption helps with simplistic calculation scenarios. However, in practice, a more comprehensive view of the problem would be required or a decision tree examined at the point where System K needed to be replaced and production alternatives introduced.

5.7 Decision trees

Many of the problems described here and elsewhere require forecast data speculation about future events. This is not avoidable and aligns with decision-making where you "use the best data available at the time the decision needed to be made." Retrospect benefits from validation of data that cannot be made in advance. In cases where there may be high sensitivity or the stakes are high some additional effort to bracket outcomes can sometimes be warranted. This is where uncertainty and risk analyses come into play, in particular, the use of *decision trees* in the assessment of probability outcomes; this topic is explored in more detail in Chapter 8, p. 223.

Multi-level decision trees are of particular value as they are a graphical representation of a problem, which can help with the organization of options and related information. Consider the situation from Figure 5.2 and the richness of the decision-making that can be accommodated with the use of probability analysis.

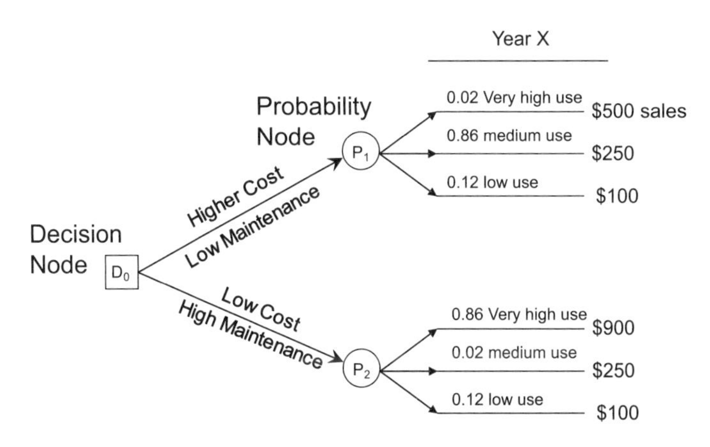

Figure 5.2: Multi-level decision tree for capital equipment selection.

Again, we will work through the details of decision tree analysis in Chapter 8, but we can make the following observations:

1. The simple utility of the decision tree can be limited if there are more than 3 or 4 options, but that is not typically an issue for most practical problems involving comparison of alternatives. At most we compare just 3 options.
2. Availability of risk data may be hard to find, but the creation of the

table may force the issue that some people may be trying to avoid. A decision tree broadens the perspective of the "upside" and "downside" and asks the analyst to be a bit more objective.

3. There is, of course, software that can be used to solve multi-level problems, but there is the risk of overlooking a concise view of the problem that can be explained in persuasive terms to senior-level management. Seasoned executives will understand the limitations of forecasts and may simply be seeking what the worst case might be and in the back of their mind consider developing contingency plans.

We can't forget that whether we like it or not, most decisions involve a lot of uncertainty and a decision tree is a handy way to outline critical path elements of the problem.

5.8 "Lease versus buy" decisions

Leasing has evolved into a $521 billion financing option in commerce today; 4 out of 5 companies are involved in some form of lease financing[7]. Although leasing was reserved primarily for real-estate transaction in the 1950s, it has now expanded to almost every asset a company needs to run its business, including leases for carpet! Knowing that – on the average – leasing tends to be *more expensive* from a cash-flow perspective in the long run, the "lease versus buy" decision must entertain factors such as interest tax savings and depreciation benefits for one to craft persuasive arguments on leasing benefits.

Many times the lease versus buy decision is reduced to numbers alone, ignoring the dynamic FASB issues surrounding leasing. We introduce some key concepts in this chapter to enlighten what is otherwise a fairly sterile discussion. M. Ameziane Lasfer and Mario Levis[8], along with Lewis and Schallheim[9] and Schallheim [10] provide ample reading on the topics of leasing models as they pertain to interest tax savings on debt capacity, lease contract tax transfers, net tax advantages, and depreciation tax savings benefit, so those discussions are not repeated here. Indeed the book by Schallheim[10] is a well-written discussion of "lease vs buy" that is more than adequate for most engineering manager concerns! For any additional information the FASB site `www.fasb.org` has several quick-guides and summaries that aim to help an average layperson understand FASB Section 13 on leasing. Simply stated, leasing avoids the initial cost of purchasing an asset in exchange for making a series of payments that generally have tax advantages.

5.8.1 FASB view of a lease

The Financial Accounting Standards Board has defined a lease as "*an agreement conveying the right to use property, plant, or equipment (land and/or depreciable assets) usually for a stated period of time*" ([11], p. 4). A lease is a *binding, legal contractual agreement* between the owner of the asset (the *lessor*) and the renter of the asset (the *lessee*). Leasing is often viewed as an important source of financing: estimates vary, but somewhere between 25% and 40% of all capital equipment is financed through lease agreements. Leases are traditionally viewed by accountants as either *operating leases* or *capital leases*:

1. **Operating leases** generally involve "short term" (cancelable) agreements, typically less than a year in duration. An asset operating lease is recognized as an expense on financial statements much in the same way you would expense equipment rental. No asset or liability is recorded. A distinct advantage of operating leases is that the lessor retains the risk of ownership, not the lessee.
2. **Capital leases**, also known as *financing leases*, are generally non-cancelable reported much differently and *are* recorded by asset and liability entries on the balance sheet. They are not expensed and are treated as if the lessee is the owner of the asset (not the lessor).

FASB provides four criteria called the "7(a)-7(d) test" ([11], p. 8) for the distinction between types of leases. A lease is a capital lease when *any one of the following criteria are met*:

1. The lessor transfers ownership of the asset to the lessee at the end of the lease term.
2. A "bargain purchase option" is given to the lessee. This is an option that allows the lessee, upon termination of the lease, to purchase the leased asset at a price significantly lower than the expected fair market value of the asset.
3. The life of the lease is equal to or greater than 75% of the economic life of the asset.
4. The present value of the minimum lease payments (MLP) is equal to or greater than 90% of the fair market value of leased property.

Application of the "7(a)-7(d) test" requires knowledge of PV and asset depreciation calculations to implement. Operating leases provide lessees the benefit of having possession of assets that do not show up on the balance sheet, and for which the cost of the lease can be expensed. The significance of categorizing a lease as a capital lease is that the lease is recorded as an asset on the balance sheet.

Traditionally lessees have preferred to negotiate operating leases since the company has the use of assets without them appearing on the balance sheet; this *off-balance-sheet financing* is viewed unfavorably by FASB since this leads to mis-representation of financial position, typically to the disadvantage of the stakeholders of the company.

5.8.2 Off-balance-sheet financing

As just mentioned, an advantage of leasing is that, for operating leases, the lessee has the advantage of the use of the asset without the asset burdening the company's balance sheet. To demonstrate, consider the sample balance sheet shown in Table 5.11.

Household Paving, Inc.
Condensed Balance Statement

Current Assets		Liabilities and OE	
Cash	$125,000	Current Liabilities ..	$75,000
Accounts Receivable	$50,000	Long-term Liabilities	$65,000
Inventory	$25,000	Owners' Equity	$60,000
	$200,000		$200,000

Table 5.11: Household Paving, Inc. condensed balance sheet.

The debt ratio for this case is simply

$$\begin{aligned}\texttt{Debit Ratio} &= \frac{\texttt{Current Liabilities} + \texttt{Long-term Liabilities}}{\texttt{Total Assets}} \\ &= \frac{\$75,000 + \$65,000}{\$200,000} = 70\%\end{aligned}$$

This is a very high debt ratio, and the company is probably unlikely or marginally able to borrow money. If a new piece of equipment were to be acquired for $58,000, and the asset had to be included on the balance sheet, then the debt ratio would reflect the equal addition of an asset and a liability:

$$\texttt{Debt Ratio} = \frac{\$75,000 + \$65,000 + \$58,000}{\$200,000 + \$58,000} = 79.2\%$$

This debt ratio is an increase of over 9% and highly unfavorable to the company. Thus it is very important to identify the correct categorization of a leased asset; operating leases *will not* enter balance sheet calculations, but capital leases *will have* a direct (normally adverse) impact on the balance sheet (from an external investor risk standpoint). As a result, most companies would prefer an off-balance-sheet situation, leading FASB to keep up with "lease games" to provide guidelines promoting transparency of financial statements by requiring capital leases to appear on the balance sheet. Still, you can imagine that a lot of massaging of data might be in action to try to recast a capital lease as an operating lease; the practice of misrepresentation keeps FASB on the move, as we shall see in the next section.

5.8.3 Proposed FASB lease classification changes

Current FASB standards were effective January 1, 1977, and have been amended numerous times. An important and dramatic set of revisions to the standards is expected to take place about the time of the printing of this textbook. What is proposed by FASB and the International Accounting Standards Board (IASB) is to recognize *all leased equipment* as an asset appearing on the balance sheet, essentially characterizing all leases as capital leases. Only in the case where a lease does not exceed 12 months in duration would the lease be considered a "rental" (and thus would not be disclosed except as a footnote in the financial statements). We recommend referencing `www.fasb.org` for more details and updates to FASB-13 "Topic 840" governing leases and related work under way to harmonize with IAS-17. Leasing is a popular way to finance equipment since a majority of leases are categorized as operating leases and are thus generally transparent to the balance sheet. Converting operating leases to capital leases significantly impacts balance sheet ratios, particularly the debt-to-equity (D/E) ratio, an integral part of credit ratings that influence loan interest rates. The proposed FASB-13 guideline changes will cause D/E to balloon, thus eroding a company's creditworthiness. An unintended consequence of the policy change might be the trend for leases to be renegotiated for 12-month terms to retain operational lease status.

5.8.4 Advantages and disadvantages of leasing

As with any financial instrument there are advantages and disadvantages to its use, and only the context of the decision-making can determine what is "best" for a given situation. Although details vary with specific situations, some general advantages of leasing are:

1. Reduction of risk of obsolescence due to technology or market changes.
2. Providing a hedge against inflation and general business risk.
3. Possibility of "No money down" and can be tailored to the lessee's needs.
4. Relaxed credit requirements for start-up and struggling companies.
5. Avoidance of loan covenants and other restrictions in debt contracts.

There are distinct disadvantages to a lease as well:

1. You generally pay more for the asset in the long run than if you were to buy it.
2. If the lease fails an operational lease test, the asset appears on the balance sheet.
3. Businesses that are not sound financially may take on more debt than they can service.

These disadvantages are generally not "showstoppers" for businesses, but when carefully managed (i.e., timely payments, ensuring utilization to create value, avoiding unreasonable residual or termination fees, etc.) a lease can provide flexibility and affordability of assets otherwise unavailable to the firm. Quite unfortunately, many consumers do not think critically about leasing and end up buying a car, for instance, "at prices they can't afford with money they don't have and for features they do not need."

5.8.5 Before-tax "lease versus buy" scenario

Darshan Engineering, LLC anticipates the need for a high-speed profilometer with digital recording capability for a multi-year quality control subcontract they believe they will win within the next 6 months. Although the system costs $43,000 there is enough time to order a system and have it installed; management is pondering whether to lease the machine for $9,000 per year (in the event they lose the contract after 3 years) or whether to buy it outright.

The company MARR is 12%. Problem data are gathered in Table 5.12. Solving for the "best" alternative involves finding the equivalent EUAW or the PV for the lease and system purchase data provided. In this particular problem the solution is well-suited for a PV approach. The PV of each option is the sum of three components: (a) the initial purchase price or down payment, (b) the PV of the operating costs, and (c) the PV of the salvage costs or any residual fees/charges.

Project Parameter	Buy	Lease
First cost, P	$43,000	$9,000
Operating costs, OE	$9,500	$9,500
Service period, n	8 years	N/A
Salvage value, SV	$7,000	N/A

Table 5.12: High-speed profilometer "lease versus buy" options.

The PV calculations have the familiar form:

$$\begin{aligned} PV_B &= P + OC\ (P/A, i\%, n) - SV\ (P/F, i\%, n) \\ &= \$43,000 + \$9,500(P/A, 12\%, 8) - \$7,000\ (P/F, 12\%, 8) \\ &= \$43,000 + \$9,500(4.9676) - \$7,000\ (0.4039) \\ &= \$93,020 \end{aligned}$$

$$PV_L = (\$9,000 + \$9,500)(P/A, 12\%, 8) = \$18,500(4.9676) = \$91,900$$

Since we are computing the present value of the *costs* of each system, then given $PV_B > PV_L$ the outcome of leasing is the preferred course of action. For the small difference of $1,120 in cost, even if the lease were to be more expensive it might still make sense to lease if the company had a short-term cash-flow problem or if the new business revenue stream remained uncertain. In this problem the outcome is independent of the operating costs, since the costs are the same whether the system is purchased or leased; we could actually have dropped that part of the calculation (but included it for the sake of completeness).

Like many of the PV problems we have performed, it was tacitly assumed payments are made at the end of each period. Generally this is not the case for a lease, where the payments are made at the *beginning* of each period. Figure 5.3 illustrates the situation. The impact on the PV can potentially be significant and influence the decision.

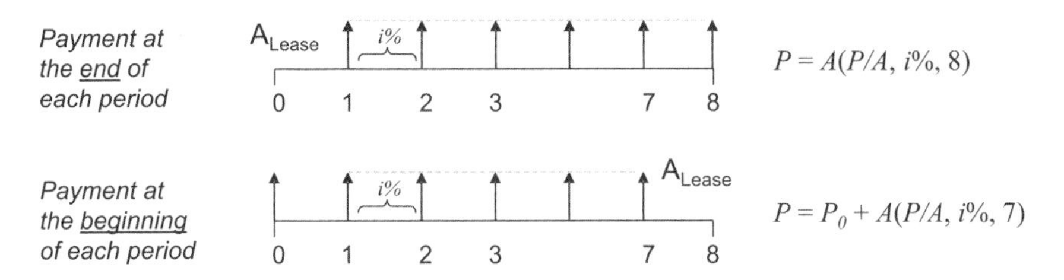

Figure 5.3: Difference between end-of-period and beginning-of-period lease payments.

The PV of the lease with beginning-of-period payments is thus

$$\begin{aligned} PV_L &= \$9,000 + \$9,000(P/A, 12\%, 7) + \$9,500(P/A, 12\%, 8) \\ &= \$9,000(1 + 4.5638) + \$9,500(4.9676) \\ &= \$50,074 + \$47,192 = \$97,266 \end{aligned}$$

We see the advantage to the lessor when payments are beginning-of-period, though the lease now has PV that is $5,366 higher than previously, adversely affecting the preference to leasing. Other items such as fees must be carefully considered as they may also affect outcomes. Let's assume the dealer is willing to finance the purchase but needs a 10% down payment of $4,300. The annual payment is easily found:

$$A_B = \$38,700(A/P, 12\%, 8) = \$38,700(0.2013) = \$7,790.31$$

Suppose your cash-flow situation is tight; the dealer may do you a favor on the purchase and fold in the fees into the loan. If we also include approximately $2,500 for taxes, title, and fees, then a $41,200 net loan with 8 payments of $7,790.31 means the effective interest rate is 13.95%. When working through options it is essential to track more than the initial purchase price and fully account for a variety of add-ons. It is not that the additional fees are invalid, but that you can start to lose track of what might be hidden or not evident that can result in net interest rates much higher than what was planned.

Returning to the $43,000 purchase, this option requires $6,800 up-front ($4,300 down payment and $2,500 in fees) and at the end of the month there is also the first payment of $7,790.31. That adds up to over $14,500 to pay during the first 30 days of ownership; in the very short run the $9,000 lease option starts to look fairly attractive.

One important fact to keep in mind is the situation you are confronted with at the end of the 8-year project period. For the purchase option you receive notice that the loan has been paid in full and along with that will come the title to the capital equipment. You have in your possession an asset worth approximately $7,000 and you might even have the option to continue to use it in production. On the other hand, the lease will require the return of the asset (or a buy-out) with potential fees for excessive wear-and-tear, among other contract clauses. For any large capital purchase it is good to have a "systems perspective."

We have illustrated that, for the baseline comparison, the \$43,000 asset has a beginning-of-period payment lease PV = \$97,266 that exceeds the purchase PV = \$93,020, thus making the purchase the best option from a PV standpoint. Although many financial decisions are easier and generally accurate using before-tax-cash-flow (BTCF) methods, the next section illustrates why after-tax cash flow (ATCF) is often warranted.

5.8.6 After-tax "lease versus buy" scenario.

To make the point clear about the impact of taxes on the lease-versus-buy decision, let's reconsider the previous problem from Table 5.12, p. 160 but to make the calculations simple, let's ignore OC. As well, assume the corporate tax rate is 42%. MARR is still 12%. The narrative and calculation methods outlined in Section 5.5 apply directly to the after-tax analysis here. Taxes pertinent to capital budgeting are discussed in Section 9.5.3 (p. 313). Let's work through each element of the sample problem in Table 5.12, beginning with the "Buy option."

1. The purchase of the system for \$43,000 is the key investment, shown in Table 5.13 under year 0. Assuming this was a business-to-business transaction, sales or excise taxes are not levied or indicated.
2. Straight-line depreciation of a \$43,000 asset with no salvage value is an annual depreciation allowance of

$$D_j = \frac{\$43,000 - \$0}{8} = \$5,375$$

   As discussed on page 150 in Section 5.5 this is a depreciation expense credit that increases net income for each of years 1 through 8, so \$5,375 is entered as the BTCF, Line 1 in Table 5.13.
3. With a corporate tax rate of 42%, the after-tax cash flow (ATCF, Line 2) of the BTCF is simply $ATCF = 0.42(\$5,375)$.
4. The present value of the ATCF is

$$PV = -\$2,258(P/A, 12\%, 8) = -\$31,786$$

This problem is greatly simplified by the assumption of straight-line depreciation, though with spreadsheets the assumption is easily modified to another depreciation method.

Calculations for the lease option follow a similar pattern, but make note of the following:

1. The lease expense on the income statement reduces net income and thus reduces the income tax liability. Recalling that operational costs are the same for the buy and lease options, we are not including them in this analysis. Line 4 in Table 5.13 for BTCF thus shows a -\$9,000 annual

payment, beginning with Year 0 and ending with Year 7. This payment is a BTCF that decreases corporate income.

2. Lease payments are at the *beginning* of each period. Year 8 end of period payments are not relevant, and in Table 5.13 we have simply indicated "NA" (not applicable) in the spreadsheet.
3. Pay close attention to the sign (+/-) on the BTCF. The depreciation allocation on Line 1 is different than the expense on Line 4.
4. With a corporate tax rate of 42%, the reduction in tax liability is

$$\texttt{Tax Credit} = 0.42(\$9,000) = \$3,780$$

from which the ATCF of the $9,000 lease reflects a decreased effective lease expense

$$ATCF = -\$9,000 + \$3,780 \; = \; -\$5,220$$

5. The present value of the ATCF involves a slightly different calculation sequence for the lease situation. Note the effect of the beginning-of-period payment

$$PV = \; -\$5,220 - \$5,220(P/A, 12\%, 7) \; = \; -\$29,043$$

	Year 0	1	2	...	7	8
Buy Option						
1. BTCF	-$43,000	$5,375	$5,375	...	$5,375	$5,375
2. ATCF	0	$2,258	$2,258	...	$2,258	$2,258
3. PV	-$31,786					
Lease Option						
4. BTCF	-$9,000	-$9,000	-$9,000	...	-$9,000	NA
5. ATCF	-$5,220	-$5,220	-$5,220	...	-$5,220	NA
6. PV	-$29,043					

Table 5.13: After-tax cash flow for profilometer problem.

To summarize, for the buy option, $PV_B = \$31,786$, and for the lease option, $PV_L = \$29,043$. As we are computing the present value of the *costs* of each system, then given $PV_B > PV_L$ the outcome of leasing is still the preferred course of action (lower expense), but by a very narrow margin of $2,743.

Some interesting variations on the problem arise by moderately small changes in problem parameters. At a lower MARR (already low) of 8%, the buy option $PV_B = \$30,027$, and for the lease option, $PV_L = \$32,397$. Now, $PV_B < PV_L$ and the *buy option* becomes favorable. A similar reversal exists if the duration of the lease grows to 10 years. This makes intuitive sense, as the buy option has more time for asset amortization to occur or occur more favorably. It is important to understand how calculation and decision outcomes can be impacted by variations in parameters; Section 8.4, p. 225 provides a more detailed look at parametric analysis. Valuable tools are presented in that chapter.

5.9 Summary

The ability to compare alternatives is a fundamental engineering manager skill. Typically calculations are iterative as decision-makers work through a series of "what-if" calculations and may end up converging on an answer that does not fit the "textbook" outcome. Items like *opportunity cost*, *credit lines with covenants* and other financial contexts may require compromise. Still, in a perfect world our objective would be to (a) maximize the expansion of wealth, or (b) improve profitability through cost-reduction decisions. We saw through several examples the foundation for these arguments. As well, we saw that very basic concepts from the time value of money (TVM) come up over and over again, and the problems at the end of the chapter provide ample opportunity to test different ways of solving them.

Taxes have been used to influence economic policy for years[5], and it is unlikely this powerful incentive to modify corporate behavior will change anytime soon. It should also be clear that numerous areas of finance still create controversy on what is ethical and regulatory and advisory bodies will continue to ferret out and refine practice areas that improve transparency of financial statements for all stakeholders.

At several points in the chapter we saw that problem data may be provided, but it made sense to perform even simple parametrics to test the "robustness" of a solution to understand what factors might change that could render the original conclusion invalid. In many cases we seek to solve problems to "gain insight, not numbers." This is an excellent time to consider Table A.1:

	Problem-solving		Engineering design process
1.	Encounter problem	1.	Define problem
2.	Collect data	2.	Collect data
3.	Analyze data to specify problem	3.	Formulate hypothesis
4.	Determine plan of action	4.	Design plan to test hypothesis
5.	Execute action plan	5.	Test hypothesis
6.	Evaluate plan for effectiveness	6.	Interpret results
		7.	Evaluate for study conclusion

We cannot get caught up in just Steps 3-4. We also need to consider how we will develop a plan for determining the effectiveness of our decision, how we might need to interpret results in rapidly changing industrial contexts, and how to draw conclusions on ways to improve our analysis for the next (inevitable) iteration!

5.10 Problems to work

Problem 5.1

A polyester powder plant power has a cyclone separation system that is beginning to look like it may not meet upcoming OSHA safety and operational standards. The system is under consideration for replacement for a 5-year project horizon. The existing separation system has a value in the market of about $5,500. If the system is kept in service, it is anticipated to have annual operating and maintenance costs of over $7,200, with zero market value at the end of the project horizon. A new system is expected to cost $18,000, will have operating and maintenance costs of $4,250, and will have a market value of $8,600 at the end of the planning horizon. Determine the preferred alternative. Assume the company MARR is 18% (before taxes).

Problem 5.2

Action Diagnostics is considering one of the following centrifuge systems:

	System D	System E
Cost	$16,000	$24,000
Savings/yr	$1,500	$2,200
IRR	6%	8%

Each machine will have a 10-year life with no salvage value. If we assume a corporate MARR of 10% is reasonable, which system do you recommend?

Problem 5.3

Construct a sensitivity analysis of the outcome in Problem 5.3, varying the IRR of System E from 5% to 10% in increments of 0.5%. How do you interpret the results and what does the outcome suggest in terms of selecting System D over System E?

Problem 5.4

SmartLabs Diagnostics is considering one of the following three microscopes with digital recording and pattern detection software.

	X	Y	Z
Cost	$20,000	$30,000	$15,000
Savings	$1,199	$2,199	$1,000
IRR	6%	9%	7%

Each machine will have a 20-year life with no salvage value. There is a lot of debate within the company as to whether these devices are really going to last that long, but they assume a MARR of 10% is reasonable. Which system do you recommend?

Problem 5.5

Orbital Milling Services is going to replace one of its new 3-D prototyping systems. High use in the prototyping shop means the systems are not expected to last more than 5 years. If MARR is 12%, use an IRR analysis to determine which system should be purchased:

	System K	System L
Cost	$14,000	$20,000
Annual OC	$3,000	$1,500
Salvage Value	$4,000	$6,000

Problem 5.6

Three years into a 10-year project, Consolidated Shipping Associates is considering an upgrade to a package conveyor line. A local supplier has two systems on sale as shown below; the QuickLine series is a close-out special (discounted 30% from list of $15,999) and the RFID Rider is built with RFID tracking as an option, though "as-is" the floor model does not have the RFID electronics ($10,000 up-charge) included. Because both systems are close-outs, there is no expectation the conveyor lines will be of much value at the end of the 8-year project period. Which system should be chosen?

Project Parameter (15% MARR)	QuickLine	RFID Rider	Difference
Initial project investment, P	$11,199	$19,995	$8,796
Expected duration of use, N	8 years	8 years	None
Estimated annual net revenue	$2,500	$5,150	$2,650

Problem 5.7

At the end of FY20xx MRI Imaging has determined they need to increase their imaging capabilities to compete in the growing sports health management arena. Their closest competitor has three times the imaging capital equipment and they run at 82% capacity. MRI's board of directors estimates they could safely add $500,000 in capacity and the regional market would still not be saturated. The funds for the project are to be obtained by allocating $350,000 in retained earnings from the prior year and using $150,000 from their line of credit, if needed. The company was expecting to use the same imaging vendor they've always used, and the $350,000 investment in System K would produce a net rate of return of 20%; the return is considered fairly good in the down service economy where the company MARR has dropped to 8%. A new imaging vendor hears about the potential sale and offers a more advanced System Z that costs $500,000, though; the vendor claims the higher net rate of return of 22% is "worth it in the long run." The argument is that the higher cost enables faster imaging and greater patient throughput. "You'll make up the difference in the capital cost in three years," exclaims the salesman. MRI is a conservative company, and the board is inclined to spend as little on capital equipment as possible. Can it be argued that the more expensive system is worth it?

Project Parameter (8% MARR)	System K	System Z	Difference
Initial project investment, P	$350,000	$500,000	$150,000
Expected duration of use, N	7 years	7 years	None
Anticipated annual net return	20%	22%	2%
Annual net revenue	$70,000	$110,000	$40,000

Problem 5.8
SMS Microsystems is considering the replacement of a confocal microscope (CFM) in the quality control department. The closest CFM will require at least \$19,500 for installation and will have an estimated economic life of 10 years; its end-of-life value is estimated to be roughly \$3,000 MV. Annual operational and maintenance expenses are estimated to average about \$14,000 per year. Due to the dynamic nature of microsystems technology, the existing system (defender) has a book value of \$7,000 but a present market value of only \$4,000. It is estimated that over the next three years the system will have the MV, BV, and expense data shown in the table below. Using a before-tax interest rate of 17% per year, make a comparison to determine whether it is economical to make the replacement now.

Year	MV	BV	Expenses
1	\$2,000	\$4,000	\$18,000
2	\$1,000	\$2,000	\$24,000
3	\$500	\$1,000	\$30,000

Problem 5.9
A detached warehouse cooling system station was destroyed by high-water flooding, and that substation will have to be scrapped and entirely re-built. If the original system is replaced the substation will cost \$50,000 and it is estimated that with maintenance of \$2,800 annually the substation will be good for 15 years. If a more modern "flood-proof" system is purchased at \$75,000, then the maintenance drops by \$1,800 to \$1,000 annually. Under the assumption the company has a 10% MARR, what is the IRR of building the flood-proof version of the substation?

Problem 5.10
Darshan Engineering, LLC anticipates the need for a high-speed profilometer with digital recording capability for a multi-year quality control subcontract they believe they will win within the next 6 months. Although the system costs \$43,000 there is enough time to order a system and have it installed; management is pondering whether to lease the machine for \$9,000 per year (in the event they lose the contract after 3 years) or whether to buy it outright. The company MARR is 12%. Problem data are gathered below.

Project Parameter	Buy	Lease
First cost, P	\$43,000	\$9,000
Operating costs, OE	\$9,500	\$9,500
Service period, n	8 years	NA
Salvage value, SV	\$7,000	NA

Problem 5.11
Darshan Engineering, LLC also is interested in an automatic board etching system to reduce labor of specialized MEMS packaging. Which machine should be purchased? The company MARR is 12%. Problem data is gathered below.

	Manual	Automatic
First cost	\$10,000	\$15,000
Operating costs	\$3,000	\$1,500
Service period	10 years	5 years
Salvage value	N/A	\$2,500

Problem 5.12
Darshan's management is pondering whether to lease the machine for \$3,000 per year or whether to buy it outright. The company MARR is 12%. Problem data is gathered below. What is your recommendation?

Project Parameter	Buy	Lease
First cost	$15,000	$3,000
Operating costs	$1,500	$1,500
Service period	5 years	N/A
Salvage value	$2,500	N/A

Problem 5.13

Darshan Engineering, LLC has decided they need to conduct an after-tax analysis for Problem 5.11 and so the estimated annual receipts and expenses are shown below. The company MARR is 12%. Which system is preferred?

	Manual	Automatic
First cost	$12,000	$15,000
Revenue	$23,000	$38,500
Operating costs	$3,000	$1,500
Service period	8 years	4 years
Salvage value	$2,500	$0

Problem 5.14

Jason anticipates investing about $75,000 in a high-end 3-D laser cutter that he estimates will generate $20,000 annually for the next decade. Expenses are reasonable and projected to be about $6,500 per year. Assuming straight-line depreciation, no salvage value, and combined taxes of 40%, what is the rate of return for this project?

Problem 5.15

Jason moves ahead with the investment in Problem 5.14, but the business turns out to be much better than expected and a competitor wants to buy Jason's system for $50,000 two years after he started. How would you handle the depreciation of the system? Create at least 2 depreciation scenarios and determine the depreciation loss or recaptured depreciation that would need to be recorded if the system were sold.

Problem 5.16

Given the interest of the competitor in his laser cutter activity, Jason is considering *not* selling but quadrupling the production facility size and investing in 3 more systems at a cost of $250,000. Given this new "critical mass," Jason believes revenues will "skyrocket" and reach $90,000 in the following year, and *increase* by $7,500 each year after that. Does this expansion make financial sense? His competitive edge will last 8 years at most, and it is estimated taxes will drop slightly from 40%.

Problem 5.17

In Problem 5.16, how much of a difference does MACRS depreciation versus straight-line depreciation change the calculation outcomes?

Problem 5.18

Jason discovers that he can take an investment tax credit in the current fiscal year on any capital equipment purchased for his expansion plans in Problem 5.15. The state credit is 33% for investments up to $500,000 (for small businesses). How much more attractive does this make the project?

Problem 5.19

In Problem 5.16, suppose Jason ends up with revenues of $45,000 and recaptured depreciation of $1,200 for the current year-end. Expenses were $23,500, and depreciation expenses were $11,575. What is Ted's taxable income?

Problem 5.20

Treadway Moving Service specializes in transferring sensitive electronic instruments. The prospect of some new 3-D dual-material prototyping machines exists but the trailer set-up is just over $125,000 to purchase. The investment is still appealing since Boston Consulting Group surveys show that the business can expect to grow for at least another 10 years. Treadway's marginal tax rate is 40% and they have a fairly good credit rating so they obtain a loan for two trailers at a 10% interest rate for a 60-month loan. The depreciation schedule for the trailer is shown below. Leasing is popular, and it turns out the dealer will lease to Treadway. The annual payment for two trailers was quoted at $45,000, and after some debate the purchase option at lease-end was set at 1/5 of cost ($25,000 for each trailer). If Treadway has to absorb all taxes, maintenance, and insurance, is leasing the best option for Treadway?

Year	% of cost
1	40%
2	35%
3	30%
4	20%
5	10%

Problem 5.21

On average it is generally much more expensive to lease an asset than to buy it outright. And, as we saw in Section 5.8.3, there are regulations that erase some other traditional advantages to leasing (off-balance-sheet financing). Why, then, has leasing continued to grow in popularity?

Problem 5.22

Treadway has gotten used to the idea of leasing and is now considering an equipment lease of a computer. The lease terms call for a lease amount of $3,280, a residual value of $750 and 18 monthly payments. The lease carries an interest rate of 12% per year. How much is Treadway's monthly payment?

Problem 5.23

A credit check on Treadway shows that the owner has co-signed on his son's student loans, and this has dropped his credit rating by 110 points. The leasing company would feel "more comfortable" if Treadway would make a 3-month payment up front. What is the monthly payment in this case?

Problem 5.24

Treadway is really having fun "getting stuff cheap" and wants to buy his brother-in-law a car for "company sales" that costs $22,765. The lease would be for a standard 36 months with a residual value of 50% cost ($11,382). If the interest rate is 12% compounded monthly, what is the monthly payment going to be?

References

[1] Tung Au. Overall rate of return as a profit measure for capital projects. *Engineering Costs and Production Economics*, (7):235–241, 1983.

[2] William R. Lasher. *Practical financial management.* South-Western CENGAGE Learning, 2011.

[3] R.F. Meigs and W.B. Meigs. *Financial accounting.* McGraw-Hill, 6th edition, 1989.

[4] B.E. Needles, M. Powers, and S.V. Crosson. *Principles of accounting.* Houghton Mifflin, 10th edition, 2008.

[5] M. Jones and J. Tanchoco. Replacement policy: The impact of technological advances. *Engineering Costs and Production Economics*, (11):79–86, 1987.

[6] Jill Jonnes. *Empires of light: Edison, Tesla, Westinghouse, and the race to electrify the world.* Random House Trade Paperbacks, 2004.

[7] ELFA. *2013 Survey of equipment finance activity.* Equipment Leasing and Finance Association, 2013. `http://www.elfaonline.org/`.

[8] M. Ameziane Lasfer and Mario Levis. The determinants of the leasing decision of small and large companies. *European Financial Management*, 4(2):159–184, 1998.

[9] C. M. Lewis and J. S. Schallheim. Are debt and leases substitutes? *Journal of Financial and Quantitative Analysis*, 27(4):497–511, 1992.

[10] J. S. Schallheim. *Lease or buy? Principles for sound decision-making.* Harvard Business School Press, 1994.

[11] FASB. *Statement of financial accounting standards No. 13.* Financial Accounting Standards Board, 1979.

Chapter 6

Intellectual Property as an Asset for Wealth Creation

6.1 Learning objectives

In his February 2004 lecture at Stanford University, Alan Greenspan observed, "The increase in the value of raw materials accounts for only a fraction of the overall growth of U.S. gross domestic product (GDP). The rest of that growth reflects the embodiment of *ideas in products and services* that consumers value. This shift of emphasis from physical materials to ideas as the core of value creation appears to have accelerated in recent decades." From a business perspective this structural shift in the value derived from business activity renews emphasis on understanding how intellectual property (IP) is used to create wealth. This chapter expands on the idea that IP is an intangible asset to be *managed for wealth creation* much the way we do for tangible assets such as buildings and industrial processes.

After reading and discussion sessions, the student should be able to:

1. Understand the basic types of intellectual property.
2. Describe the pros and cons of trade secrets over patents.
3. Differentiate "real" markets from "financial" market dynamics.
4. Identify a key mechanism by which IP is linked to wealth creation.
5. Describe the impact of the Bayh-Dole and Sarbanes-Oxley Acts on management.
6. Describe the impact of the "America Invents Act of 2010" on IP management.
7. Identify and describe the three basic methods for valuation and the preferred method for IP.

6.2 Roadmap for wealth creation

While intellectual property (IP) involves topics that we have discussed previously (investment choices, risk, the time value of money, and asset maintenance), IP is intrinsically interdisciplinary, involving three fundamentally different subjects: business, technology, and law. While most students emerge from engineering school with a general understanding of what IP *is*, less often is there an appreciation of *how IP creates wealth.* A simple reason for this is that although engineering students will take some type of course in Engineering Economy focused on *internal, project* activities, less commonly do such courses wander into the topic of *real versus financial markets* or look at *market valuation of intangible assets.* Understanding "real" versus "financial" markets involves a few subtle changes in perspective that the decision-maker must appreciate and undertake, not the least of which is the need to assess *intangibles* in the *context* of the financial (external to the company) markets.

Greenspan's observation and comment that knowledge is an important factor of production is not entirely new. This notion captured the attention of writers long ago, at the end of the industrial revolution. The structural shift from an agrarian to a manufacturing economy in the late 1800s prompted researchers such as Marshall[1] to wonder what had happened. We're in the midst of another paradigm shift, this time to a "knowledge economy," and on any given day the business press promotes "innovation" as the key to economic growth. There is no question that IP is an important piece of the wealth creation puzzle, and this chapter is intended to clarify the what, why and how!

To "clarify" often means to "simplify," and thus to guide the thinking that will unfold in this chapter, Figure 6.1 provides a simple roadmap of three major topics to cover, where each major topic has associated with it four steps in the overall wealth-creation process. The remainder of the chapter works through each of the 12 steps in one way or another.

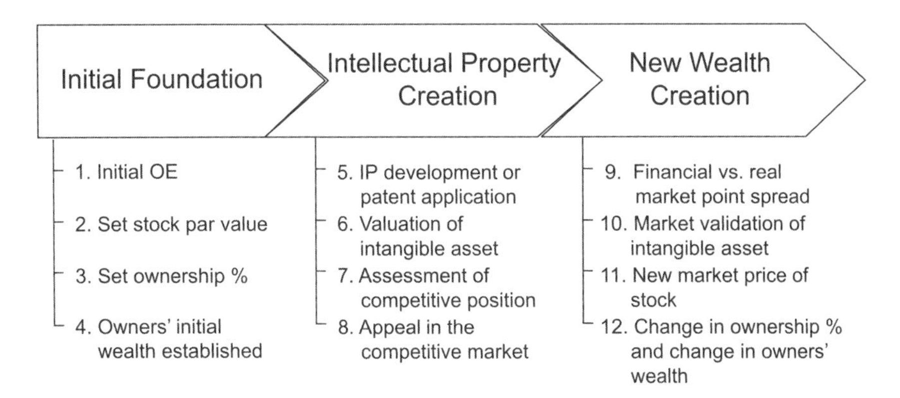

Figure 6.1: Twelve steps for wealth creation through intellectual property creation.

The scope of this chapter is a bit difficult to limit. We don't want to repeat what can be found elsewhere, but, at the same time, *do* need a basic understanding of fundamentals so that concept subtleties can be appreciated and understood. Thus, in the spirit of a *Field Guide*, it is worthwhile to pause and make a few comments about references one might consider as we move forward or later when reflecting or refining thoughts about wealth creation through IP.

6.2.1 Books to explore

There are many books and websites suitable for self-study on the topic of IP. The choice below might seem a bit eclectic, but, at a minimum, the recommendations are simply a starting point for the student to build his or her own collection:

1. Steven L. Oberholtzer[2] *The Basics of Intellectual Property Law* is a *very* brief monograph published by Brinks, Hofer, Gilson, and Lione, and is available by request from their resource center (`http://www.brinkshofer.com/resource_center/8`). This concise document gets to the important points very quickly and summarizes the essential aspects of patents for the layperson. The book is no more than a two-hour read and is worth having as a handy reference (and when you are done, to pass along to others). You'll come up to speed very quickly and at least begin to know some of the questions to ask about patents and patent law.
2. Weston Anson and Donna Suchy[3] do a commendable job as stated in the preface to "strike a balance between complexity and simplicity" with their reference book *Intellectual Property Valuation: A Primer for Identifying and Determining Value*. The focus of the textbook is on *valuation*, but the authors also discuss related topics (what to do in bankruptcy, tax issues, and FASB concerns) and supplement many chapters with glos-

saries. In particular, Chapter 8 (p. 74) on "patents" is a great follow-on read to the monograph by Oberholtzer[2].

3. Howard B. Rockman's[4] *Intellectual Property Law for Engineers and Scientists* is a comprehensive reference book with useful definitions, historical comments, examples, and patent case studies. Developed as a textbook for a graduate course in IP by an intellectual property attorney, this book covers most of the topics of concerns to engineering managers. Chapter 1 and Chapter 4 are stand-alone chapters that introduce the subject of IP in a slightly different way than Anson and Suchy[3]; both are worth time spent reviewing.
4. The Association of University Technology Managers (AUTM) produces a series of educational pamphlets that are very easy to read and readily available; Lisa Mueller and Jill Sorenson's monograph[5] *An Inventors Guide to Patents and Patenting* has clarity similar to Oberholtzer[2]. Somewhat less of engineering interest (it seems) but equally important is the easy-to-follow and quite comprehensive monograph by Charles Valauskas and Catherine Innes[6], *Copyright Protection of Software, Multimedia, and Other Works: An Author's Guide* – their focus on the "how" versus "what" is helpful to those seeking practical information. In situations involving biomedical materials, Brian Leslie[7], *Material Transfer Agreements* can help with an understanding of a variety of unique biomedical aspects as well as confidentiality, warranty, and sample agreements.
5. For biomedical engineers, the work of Stefanos Zenios, Josh Makower and Paul Yock[8], "BioDesign: The Process of Innovating Medical Technologies" is quickly becoming a "go-to" reference. Chapter 4, Section 1 on "Intellectual Property Basics" (pp. 210-272) is fairly inclusive and contains useful checklists, a sample provisional patent, and sample agreements. The section is not overburdened with medical terms and concepts, so the average engineer will find this a good reference and an easy read.
6. Robert Shearer's[9], *Business Power: Creating New Wealth From IP Assets* has a few sections that are of particular interest. Chapter 4 (pp. 45-55) is somewhat short, but has a clear overview of SEC and FASB concerns; Chapter 7, "Capturing the Value of Trade Secrets" has a simplified, but clear, narrative on what needs to be considered for valuation of trade secrets. FASB concerns are touched on in several areas, including Chapter 9 on "Valuation."
7. The IEEE report *Intellectual Property and the Employee Engineer* authored by Orin Laney[10] is a brief 21-page document, but has very useful information on the implications of employment agreements and negotiation strategies. While this sounds more like a teaching in contract law than intellectual property, it is one of the few documents readily accessible that can help understand IP risks associated with entering/exiting employment with a company.
8. A reading list would not be complete without a discussion of global and

international affairs, and for this the World Intellectual Property Organization (WIPO) has published *World Intellectual Property Report: The Changing Face of Innovation* [11]. Within this comprehensive document, Chapter 3, "Balancing Collaboration and Competition," adds a useful perspective to the strategic management of IP.

6.2.2 Three links to explore

As with just about any popular topic, an Internet search of "IP valuation" turns up millions of websites to explore. To counter this overwhelming volume of information, I am providing the top three sites that I strongly recommend for *regular* visitation and exploration!

First: The United States Patent and Trademark Office (USPTO), `http://www.uspto.gov`, has *the* database for patent searches, along with current information on the progress of patent reform, most notably the America Invents Act.

Second: World Intellectual Property Organization (WIPO), `wipo.int`. Valuation is covered in good detail.

Third: Federal Reserve speeches are educational. There is always something new and *very* informative on current events. My favorite is Alan Greenspan's Stanford lecture: `http://federalreserve.gov/boarddocs/speeches/2004/200402272/default.htm`. Check it out!

6.2.3 Politics to explore

Our roadmap for this chapter would not be complete without underscoring three fundamental facets of wealth creation: (a) The concepts in this chapter have perishable aspects related to current and anticipated changes in the law, (b) new administrations may choose to be more or less vigilant about enforcing that laws are already on the books, and (c) if you create a lot of wealth through intellectual property, expect and plan on litigation. We often hear that "perception is reality." This, and "you get what you negotiate, not what is fair," apply to just about every intangible asset transaction!

Figure 6.2 illustrates that laws generally do not arise from engineering "first principles" but rather involve a political, negotiated process. Because this process is not perfect, unintended consequences are common! The intellectual property field is not exempt and continues to undergo a radical transformation related to the passage of S.23, the Leahy-Smith America Invents Act. Many "truths" prior to the Act are affected and evolve in the years ahead.

Sullivan's *How our laws are made*[12] sheds light on how you can improve the process. "Jump in, the water is fine."

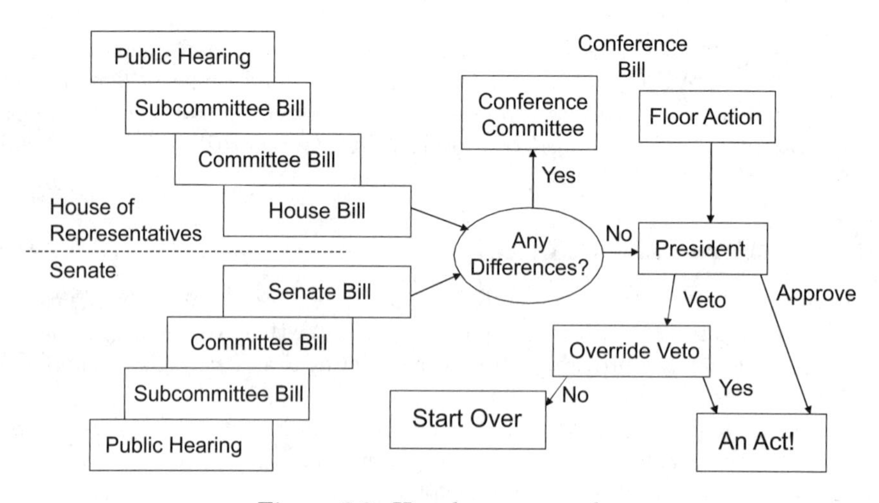

Figure 6.2: How laws are made.

6.3 Intellectual property principles

The field of intellectual property has been around in some form since the mid-fifteenth century, and many good books, articles, and websites are available for current as well as historical information. As odd as it might seem, a historical perspective is relevant to acquire as it surprises engineers sometimes that the underlying purpose of patents is for *economic development* and thus the crafting of rules and laws is inherently a *political, not scientific*, endeavor. Engineers and other scientists accustomed to scientific axioms defining disciplined problem-solving find the interdisciplinary nature of IP treatment a bit convoluted, thus frustrating those who prefer a singular answer to IP strategy. The legal framework for IP has evolved slowly and is crafted primarily by common law (precedents for action set by the courts). Legislative acts are significant but infrequent. You don't ever necessarily reach a "terminal" understanding of IP; you just master a snapshot of the field at a specific moment in time.

Within the last two decades, the landmark General Agreement on Tariffs and Trade (GATT) influenced our view of IP in three fundamental ways: (a) strengthening the law by enabling infringement remedies in federal district court practice, (b) seeking to harmonize the US "first to invent" with the European "first to file" ownership claim, and (c) expanding the patent term from 14 to 20 years for those patents filed after 1995[13][14]. More recently, the 2010 Leahy-Smith America Invents Act Implementation mentioned in Section 6.2.1 has the potential to render obsolete much of what is "prevailing wisdom" today – reference information and advice written before 2012 will need to be critically reexamined. No attempt is made to try to trace the scope and depth of existing legislation in this book. Any ambiguities that the America Invents Act produces will most certainly be worked through the courts.

This *Field Guide* is about financial decision-making, so we try to focus on concepts that have endured the test of time, especially principles related to *wealth creation.* In Section 6.3.1 we highlight enough of the basics of IP to enable financial decision-making to be meaningful. The nuances of governing laws and other legal matters are actually quite fascinating, but we only have the space to venture where topics are relevant to decision-making. The concepts of a *limited monopoly* and *protectable property* appeal to the business venture, though I imagine some (maybe) are swept up in the altruistic objective of "promoting science and the useful arts."

The powerful influence of IP on investment and wealth creation is at the core of preferential treatment (issuance of patents) to a special group of individuals. What right does the government have to do this? Why do a "chosen few" have privilege over the majority? The answer is in the *greater good.* It is in the government's interest to bless an assured monopoly to innovators so that they might educate the population for the "greater good" of national economic development. Absent the "greater good" argument, the existence of a patent might otherwise be viewed as an injustice.

A historical reasoning for an assured (but time-limited) monopoly is quite clearly expressed by Mueller and Sorenson[5]:

> *In England, letters patents were used as inducement for foreign artisans to introduce continental technologies ... it was believed that, by luring skilled artisans with the reward of a limited monopoly, the community would gain from the fruits of skilled labor.*

In the United States, similar economic development objectives for the "greater good" made their way into the Constitution, Article 1, Section 8, Clause 8 (see Bernstein[15], p. 44):

> *To promote the Progress of Science and the useful Arts, by securing for a limited Times to Authors and Inventors the exclusive right to their respective Writings and Discoveries.*

Constitutional foundations are appealing and persuasive from a public policy perspective, but converting this into an understanding of investment cause-and-effect is a bit of a challenge. Here, we benefit from the four theories put forth by R. Mazzoleni and R. Nelson[16]. These principles help identify the foundation for wealth creation through IP investments:

- *Invention inducement Theory:* Patents motivate innovation and invention – absent a robust legal system there will be no invention.
- *Disclosure Theory:* Public use and knowledge of inventions derive from inventors induced to disclose information otherwise kept in secret.
- *Development and Commercialization Theory:* Investments needed to bring products and services to market (commercialization) are promoted through a limited monopoly.
- *Prospect Development Theory:* Research and development investments will occur within areas of the patent claim and foster commercialization objectives.

Of particular interest is the *invention inducement theory* as we believe this most pertinent to decision-making: that public policy can be shaped or changes in the legal framework crafted for favorable (or unfavorable) patent protection that will foster (or discourage) investment in IP. As mentioned earlier, the monopoly on protectable property is central to the financial attraction of IP. Faced with choosing between an investment in product production machinery *or* in new product ideas, Table 6.1 reveals three critical differences to consider between physical and intellectual property. The strategic advantage of intellectual (over *physical*) property is quite evident, particularly as IP is *extensible and does not depreciate* over time, that is, of course, *if* your IP is positioned as strong in the first place.

In this chapter, we are unable to dive extensively into the fundamental philosophical principles underpinning the role of government providing preferential treatment to innovators. There is no question that lots of fascinating topics would surface if we did, but we can only encourage readers to explore this topic on their own. For now, we must forge ahead and turn our attention in the next section to a more detailed look at patents themselves. This then paves the way to dig in and ultimately reach our objective of clarifying IP as a source of wealth.

Factor	Physical	Intellectual
Multi-use	Use by one firm precludes simultaneous use by another	Use by one firm does not preclude simultaneous use by another
Depreciation	Depreciates, wears out	Does not wear out (but might become obsolete)
Protection	Generally can enforce and protect ownership	Difficult and expensive to enforce and protect ownership

Table 6.1: Comparison between physical and intellectual property.

6.3.1 Categories and purpose

The term *intellectual property* (IP) is often viewed as synonymous with patents, and certainly patents are *one* form of IP. However, we should adopt a broader concept of "intellectual creative efforts" for which different characterizations lead to different categories of protectable intangible intellectual property. Table 6.2 clarifies the intention of each category. Different venues have different state and federal laws to help innovators protect their efforts.

Category	Purpose
Patent	Protect the idea
Copyright	Protect the unique expression of the idea
Trademark	Protect the reputation of a product
Trade secret	Protect the sharing of an idea

Table 6.2: Types of IP protection and purpose.

It is common to default to talking about patents when speaking of IP since patents have been the traditional visible investor force for widespread wealth generation over the past few decades. However, the backlog of patents in the US Trademark and Patent Office (USTPO) queue (about 650,000) awaiting first action by an USTPO examiner *and* the expected 4-year wait for a first action to occur is a concern for many investors; the time to return on investment (ROI) and a declined patent application represent risk. With the protracted time to be awarded a patent there is growing sense that other forms of IP (trade secrets) might be a more expedient way to protect intellectual property. Patents *and* trade secrets therefore become of primary interest in the discussion to follow.

As mentioned earlier, state and federal laws exist to enforce IP rights. These days, language for federal statutes is readily available online from a variety of websites. For instance the Cornell University school of law maintains a complete listing of the United States Code (USC) at `http://www.law.cornell.edu/uscode/text`. Alternatively, you can track specific US legislation that possible affects IP through the Library of Congress website `http://thomas.loc.gov/home/thomas.php`.

The Cornell University website illustrates that Section 17 of the USC covers Copyrights, and Section 35 of the USC covers Patents. Trademarks are part of Commerce and Trade, Title 15, Chapter 22, Subchapter I, §1052. Trade secrets are covered by state law, not federal law.

6.3.2 Patents

Patents are popular to talk about in engineering circles because of their central and visible role in creating competitive advantage; we also hear a lot about patents and patent lawsuits in the news. Early patents covered products and technology that might surprise us today. Think of the wealth associated with having a monopoly on the making of glass! Reuters[17] notes:

> The first recorded patent of invention was granted to John of Utynam. In 1449, he was awarded a 20-year monopoly for a glass-making process previously unknown in England (subsequently, he supplied glass for the windows of Eton College Chapel, UK). In return for his monopoly, John of Utynam was required to teach his process to native Englishmen. That same function of passing on information is now fulfilled by the publication of a patent specification.

It is entertaining to imagine royalty writing a letter to you conferring rights and privileges as an inventor. Sounds like fun to get a letter like that! With a royal seal the letters served as proof of those rights. Letters closed by a royal seal were called "litterae clausae." Letters that carried the royal seal, but were *open to inspection* were "litterae patentes," meaning an open letter. Today, we just call these conferred rights "patents." The first U.S. patent was granted in 1790 to Samuel Hopkins of Philadelphia for "making pot and pearl ashes" - a cleaning formula used in making soap. Activities of daily living are improved through the ability to profit from patents.

A common myth is that a "successful" patent will make the inventor rich. Table 6.3, adapted from the work of Dorf[18] illustrates, there are several well-know patents whose inventors spent as much in litigation as they earned in royalties. Eli Whitney actually had to use the proceeds from his patent on interchangeable parts to support the on-going litigation for the cotton gin. In his case the published patent educated the competition so well they knew exactly how to work around his invention claims.

Unlike tangible assets, we can't "see" or "touch" intellectual property, so protection requires some tangible form or description of the creation in order for ownership to be established. Let's explore the characteristics and document components needed to develop and submit a patent.

Patents are generally granted on "new and useful processes, machines, manufactures or compositions of matter, or any new and useful improvement thereof" (35 USC §101); within this framework are three general types of patents:

Invention	Inventor	Date
Cotton gin	Eli Whitney	1794
Sewing machine	Elias Howe	1846
Barbed wire	Joseph Glidden	1874
Telephone	Alexander Graham Bell	1876
Lightbulb	Thomas Edison	1880
Airplane	Orville and Wilbur Wright	1906
Gyroscope	Elmer A. Sperry	1916
Television	Philo T. Farnsworth	1927
Xerography	Chester Carlton	1942
Transistor	J. Bardeen, W. Brattan, W. Shockley	1950

Table 6.3: Examples of patents difficult to enforce; in boldface are inventions involving significant litigation cost. Adapted from Dorf[18]

1. *Utility patents* - Encompass the function or operation of a machine or device.
2. *Design patents* - Protect "new original decorative" appearance of an object (not the function of the device).
3. *Plant patents* - Highly specialized protection for a certain category of asexually reproduced plants.

Patents themselves are documents composed of three key parts:

1. *Specification* - A written description of the invention, including the "best mode" of practicing the invention and adequate detail for those "skilled in the art" to duplicate the invention.
2. *Claims* - Definition of the extent and rights for which the application is seeking protection.
3. *Drawings* - Illustrations of examples of the concept.

To be patented, an invention must fulfill four requirements of "patentability":

1. *Novelty* - The invention must be *new.* If it has been in the public domain (known, used, sold, or described in printed material) in the year prior to your application, your invention cannot be patented (35 USC §102).
2. *Utility* - Being "new" is not adequate if the invention is not useful (35 USC §101).
3. *Nonobvious* - Must give new and nonobvious results compared to known approaches (35 USC §103).
4. *Enablement* - The application must provide sufficient description to enable one of ordinary skill in the art to practice the invention (35 USC §112).

Although the patent application process seems to have a lot of folklore associated with it, Figure 6.3 is a simplified flowchart to help identify major steps in the process. An experienced patent attorney is recommended – but not required – to navigate the process. Terminology is important. A patent agent is quite often a lawyer who specializes in patent law, but not always. A *patent agent* is someone admitted to practice before the U.S. Patent Office, but is not required to have the same legal credentials as someone who practices in a court of law. A *patent attorney* is someone who is admitted to practice before the courts *and* who is also admitted to practice before the U.S. Patent Office.

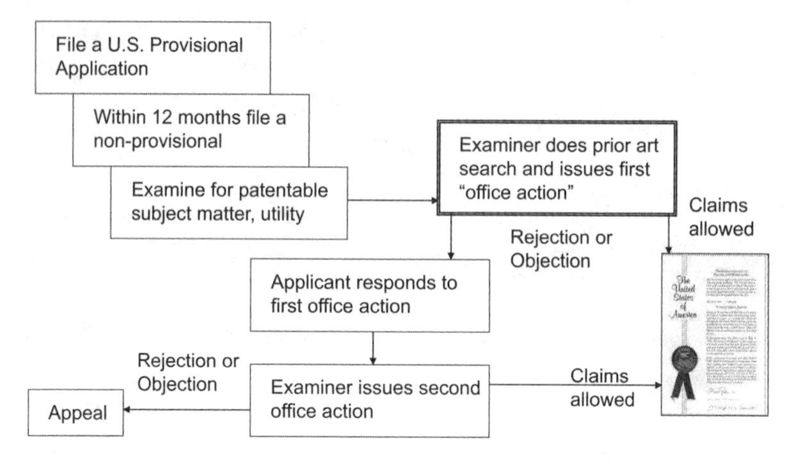

Figure 6.3: Summary of the patent application process.

We often hear that it costs a lot of money to "get" a patent. The *filing fee* is actually quite reasonable and ranges between \$95 and \$280 depending on the size of your firm and how it is filed. Legal preparation fees are a separate matter, and these range from \$5,000 to \$20,000.

A few interesting ideas about patents are that:

1. A patent is not required to manufacture or use a new product or invention. Many products are on the market that are sold without patents.
2. An invention does not need to be perfected (do not need to reduce to practice to apply or be granted a patent).
3. Most patents are not worth much money.
4. It does not cost a significant amount to file (mentioned above).
5. A patentable invention may be a combination of existing patents. The challenge will be the freedom to exercise your patent if contingent on another.

Figure 6.3 suggests there is a difference between a *provisional* and *non-provisional* patent application. Essentially the provisional is not all *that* rigorous and serves to establish the filing date of your patent application as soon as you have your idea – it "stops the clock" and indicates you believe that you are the "first to invent" and, if you practice (use) the patent, you can claim your product is *patent pending.* A non-provisional patent is the "real thing" and is queued up for full consideration by a patent examiner.

6.3.3 Trade secrets

In light of the recent legislative changes suggested by the America Invents Act[19], trade secrets are getting renewed attention as a venue for protecting intellectual property. Descriptions vary, but trade secrets represent any confidential information deemed adequately valuable by a company to provide the owner a real or potential competitive advantage in the marketplace. Part of the value of trade secret assets is the secrecy of the assets. Unlike a patent, trade secrets do not need to meet the criteria of novelty, non-obviousness, and usefulness, so from one perspective, trade secrets cast a wider net in terms of the scope of intellectual property that can be protected. As well, trade secrets do not have a "sunset" to their life, after which the information defaults to the public domain.

Many people do not realize that trade secrets do not benefit from the laws of the federal government but are instead governed by state laws. And state laws vary widely, so the National Conference of Commissioners on Uniform State Laws (NCCUSL) seeks to promote harmonization of state laws through the Uniform Trade Secrets Act (UTSA). The UTSA provides a precise, yet fairly simple, definition[20] of a trade secret as

> *Information, including a formula, pattern, compilation, program, device, method, technique, or process, that derives independent economic value, actual or potential, from not being generally known to or readily ascertainable through appropriate means by other persons who might obtain economic value from its disclosure or use; and is the subject of efforts that are reasonable under the circumstances to maintain its secrecy.*

A majority of the states have enacted provisions of the UTSA, and its website `http://uniformlaws.org` provides a cross reference for those interested in specific details. Previously, trade secret law was traditionally set through common law (precedent via the courts), until the NCCUSL led the Restatement of Torts (1939)[20] wherein §757 and §758 set forth the basic principles of trade secret law as we generally know them today. In particular, §757 lists six factors frequently cited in textbooks in checklist form for determining whether information constitutes a trade secret:

1. How much of this information is already known outside the company?
2. To what extent is the information known *inside* the company?
3. What measures have been taken to keep the secret a secret?
4. How valuable is the information to the company and to its competitors?
5. What were the cost and time involved to develop the information?
6. How difficult or easy would it be for someone else to create the information?

The World Intellectual Property Organization definition recognizes that trade secrets are more than just inventions or manufacturing processes and that many items that do not meet normal patentability criteria might be considered a trade secret[21]:

> *The subject matter of trade secrets is usually defined in broad terms and includes sales methods, distribution methods, consumer profiles, advertising strategies, lists of suppliers and clients, and manufacturing processes. While a final determination of what information constitutes a trade secret will depend on the circumstances of each individual case, clearly unfair practices in respect of secret information include industrial or commercial espionage, breach of contract and breach of confidence.*

Trade secrets encompass a wide variety of customer information, databases, know-how, and many other general industrial secrets as long as *they provide the owner a real or potential competitive advantage in the marketplace.* The unauthorized use of such information is a violation of the trade secret and the owner can seek remedy in court. Some advantages of trade secrets include:

1. Trade secret protection is indefinite as long as the secret remains a secret.
2. There are no registration or filing costs.
3. The protection takes effect immediately.
4. Ongoing protection does not require disclosure.

Confidential business information qualifies as a trade secret, too. But there can be complications if the information also meets the criteria for patentability:

1. Trade secrets manifest in a product might be subject to being "reverse engineered" and in most states the third party would be entitled to use it. Trade secret protection is not an exclusive right and you may not be able to prevent use by third parties.
2. Trade secrets made public are no longer secrets and the value is diminished.
3. Patents are easier to enforce than a trade secret, but with litigation costs these days, that is not saying much.
4. Someone else can patent an idea developed by legitimate means that might also be the subject of your trade secret.

6.3.4 Copyrights and trademarks

The subject of copyrights and trademarks seems less at the forefront of engineering project discussions, as patents and trade secrets dominate thought about investment and financial returns. For this reason we only briefly discuss copyrights and trademarks, but note that software is a special case of IP that *might* be covered by a patent, copyright, or both, but to the extent the software is original would always be covered by copyright. Software programs are considered "original works of authorship" under the copyright statute. We don't need to look much further than the on-going Yahoo! versus Facebook suits and counter-suits to understand the potential value of software and related infringement risk.

Copyrights

Table 6.2 identified a copyright as protecting the *expression of an idea* and does not protect the idea *itself.* Generally this protects *original* works of authorship expressed in a tangible way such as a book or Internet site. Copyrights cover many mediums of expression and do not necessarily need to be a published work – music, dance, and software programs all qualify for copyright protection.

For instance, this chapter discusses the idea of intellectual property. The idea is not new, but many writers have expressed unique perspectives about intellectual property in papers, journals, and books, and their work in these "tangible" media are protectable. Anyone who uses these ideas must indicate use and acknowledge the source. So while the current book chapter has many original thoughts, we have acknowledged the use of other works that contribute to the understanding and interpretation of intellectual property. As is clear even in the present book, we extensively use citations to acknowledge the person, the specific work, and the date of the work that was used to help shape our thinking. This lets the reader clearly know what is original in the book and what is not.

In the US, submissions of work for copyright protection are sent to the Register of Copyrights within the Library of Congress. The process is fairly simple and does not carry with it an examination for novelty, utility, or the potential for confusion with other products. Software is unique in that it might qualify for copyright and patent protection.

The terms of protection for copyrights are considerably longer than for patents. Today, the copyright term is the *life of the author, plus 50 years.* Works performed for a fee have protection for 100 years from creation or 75 years from publication, whichever is shorter.

Copyrights have influenced today's "wired world" in more ways than most people realize. Were it not for strong copyright protection, the success of the Open Source movement would have been limited, and the development of the Linux operating system and other core technologies that drive the internet may never have proliferated.

Trademarks

Trademarks protect the reputation or "goodwill" of the manufacturer or producer of goods. They are a non-functional word, logo, slogan, symbol, design, or any combination that distinguishes a product or service or source of a product or service from its competitor. The key objective of registering a trademark is to establish the rights of the "mark" so that the consumer is not confused about the source of the goods. The mark itself is not of value as a stand-alone entity, and is only valid in the context of or adjunct to the product or service it denotes.

It seems that "trade-marking" has some origins with the guilds (trades) of the Middle Ages in which craftsmen would affix marks to goods so that the superior nature of a guild product could be easily identified. In this way the mark of a trade guild would distinguish their products from inferior competitors and instill within consumers a trust and loyalty to a guild product. It is easy to imagine how hard it would be for the common person to assess goods that might be short in weight, or built with poor quality materials and inferior craftsmanship, and the trademark emerged as a way to stand out. The notion of guilds was eventually replaced by capitalism after the Industrial Revolution, and trademarks simply were to identify the *source* of goods.

The challenge of forgery and counterfeiting led to laws for remedy, and today most every country has a process for registering a trademark or service mark. In some situations you do not have to register as long as there has been a history of usage by a producer, and it can be proven that the public has been made aware of the product and it is distinctive in the mind of the public. Again the idea is to avoid confusion in the marketplace of the source of the goods; registry does not affirm or refute quality.

As long as a trademark is used properly, it can have an indefinite life. But abandoned or incorrectly used the trademark can be lost. Another hazard is when the product is so successful that it becomes the generic name for the product it represents, as in the case of aspirin and escalator.

6.3.5 Intellectual property as an intangible asset

Chapter 2 helped identify categorization of assets and their role in enabling management to create value for company stakeholders. For simplicity, Chapter 2 centered on the accounting of *tangible* assets such as property, plant, or equipment. A tangible asset has the "capability of being perceived or precisely identified" (Morris[22]) and there have emerged guidelines and standard approaches (Yegge[23]; Siegel and Shin[24]; FASB) to account for or assess the value of these assets. However, the so-called "knowledge economy" alluded to in Greenspan's speech[25] implies the "new economy" is now based more on *intangible* assets. Let's explore these type of assets in more detail and related thoughts about the ways in which intangible assets create value.

The valuation of intangible assets can be frustrating for two simple reasons:

1. The context of the valuation has an impact on the valuation.
2. The concept of intangible assets is often used interchangeably with intellectual property.
3. There are several categories of intangible assets and they overlap.

It helps to try to differentiate between *goodwill, intangible assets*, and *intellectual property*. There is no one definition for each of these terms – and they are somewhat intertwined – so be thoughtful and patient as you read!

Intangible asset: An asset that is not monetary or physical in nature but that (a) is a quantifiable asset that has similarities to other intangible assets in the marketplace, (b) has some proof of existence, (c) is something that can be legally owned, and (d) is something that can be identified or associated with a particular company.

Intellectual property: A subset of intangible assets that have been granted certain legal recognition, protections, and rights. Hopefully, most engineers could easily recite that these commonly take the form of patents, trademarks, copyrights, and trade secrets.

Goodwill: Recorded value on a company's balance sheet in excess of the value of all tangible and uniquely identifiable intangible assets.

The pathway to analysis is important here. With no precedent to valuation, you can undertake an "If A, then B, else C" approach to deducing the value of independent, uniquely identifiable intangible assets. For instance, in the case of design technology, proprietary test results, process technology, or operating platforms there may be identifiable fair market value, tax value, or licensing value that can (among other methods) be used to establish a monetary value of the intangible asset. Of course, the *context* of the method becomes critical in reaching a specific value, but this simply sets the stage for a higher variance in the outcome. The context itself does not mean the deductive process is absent or inapplicable. Other intangible assets have a more inductive origin, that is, we might have some general observations about an asset and then through a series of observations conclude what the value of that asset is. For instance, we might have an Internet-based company that has a website, e-commerce capability, a 1-800 number, a customer base, and specialty software and hardware that *bundled together* could be monetized for tax purposes based on similar entity comparable value. It might be the case, though, that the monetized value of a particular company's unique bundle of assets has a *market value* much higher than the monetized value, the difference in value being *goodwill*. So again, through a series of conclusions about the value of unique intangible assets and the perceived market value of a company we can compute goodwill:

$$Goodwill = \left(Company\ Market\ Value - \sum Assets\right) \tag{6.1}$$

The value of goodwill tends to be time-dependent and largely a function of perception of value in the market and is thus highly variable (check out the wild fluctuations of the value of Facebook), so, although somewhat elusive, "goodwill" *can* be calculated and a value established (or negotiated). Goodwill is quite controversial because its value can significantly affect the reporting of earnings and assets, and is thus often the topic of FASB and SEC guidance to companies (Marshall et al.[26], p. 206; Anson[3], p. 55, p. 66).

Without doubt intangible assets enable a company to produce products and services of measurable value, but, in contrast to the very deterministic flavor of Chapter 2, the simplistic internal perspective of value is now giving way to the influence of external issues and stakeholder pressures. This new-found ambiguity in financial affairs is the price we pay when embracing a knowledge economy.

Figure 6.4 illustrates the perspective that all assets are the sum of tangible and intangible assets.

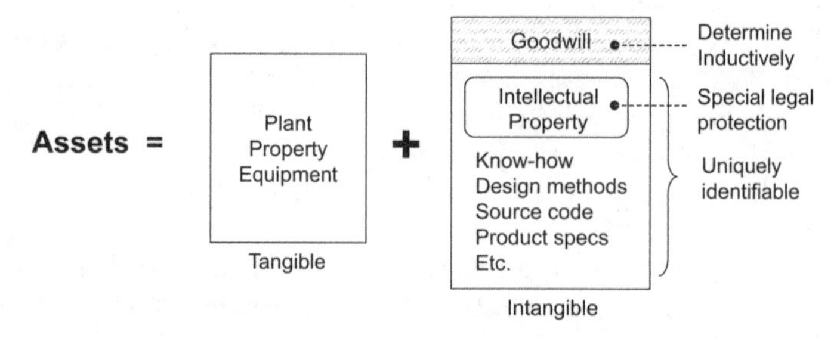

Figure 6.4: Examples of tangible and intangible assets.

A slightly different form of this model that links patents to performance objectives is presented by Carlucci and Schiuma[27], and takes advantage of the power of concept maps; this is discussed further Chapter 7.

6.3.6 Valuation of intellectual property

To some extent the approaches to valuation of intellectual property (patents, trade secrets) are similar to valuation of tangible assets (property, plant, equipment): there is precedent for application of *cost, market*, and *income* approaches to valuation, and recommended practices and guidelines have emerged over time [28][29][3]. Still, some of the details, like market comparables or replacement value, might be needed and hard to find for intellectual property, so extraordinary skill and soliciting multiple perspectives is essential. References such as Copeland et al. [30] and Yegge[23] provide excellent guidance for the analyst and have informed the present work. Table 6.4 provides a high-level comparison of factor differences.

Factor	Real Property	Intellectual Property
Categorization	Easy	Difficult
Context	Independent	Dependent
Focus	Specific Details	Specific Concepts
Methodology	Objective	Subjective

Table 6.4: Difference in key valuation factors between real property and intellectual property.

The specific valuation approach (market, cost, or income) to use in each situation requires a pit of prodding thought, and judgment before a specific strategy can be recommended. It is a lot easier to say, "Find the approach that best indicates the value of the property," than it is to reduce to practice for a specific (often unique) IP situation. A strategy calling for "fair market" will have upper and lower bounds to value, and committing to a final number requires experience and practice. We pass along the following definitions of market, cost, and income methods, directly quoting IRS guidance (IRS Revenue Manual §4-48-5 [29]):

- *Market-based methods rely on identification of similar intellectual property sold under similar conditions. This method requires the existence of an active market involving comparable property.*
- *Cost-based methods estimate the cost to reproduce/replace or to pay to purchase the subject intellectual property (the "make or buy" decision), using historical costs or reproduction costs. Fair market value may be difficult to estimate due to quantification problems related to economic obsolescence or future income potential.*
- *Income-based methods focus on the income-producing capacity of intellectual property. The present value of the net economic benefit to be received over the life of the asset (cash receipts less cash outlays) can estimate the value. The income approach usually computes the net present value (NPV) of the intellectual property by use of the discounted cash flow (DCF) method.*

The experience of the Chartered Institute of Management Accountants [28] suggests the following priority of techniques.

Asset	Primary	Secondary	Tertiary
Patents, Technology	Income	Market	Cost
Copyrights	Income	Market	Cost
Software	Cost	Market	Income
Customer Relations	Income	Cost	Market

Table 6.5: Intangible Valuation Approach Summary; adapted from CIMA[28], Table 1.

Chapter 3 from the work of Anson [3] examines the three categories of valuation in much more detail. Overall, the author encourages the use of the income approach for valuation. This carries with it good news for the student since our friend *net present value* from Section 3.6, Equation 3.9 is back with us and ready to help.

$$P = A\left\{\frac{(1+i)^N - 1}{i(1+i)^N}\right\}, i \neq 0; \quad = A\left(\frac{P}{A}, i\%, N\right)$$

How rewarding that we've got another use for the equation! Let's take a closer look.

Income approach

The fundamental idea in the cost approach is to estimate what it would cost to replace the asset. Recall from Section 3.6 that *present worth* (PW) calculations were just a way of collapsing a future earnings stream into a single number. Figure 6.5 provides a cashflow diagram representing the basic income approach. The situation is somewhat complicated by the fact we have to forecast earnings and select an appropriate interest rate, but that's OK; we've done that before.

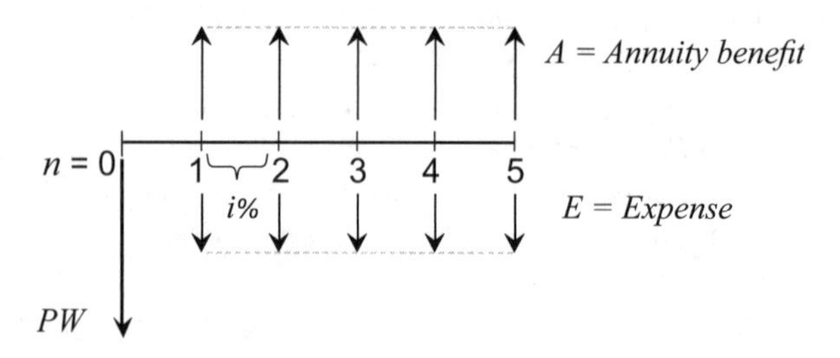

Figure 6.5: Net present worth of a forecast of future earnings.

A few decisions need to be made to proceed with the PW calculation. Let's work through the four major issues one by one.

1. A critical component of time-value of money calculation is the interest rate i to use. We know the interest rate is the opportunity cost of capital, and this could be prevailing investment rates of return or the company's minimum acceptable rate of return (MARR). For the valuation exercise, assume management is bullish on innovation and sets $MARR = 12\%$.
2. The cash forecast requires an estimate of the expected timeframe of the benefit of IP, and in this case assume we are examining a moderately perishable technology in a highly competitive market, so the estimated life of the asset is $N = 5$ years.
3. The benefit of IP is based on the enhanced revenue to our business. If we assume that the use of IP has created the ability to produce a preferred product with an 8% price premium relative to our competitors, this would mean a $100,000 benefit on a business with $1,250,000 in revenue. Here, $A = \$100,000$.
4. Lastly, there might be a special *expense* associated with the IP. This might be a sub-royalty for a related patent needed to enable complete freedom to operate, or some other expense unique to the IP. Assume this is a small portion of the incremental revenue, say, 2.5%. In this case, $E = \$2,500$.

With these four parameters in hand we can proceed to calculate the PW as a measure of value of the IP.

$$
\begin{aligned}
PW &= (A-E)\left(\frac{P}{A}, i\%, N\right) \\
&= (\$100,000 - \$2,500)\left(\frac{P}{A}, 12\%, 5\right) \\
&= \$97,500\left(\frac{P}{A}, 12\%, 5\right) \\
&= \$97,500(3.6048) \\
&= \$351,468
\end{aligned}
$$

Suppose that a $350K PW is not that exciting to management. Since everything is just an estimate anyway, re-consider that the IP has an expected life of 20 years, there are no special expenses associated with the intellectual property, and management decides to drop the MARR to match more closely conventional investment returns of 10%. The *PW* calculation is now:

$$
\begin{aligned}
PW &= \$100,000\left(\frac{P}{A}, 10\%, 20\right) \\
&= \$100,000(8.5136) \\
&= \$851,360
\end{aligned}
$$

This is a much higher valuation, almost 2.5x our original estimate! This, however, points out a critical issue with valuation. As we said, the "devil in in the concept, not the details." The reputation of the analyst performing the valuation has a tremendous influence on the valuation outcome.

It should be evident one that could take considerable latitude in identifying revenue versus expense relative to the IP, if it is even possible to separate out revenue in this way. Further the expected useful life of the IP is speculative, but there are probably reasonable industry-specific trends (consumer electronics versus appliances). And, of course, we know from Chapter 3 that selecting the interest rate requires care, too.

In closing, it is worthwhile to mention the *relief from royalty method*, where based on comparable technology we estimate the third-party royalty rate that would be required. It does tend to be a bit more conservative and less subject to PW parameter speculation primarily since market comparables anchor the calculation.

6.4 Wealth creation

Earlier in Chapter 2 we noted that Owners' Equity (OE) often increases after the founders' original investment in the company. Many times other people (family members, friends, angel investors, or venture capitalists) are subsequently invited to invest in the company. On the balance sheet, OE funds are matched on the asset side by an increase in cash. As illustrated earlier in

Figure 1.8, management employs these new assets to grow the business. In this section we describe a secondary benefit of the investment as it pertains to potential increases in stock price and thus wealth generation for the founders.

We'll use the example of Bethesda Imaging, Inc. to illustrate one method of wealth generation through the development of IP. Let's assume that Bethesda Imaging, Inc. is owned by Martha and Jake Bethesda. Jake had hoped to become an actor after leaving college with a degree in biomedical engineering but, while waiting tables to earn a living, he stumbled upon an idea for speeding up the identification of arterial plaque growth by automating MRI pattern analysis. Martha was a computer science graduate currently doing ICD-10 conversions for a local insurance company. The two of them were married a year ago and decided to start their company part-time as a way to fulfill professional ambition. The company was registered in the State of Ohio on January 30, 2010; they felt no need to get an attorney and so simply paid the $175 application fee for Form OH-532 and an additional $500 filing fee for the issuance of 10,000 shares of stock. Table 6.6 illustrates the stock distribution:

	Initial Incorporation		
	Shares	Stake	Value
Stock Value Profile			
Corporate Valuation			$2,000
Shares Authorized	10,000		
Shares Outstanding	2,000		
Shares Remaining	8,000		
Stock Price			$1.00
Initial Assignments			
President - Martha	1,000	50.0%	$1,000
CEO - Jake	1,000	50.0%	$1,000
Executive - C Option	0	0.0%	$0
Executive - D Option	0	0.0%	$0
Executive - E Option	0	0.0%	$0
Follow-on Funding			
Partner - X			
Partner - Y			
Partner - Z			
	2,000	100.0%	**$2,000**

Table 6.6: Initial stock issuance of Bethesda Imaging, Inc.

From Table 6.6 it is evident that only 2,000 shares were *distributed* (1,000 each to Martha and Jake) of the original 10,000 shares *authorized* in the article of incorporation, leaving 8,000 shares remaining. Recalling the balance sheet discussion of Chapter 2, we quickly recognize that if 2,000 common shares have a declared par value of $1/share, then the initial "value" of the company is \$2,000. This means that Jake and Martha each contributed \$1,000 to start up the company, and in exchange for that investment they each own 50% of the company. Of course, it cost $\$175 + \$500 = \$675$ to start the company, so they have $\$2,000 - \$675 = \$1,325$ left in the company. And now the company is now only worth \$1,325, but still, Martha and Jake share equally in the ownership of the company.

Bethesda Imaging, Inc.
Balance Sheet
January 31, 2010

Assets		**Liabilities**	
Cash	\$1,325	Notes payable	\$0
Capital Equipment	\$0	Equity	
Furniture	\$0	Common Stock	\$2,000
		Retained Earnings	(\$675)
Total Assets	\$1,325	Liabilities and OE	\$1,325

Table 6.7: Simplified balance sheet for Bethesda Imaging, Inc.

Well, great, 50% of *something* is still better than 50% of *nothing*, but what happened to the great idea that was the catalyst for starting the company in the first place? And what happened to the value that the IP was supposed to create?

The fact is that many start-ups such as Bethesda languish for quite a while. Despite the creative and innovative forces within the Bethesda team, no realizable value has been created for two reasons:

1. The team has not reached the point where the idea is an opportunity. Martha and Jake need to know that (a) there is a need for the product or service in the market, (b) that once in the market a business model exists that enables profit to be made, and (c) they are the business team that can execute on a plan to deliver potential value; these three items are essential for an idea to be considered an opportunity.
2. The idea must be protectable in some way, whether through the patent system or if possible to keep as a trade secret.

The *protectability* of an idea *properly positioned* as a revenue opportunity is the foundation of value in the market. If Bethesda believes that the \$1,375 they still have in the bank is adequate to get the software to market, fine! The story ends there. They produce the software and then license it to customers in the market and collect revenue for years to come. It is unlikely, however, that the duo of Jake and Martha are by themselves adequate to develop, test, and market the product; they will need additional resources. The most likely step is for Martha and Jake to consider selling some of the 8,000 remaining shares of company common stock to an outsider. This is not an easy step to take for a couple of reasons; let's take a closer look.

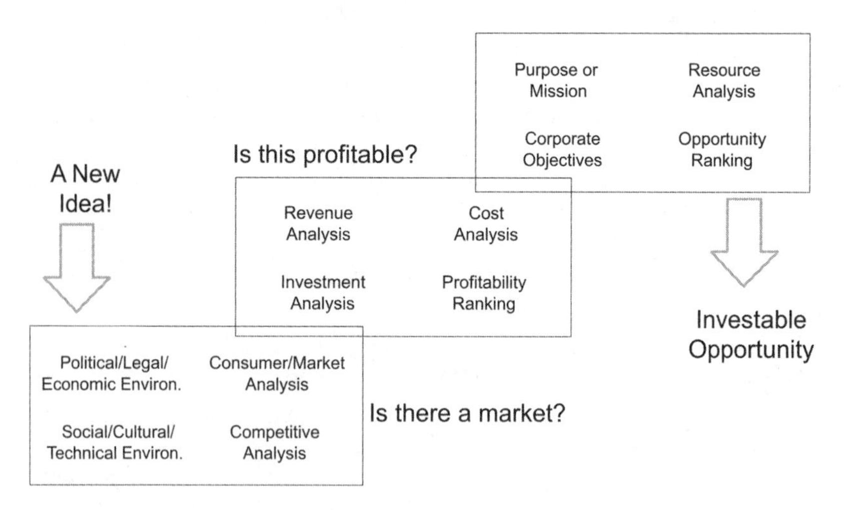

Figure 6.6: Steps to distinguish between an idea and an investable opportunity.

6.4.1 Stock valuation: Perception is reality

There are two challenges (well, more, really, but these are the top two) confronting Bethesda Imaging, Inc.:

1. What price should the stock be sold for?
2. How much stock should they sell?

Although the two questions are somewhat related, the first concerns the perceived value of the company and the second establishes how much control the new stockholder (owner) will have.

Stock price is computed as the quotient of the *value of the company* and the current *number of shares* of common stock distributed. While determining the number of shares of stock distributed is easy to find, estimating the value of the company can be difficult. At the outset the nominal price of $1/share leads to a nominal company valuation of $2,000. However, let's suppose that Jake had been doing a lot of productive research and development over the past year – through "sweat equity" the firm has not languished at all and much progress has been made. Value would accrue from (a) advances in the imaging software development, (b) an understanding of the market need and price points, and (c) customer relationships to enable product sales. Ideally, unique software capabilities have been invented that solve specific customer imaging problems, and enablements meet conditions of patentability. Let's further suppose that Bethesda has taken the step of filing a few non-provisional patents over the course of the year. All this arduous work over the past year has created value in the minds of Bethesda's founders. Absent any recordable transactions for tangible asset purchase, the internal value is minimal; some accounts simply recommend placing a balance sheet entry of $1 just to communicate the idea that intangible assets are a part of the company (Marshall[26], p. 206). Bethesda knows they "have something" worth a lot more than $1/share, but how much? This is the job of *valuation*. Glance back at Section 6.3.6.

Copeland et al. [30] and Yegge[23] were two references we used in Section 6.3.6 when discussing *intangible asset* valuation. Common ways to value a company were highlighted, but Yegge[23] in particular is useful and persuasively underscores that valuation is more of an art than a science, though in the end we find comfort in the *income approach* (Copeland [30] provides good examples, too). Some people might say, "The value of a company is simply the price someone offers," and allowing this is the case, then the *perception* of value is what dominates the transaction. One might reasonably expect value in the minds of the founders to be higher than that of a buyer, but once a mutually agreeable price point has been set, you might say this sets the market value of the company. Yegge[23] works though 13 detailed cases of a wide variety of businesses, and Chapter 9 of Yegge is a concise summary of six popular methods, including DCF from two perspectives. The recommendation, of course, is the *income approach* outlined in Section 6.3.6.

Imagine that a stock transaction price of $500/share has been negotiated between an interested Angel Investor and Bethesda. This might seem like a high stock price to the layman, but this is typical of many *private placements* for an initial investments round. The message this placement sends is that the investor has high hopes for the future earning potential of the company. And, we are not talking about a public company (yet) traded in the stock market (where it seems customary to try to keep share price below $100), so even a price of $1,250/share is not really a concern. Table 6.8 illustrates the impact of the angel investor purchasing 1,500 shares of Bethesda at the negotiated $500/share price. The column marked "Round I" shows that the company is now worth a total of $1,750,000 based on an increase in the value of company stock from $1/share to $500/share.

Examine Table 6.8 closely to notice the impact of the stock purchase. First, this was just a private equity investment, not an initial public offering as described a bit later in Section 6.4.2. In this privately negotiated deal, the angel investor has made a $750,000 investment to acquire a 35.3% equity position in the company. While Bethesda's President and CEO each see their stake in the company drop from 50% to 28.6%, together Jake and Martha still own a majority share of the company (a combined 57.2%) *and* the value of their stake in the company just went from $1,000 each to $500,000. Combined, Jake and Martha are now millionaires!

Consider carefully what just happened. If we examine the balance sheet, the par value of the stock has not changed. This touches on the significant difference between the *financial market* and the *real market*. Up to the point of angel investor involvement, we were discussing transactions in a so-called "real" market; all of what is recorded on the books is based on the activities of the company on a day-to-day basis. The real activities of the company dominate.

However, when a company decides to enter into the *financial market*, then the interaction between the owners and the investors circles around what the potential investors *perceive* the shares of the company are worth. Investors

	Initial Incorporation			Round I		
	Shares	Stake	Value	Shares	Stake	Value
Stock Value Profile						
Corporate Valuation			$2,000			$1,750,000
Shares Authorized	10,000					
Shares Outstanding	2,000			3,500		
Shares Remaining	8,000			6,500		
Stock Price			$1.00			$500.00
Initial Assignments						
President - Martha	1,000	50.0%	$1,000	1,000	28.6%	$500,000
CEO - Jake	1,000	50.0%	$1,000	1,000	28.6%	$500,000
Executive - C option	0	0.0%	$0	0	0.0%	$0
Executive - D option	0	0.0%	$0	0	0.0%	$0
Executive - E option	0	0.0%	$0	0	0.0%	$0
Angel Investor - A			$0	1,500	35.3%	$750,000
Angel Investor - B			$0	0	0.0%	$0
Angel Investor - C			$0	0	0.0%	$0
	2,000	100.0%	$2,000	3,500	100.0%	$1,750,000

Table 6.8: Private investor purchase of stock from Bethesda Imaging, Inc.

then purchase or trade stocks based on what they *think* the dividends or share price will provide in the future. In essence, the financial market brings into play equity investors in a company based on future expectations of the company, and not necessarily based on what is really happening today or in the recent past. And in more cases than not, the presence of a unique position in the market – that provided by intellectual property – seems to be the driver of future expected performance and wealth.

So, if "pre-money" (that is, before the investment is made) Bethesda is worth $1/share but the IP developed creates the expectation of a company worth $500/share, then what the financial market is telling you is that if you proceed as planned and successfully execute on company plans, then the financial market is willing to pay a *premium* of $499/share.

Managing the company in the real and financial markets is not easy. You have to execute day-to-day in the real market, but with an eye to satisfying the financial market expectations. Bethesda may have convinced the angel investor that the performance potential of the company has a significant upside, but now they have to deliver, and this has an irreversible effect on the company and its operational context.

As a final conceptual step, let's look at what happens when another investor jumps in.

What just happened? It looks like the co-founders' wealth has just increased dramatically, but now they own less than a majority share of stock. Are they in control anymore? Actually, who *is* in control? Some ways to think about this problem are:

1. The stock price rose; why?

	Initial Incorporation			Round I			Round II		
	Shares	Stake	Value	Shares	Stake	Value	Shares	Stake	Value
Stock Value Profile									
Corporate Valuation			$2,000			$1,750,000			$5,312,500
Shares Authorized	10,000								
Shares Outstanding	2,000			3,500			4,250		
Shares Remaining	8,000			6,500			5,750		
Stock Price			$1.00			$500.00			$1,250.00
Initial Assignments									
President - Martha	1,000	50.0%	$1,000	1,000	28.6%	$500,000	1,000	23.5%	$1,250,000
CEO - Jake	1,000	50.0%	$1,000	1,000	28.6%	$500,000	1,000	23.5%	$1,250,000
Executive - C option	0	0.0%	$0	0	0.0%	$0	0	0.0%	$0
Executive - D option	0	0.0%	$0	0	0.0%	$0	0	0.0%	$0
Executive - E option	0	0.0%	$0	0	0.0%	$0	0	0.0%	$0
Angel Investor - A			$0	1,500	35.3%	$750,000	1,500	35.3%	$1,875,000
Angel Investor - B			$0	0	0.0%	$0	0	0.0%	$0
Angel Investor - C			$0	0	0.0%	$0	0	0.0%	$0
	2,000	100.0%	$2,000	3,500	100.0%	$1,750,000	3,500	82.4%	$4,375,000
Round I Dilution									
Early Stage Fund I							700	16.5%	$875,000
Early Stage Fund II							50	1.2%	$62,500
							750	17.6%	$937,500

Table 6.9: Follow-on investor purchase of stock from Bethesda Imaging, Inc.

2. Do you think the fact that two early stage firms invested is significant?
3. Is the investment more likely to help the firm achieve a milestone? Another reason?
4. Does it seem odd that the early stage investor does not have majority equity "control"?

Think about these issues and how they relate to wealth creation through IP.

As with anything else in business, there are risks, and Bethesda Imaging will have its share. The "big three" are:

Liquidity risk: You buy the company with the intention to restructure for greater efficiency and sell for a premium. Will there actually be someone else in the market that wants to buy your company when you are ready to sell in 5 years?

Market risk: Will there be structural changes in the economy (taxes, economic incentives, consumer preferences) that will cause companies in your industry to be favored (gain value) or shunned (lose value)?

Business risk: Will competition, disruptive technology, or lawsuits impact your valuation?

So, we see that there is a very theoretical element to many of our predictions, in that all values and estimates are speculative until a transaction takes place. The process is, furthermore, elusive in that what is "fair" or "reasonable" sits entirely within the minds of the negotiating parties, whose perceptions can be quite different. The idea of the "truth" is quite elusive, and the moderately short paper by Luehrman[31], "What's it worth?" is really a must-read for the clarity and historical flavor that it provides an engineer. Indeed, the way Luehrman depicts the difference between scenarios for assets-in-place and opportunities plays strongly into the relevance of decision trees in valuation exercises; we will return to this as part of the discussion in

Chapter 9. Valuation that accounts for present value and growth opportunity (management making decisions based on business events) adds a very strategic component to valuation, and that too will be explored in a bit more detail in Chapter 9, in the work of Smit and Trigeorgis[32].

6.4.2 Exit stage right: IPO equity

Companies can raise money from the public by issuing debt (bonds) or equity (stock). Sometimes there is the incorrect assumption that "selling stock" is the same as going public with an *initial public offering* (IPO). The IPO is the *first* public sale of stock and is sometimes referred to as "taking a company public." It is true that IPOs can raise cash for new companies, but so can the sale of stock ("securities") to a "small" number of private investors. The latter is commonly referred to as a *private placement* and is the assumed nature of the transactions described in the previous section; an IPO requires registration with the Securities and Exchange Commission (SEC) and often is best done with assistance of an underwriting firm. An underwriting firm helps with the timing, offering price, and whether common or preferred stock should be issued.

Taking a company public can be an exciting prospect that validates business success, but there are several advantages and disadvantages.

Advantages	Disadvantages
Increase in working capital	Reduced management flexibility
Increased market value	Loss of control
Stakeholder incentive	Performance pressure
Merger and acquisitions	Volatility

Table 6.10: Pros and cons of an initial public offering (IPO).

The *increase in working capital* that results from an IPO can be used to further R&D, retire angel/VC investor debt, or simply invest in capital equipment to foster growth of the business. An IPO increases the net worth of the corporation and might improve the debt-to-equity ratio of the company. Often the due diligence process and public documentation associated with an IPO will cause an *increase in market value* since much of the affairs of the company are transparent and liquidity of public stock is viewed favorable. Besides the general prestige (assumed success) that comes with an IPO, an interesting side effect is the likelihood that once key vendors and customers become shareholders, they are more likely to be aligned with the company; the *stakeholder incentive* will make them keener to see the company successful.

Arguments promoting an IPO have opposing factors, too. IPOs can amplify management activity that might previously have been minimized or peripheral to corporate life. There are a range of disclosures mandated by the SEC that result in a *loss of privacy*, and items such as executive pay might now be subject to extensive scrutiny or public (stakeholder) debate. Management may have been accustomed to significant autonomy in making decisions that might now require board of directors' approval as *investor relations* now take on a new meaning (read: quarterly reporting and ROI pressure). Further, management might begin to be distracted by external business factors (the general state of the economy, international affairs) that might be loosely coupled to the business but affect all companies in a broad industry segment ("technology stocks" or "global entities"). There are also expenses associated with an IPO, some short-term (legal and accounting) and others long-term (additional personnel for investor relations management).

6.5 Acts to follow

6.5.1 Unlocking federally funded research: The Bayh-Dole Act

In 1996 the so-called Bayh-Dole Act (Public Law 96-517) was enacted. This legislation was profound for three reasons:

1. A significant portion of research conducted at US universities is funded by the federal government. The act enables research institutions to retain the title to any inventions or discoveries made while doing federally funded research; this shift (from ownership defaulting to the government) was to provide an incentive for Universities to commercialize their innovations. Over time the desired collaboration between universities and industry has taken place, with the idea that dissemination and commercial development would be more efficient.
2. Licensing revenue from industry would provide additional revenue streams back to the university, and "supplementing" university research by the private sector, thus leveraging federal funding.
3. Incentivized to commercialize technology, university research that might otherwise sit in the lab on the shelf is now translated into practice creating new products and services that spur economic growth.

Essentially, the act had three systemic effects:

1. System efficiency: Rather than the government deciding what to license, the organization closest to the customer can decide for itself what is to be licensed.
2. Harmonization: Rather than a university having to navigate a bewildering array of different federal-agency guidelines, a single university policy prevails.
3. Market incentives: With the prospect of earning royalties on licensed technology, the university now has a revenue stream to underwrite the cost of implementing a technology transfer policy.

An Internet search of "Public Law 96-517" produces all sorts of reports, studies, and interpretations, but a few to note are:

1. Websites, such as those supported by AUTM, that provide considerable

background information as well as updates on relevant legislation, http://www.autm.net/Bayh_Dole_Act.htm.
2. Cornell University law school has http://www.law.cornell.edu/uscode/text/35/part-II/chapter-18.
3. Track legislation from the Library of Congress at http://thomas.loc.gov/home/bills_res.html.

6.5.2 Hope for harmonization: The America Invents Act

We have mentioned the AIA at various points throughout this chapter, and possibly the best way to know more about the act is to download it from the Library of Congress (http://thomas.loc.gov/cgi-bin/query/z?c112:S.23:) and review it. The Act is best known for the switch from *first to invent* to *first to file* but that is just a small portion of the potential impact. Several significant procedural changes have been made.

1. From first-to-invent to first-to-file
2. Preissuance submissions by third parties
3. Adds post-grant review process
4. Expands *inter partes* reexamination process
5. Marking language expanded
6. Damages and willful infringement impacted

The idea of *first to file* is illustrated in Figure 6.7. A practical outcome of this new filing framework is that *if* there is IP an organization wishes to have patented, it must act promptly. Pre-AIA, your first-to-invent claim did not require you to file until you were ready. In the new system you have to act within a year.

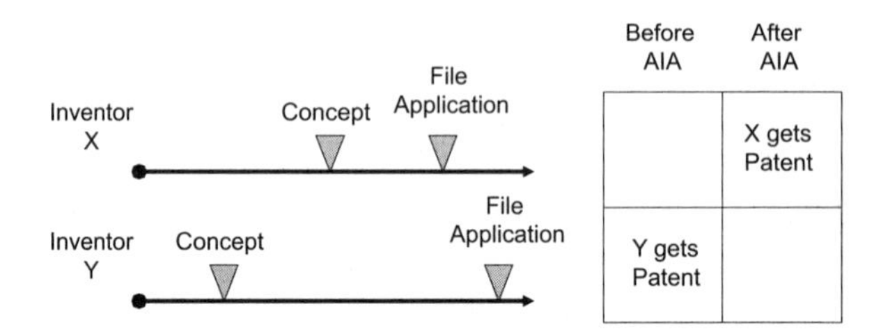

Figure 6.7: First-to-file versus first-to-invent pre- and post-AIA

Many of these changes will accomplish the objective of moving patent disputes out of the courtroom, but the procedural changes leave some trial attorneys wondering how this will affect the ability to expedite (but more likely delay) movement of a patent through the system. One line of thinking is that trade secrets will be a preferred protection of intellectual property, thus eroding a constitutional right to patent. The logic of this argument goes as follows:

1. If you chose trade secret, then essentially you abandon the right to a patent.

2. Today, you cannot receive a patent if you abandon the right (35 U.S.C. §102(c)).
3. But, as a result of the act, 35 U.S.C. §102(c) is now eliminated!
4. So, why not just keep something a trade secret until it seems time to patent?
5. Well, there is a remote possibility that the trade secret owner is prior user.
6. Thus, a patent owner holding a patent can try to sue a trade secret user to receive compensation or injunction.
7. But now, a trade secret owner can assert prior use defense and reduce the value of the patent.

Quite a scenario, eh? But the reality is that this is just one of several scenarios that will play out over time, and maybe not end up being an issue. Today, it is hard to predict what *will* happen, but it is good risk management to explore what those options might possibly be.

6.5.3 Serious about accountability: Sarbanes-Oxley

In the wake of some visible corporate scandals (Enron, WorldCom, Tyco, and Adelphia) the Sarbanes-Oxley Act of 2002 was passed to improve corporate accounting standards. Many of the officers of the companies involved in the scandals had enriched themselves at the expense of others by manipulating accounting reports. This act created *criminal* penalties for some violations which previously might have been handled through a less severe *civil* lawsuit. At the core of the act are the requirements placed upon *principal* corporate executives in three areas:

1. Certification that financial reports do not contain any untrue statements of facts (or omit facts) that might make reports misleading.
2. Executives must certify that the reports "fairly present in all material respects the financial condition and results of operations."
3. Assessment of internal controls.

Earlier, we discussed several inherent difficulties and complications associated with valuation of intangible assets; an inaccurate valuation or disclosure creates the risk that executives could unknowingly violate the act. A variety of risks must be reported that might impact the financial position of the company: impact of IP litigation, expiration of IP (loss of exclusivity), and possible USPTO reexamination proceedings.

Of course, having implied a negative tone to SARBOX (as slang has it!), there is an upside too. A trend in industry seems to be that the elevated visibility of IP and other intangibles is getting attention, providing companies the opportunity to tackle a tough subject that was previously ignored. Executives are more aware and are appraised of IP activities, and this could result in a better understanding of the IP development process (particularly for external oard members). FASB and the SEC seem to struggle with non-financial measures of value, and SARBOX appears to be somewhat of a catalyst for new activity in this area.

6.6 Problems to explore

Problem 6.1
In Section 6.2.3, we made the comment "Perception is reality." How does this play in the valuation of a *tangible* versus *intangible* asset?

Problem 6.2
Table 6.9 provided the outcome of an early stage investment in Bethesda Imaging, Inc. (see Section 6.4.1). The equity stake of the co-founders has dropped but their wealth has increased. How can that happen?

Problem 6.3
Trade secrets do not have to be novel, be non-obvious, or have utility. If that is the case, then what is the economic advantage to a trade secret?

Problem 6.4
Valuation of an intangible asset can be a challenging task. Of the three valuation methods described in Section 6.3.6, why was the *income approach* recommended? Do you agree? Why?

Problem 6.5
What is *goodwill* and how is it distinctively different from other intangible assets?

Problem 6.6
Stock valuation for Bethesda Imaging, Inc. was described in the context of "real" versus "financial" markets. What is the significance of the different markets and how does it affect stock price?

Problem 6.7
It has been proposed that patent protection is a limited monopoly provided by governments to meet economic development objectives for the "greater good" of a nation's economy. Explain, being sure to comment on the *invention-inducement theory*.

Problem 6.8
There has been a scenario depicted whereby, post-AIA, there will be a preference for trade secrets over patents. Why?

Problem 6.9
Is "goodwill" a legitimate entry on a balance sheet? Under what conditions does it arise? Provide an example from the current press where goodwill has been enhanced or impaired.

Problem 6.10
Despite IP representing an intangible asset, several approaches exist to valuation: market, cost-based, and income-based methods are common. Compare and contrast the methods and suggest reasons why the income-based approaches seem so popular.

Problem 6.11
How is stock price set at the time of company formation? What are the key factors that influence stock price pre-IPO? What dominates stock price post-IPO? How are changes in the market price of stock reflected on the balance sheet?

Problem 6.12
Explain the "bottom line" on the meaning of the Bayh-Dole Act to research institutions. In what ways do you think the Act influenced the spirit of entrepreneurship at college campuses?

Problem 6.13
Has the Sarbanes-Oxley Act "fixed" the problems with transparency in providing financial statements to the general public? Why or why not? Analyze a situation from the current press where the act has been effective. Is the concern that the Sarbanes-Oxley Act would unfairly burden small enterprises warranted?

Problem 6.14
Consider the initial stock allocations for a start-up company shown in Table 6.11, below. An early-stage investor want to invest $1,250,000 to secure 30% ownership of the company. Determine the Round I stock allocation portfolio and ownership percentages.

		Initial Incorporation		
		Shares	Stake	Value
Stock Value Profile				
Corporate Valuation				$4,500
Shares Authorized		10,000		
Shares Outstanding		4,500		
Shares Remaining		5,500		
Stock Price				$1.00
Initial Assignments				
Executive - A		1,250	27.8%	$1,250
Executive - B		1,250	27.8%	$1,250
Executive - C		1,000	22.2%	$1,000
Executive - D		500	11.1%	$500
Executive - E		500	11.1%	$500
Follow-on Funding				
Partner - X	50%			
Partner - Y	30%			
Partner - Z	20%			
		4,500	100.0%	**$4,500**

Table 6.11: Follow-on investor purchase of stock (Problem 6.14)

References

[1] A. Marshall. *Principles of economics.* McMillan, 1890.

[2] Steven L. Oberholtzer. *The basics of intellectual property law.* Brinks, Hofer, Gilson, and Lione, 2005.

[3] Weston Anson. *Intellectual property valuation.* American Bar Association, 2005.

[4] Howard B. Rockman. *Intellectual property law for engineers and scientists.* Wiley-Interscience IEE, 2004.

[5] Lisa von Bargen Mueller and Jill T. Sorenson. *An inventor's guide to patents and patenting.* AUTM Press, 2002.

[6] Charles C. Valauskas and Catherine Innes. *Copyright protection of software, multimedia, and other works: An author's guide.* AUTM Press, 1999. Educational Series No. 4.

[7] Brian Leslie. *Material transfer agreements.* AUTM Press, 1998. Educational Series No. 3.

[8] Stefanos Zenios, Josh Makower, and Paul Yock. *BioDesign: The process of innovating medical technologies.* Cambridge University Press, 2010.

[9] Robert Shearer. *Business power: Creating new wealth from IP assets.* Wiley, 2007.

[10] Orin E. Laney. *Intellectual property and the employee engineer.* IEEE, 2001.

[11] WIPO. *The changing face of innovation.* World Intellectual Property Organization, 2011. Publication 944E, ISBN 978-92-805-2160-3.

[12] John V. Sullivan. *How our laws are made.* United States Congress, 2007. `http://thomas.loc.gov/home/lawsmade.toc.html`.

[13] GATT. *United States - Section 337 of the Tariff Act OF 1930; Report by the Panel adopted on 7 November 1989.* Federal Reserve, 1989. `www.worldtradelaw.net/reports/gattpanels/sec337.pdf`.

[14] Robert G. Krupka, Phillip C. Swain, and Russell E. Levine. Section 337 and the gatt: The problem or the solution? *American Law Review*, 42:779–867, 1993.

[15] R. B. Bernstein. *The Constitution of the United States with the Declaration of Independence and the Articles of Confederation.* Fall River Press, 2002.

[16] R. Mazzoleni and R. Nelson. The benefits and costs of strong patent protection: A contribution to the current debate. *Research Policy*, 27:273–

284, 1998.

[17] Reuters. *History of Patents.* Thompson Reuters, 2005. ip-science.thomsonreuters.com.

[18] Richard C. Dorf. *The technology management handbook.* CRC Press, 1998.

[19] USPTO. *Leahy-Smith America Invents Act implementation.* United States Patent and Trademark Office, 2012. http://www.uspto.gov/aia_implementation/index.jsp.

[20] NCCUSL. *Uniform Trade Secrets Act.* National Conference of Commissioners on Uniform State Laws, 2012. http://uniformlaws.org/Act.aspx?title=Trade%20Secrets%20Act.

[21] WIPO. *What is a trade secret?* World Intellectual Property Organization, 2012. www.wipo.int/sme/en/ip_business/trade_secrets/trade_secrets.htm.

[22] William Morris. *The American Heritage Dictionary.* Houghton Mifflin, 1978.

[23] Wilbur M. Yegge. *A basic guide for valuing a company.* John Wiley and Sons, 2002.

[24] Joel G. Siegel and Jae K. Shim. *Accounting handbook.* Barrons, 4th edition, 2006.

[25] Alan Greenspan. *Stanford lecture.* Federal Reserve, 2004. federalreserve.gov/boarddocs/speeches/2004/200402272/default.htm.

[26] D.H. Marshall, W.W. McManus, and D.F. Viele. *Accounting: What the numbers mean.* McGraw-Hill, 7th edition, 2007.

[27] Daniela Carlucci and Giovanni Schiuma. Knowledge assets value map creation map. *Expert Systems with Valuations*, 32:815, 2006.

[28] CIMA. *Three Approaches to Valuing Intangible Assets.* Chartered Institute of Management Accountants, 2012. www.cgma.org/Resources/Tools/valuing-intangible-assets.pdf.

[29] IRS. *Intangible Property Valuation Guidelines: 4-48-5.* Internal Revenue Service, 2012. http://www.irs.gov/irm/part4/irm_04-048-005.html.

[30] Tom Copeland, Tim Koller, and Jack Murrin. *Valuation: Measuring and managing the value of companies.* McKinsey, 2000.

[31] Timothy A. Luehrman. *What's it worth: A general managers guide to valuation.* Harvard Business Review, 1997.

[32] Han T. Smit and Lenos Trigeorgis. Strategic planning: Valuing and managing portfolios of real options. *Research and Development Management*, 36(4):403, 2006.

Chapter 7

Concept Maps

7.1 Pausing to reflect

At the outset of this book we identified several challenges with the fairly broad topic of financial decision-making for engineers. Engineers seeking to become leaders clearly have the challenge of "learning how to learn." The purpose of this chapter is to provide a tool to enhance self-directed learning competencies.

It may seem odd to find this chapter on concept maps inserted between Chapter 6 on intellectual property and Chapter 8 on risk and uncertainty. This is quite deliberate as we move from problem solutions suggesting single numeric answers (Chapters 2-5) to those problems where an array of candidate answers are typically created (Chapter 8). Pausing to frame your knowledge in the form of a concept map can integrate your knowledge and test your ability at synthesis. If your didactic knowledge is established at this juncture, then it should be straightforward to create concept maps for each of the many concepts discussed to this point. In this chapter we'll explain what concept maps are, see how they are used, and work several illustrative examples.

Learning objectives for this chapter are modest:

1. Understand how concept maps integrate subject matter topics and ideas.
2. Be able to describe the seven steps to creating a concept map.
3. Develop a *focus question* to set the context of concept map development.
4. Demonstrate the use of a concept map to address a focus question.

7.2 Impetus for change

It is popular to accuse many accounting and finance instructional methods as leading to rote learning in which the learner's role is passive. Such instruction is "receptive" in the sense that information is presented to the student in final form; there are efficiencies to this less labor-intensive approach for both the instructor and the student. Figure 7.1 (adapted from the work of Novak [1]) illustrates in "quadrant A" being a recipient of information; not much engagement is required beyond reciting back the same information.

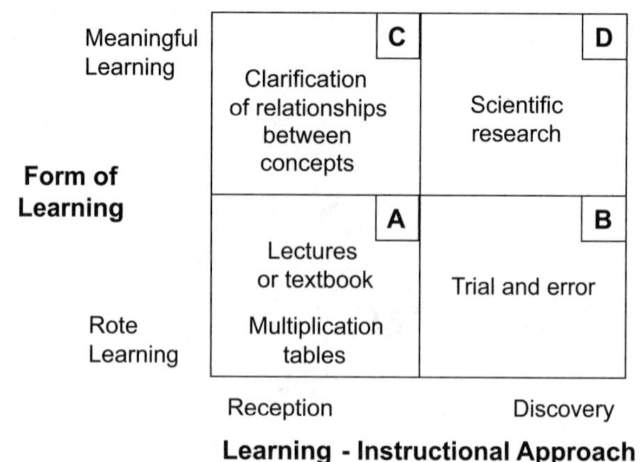

Figure 7.1: Learning matrix to distinguish between types of learning and teaching methods (adapted from the work of Novak [1], Fig 4.1, p. 101).

Quadrant "A" rote learning is hard for learners to "scale up" beyond a certain point to accommodate more and more information. Learners become overwhelmed and discouraged, and meaningful knowledge is not retained (or even learned!). Surely the reader of this book has already experienced this in the transition from Chapter 5 on comparing alternatives to Chapter 6 on intellectual property. Diminishing returns on your effort occur when

1. There is a large volume of information to be covered,
2. The nature of the topics is highly integrated, and
3. Causal skills are essential to understanding the material.

Sounds like everyday life for a university student! Medical knowledge is characterized by every item on this list and is possibly why clinical training frequently features concept maps as a learning tool (West et al. [2], Henige[3], Torre et al. [4], Luckie et al. [5], Schuster[6], and Smeltzer et al. [7]). Financial management has emerged this way, too.

Introducing new instructional methods is not easy and should be evidence-based, particularly if the industry status quo and competencies are based on, say, standardized testing scores, and for those subjects where it is essential to absorb a large amount of information in a short semester. Section 1.1 on page 1 highlighted similar challenges for the reader working through this (highly condensed) book. The situation is exacerbated in the present work since we've added several interdisciplinary elements (again, Chapter 6, on intellectual property, for example) not ordinarily found in a finance or engineering economy text.

But if the list above applies, it is not enough to simply try harder and work longer hours. Researchers have identified that many current instructional strategies are inadequate and new strategies are needed to keep pace with the increasing complexities of professional management in the post-Enron financial era (Irvine [8], Maas and Leauby [9]). Finance joins the fields of engineering (ABET[10]), psychology (Buehl and Fives[11]), and others where critical thinking skills and self-directed learning are being sought. In short, there is a clamor for learning methods to enable graduates to solve the problems of tomorrow that we do not even know exist today. Students must be empowered to go beyond *just* acquiring basic subject competencies; they also need to "learn how to learn." This requires skill sets embodied in the *meaningful learning* of quadrants "C" and "D" of Figure 7.1. Let's turn our attention to a bit of theory supporting how concept maps are intended to help.

7.2.1 Concept map theory: Ausubel, Novak, and constructivist teaching

Concept maps were developed in the early 1970s by Joseph D. Novak during a sabbatical at Harvard University. The work was originally intended to understand a child's knowledge of science but it soon became clear the method had much broader use. David Ausubel's work on cognitive psychology (Ausubel[12]; Ausubel et al.[13]) was an inspiration for Novak's work on concept maps. One of Ausubel's key ideas was that learning does not take place by random storage of information, but by the assimilation of new concepts into existing concepts. Further, each individual had a cognitive structure as a reference for new concepts and propositions. Concept mapping is a visual representation of that knowledge structure – how information is organized in memory and the potential for meaningful learning (as characterized by quadrants "C" and "D" in Figure 7.1). Constructivist teaching is not new. As early as 1910 John Dewey observed[14] that knowledge is constructed from prior knowledge interactively:

> *Science has been taught too much as an accumulation of ready-made material with which students are to be made familiar, not enough as a matter of thinking, an attitude of mind, after the pattern of which mental habits are to be transformed.*

In short, people develop unique knowledge structures, and different people can view the same events differently. If students externalize their knowledge structure through a concept map, the learners and instructor can compare them and identify alternate conceptions that could be a barrier to further learning. Discovery learning is thus harder as it must be mapped against existing knowledge. Very early habits of mind can be hard to shake.

Concept map theory has successfully found application in many subject areas, often providing a structured approach to organizing, understanding, and building on complex bodies of knowledge. As mentioned at the outset, the present chapter is a brief thematic departure from the prior six chapters. The purpose is to pause for reflection and to introduce the concept map analysis techniques so as to help integrate concepts for a richer understanding of the topics presented thus far. Concept maps are even a way to facilitate life-long learning for topics beyond the scope of this book. This is not a course on cognitive psychology, so let's cut to the chase on discovery learning that is a basis for concept maps:

1. Learning must be meaningful and absent nonsense.
2. Students must put forth the effort and be inclined to learn in a meaningful fashion.
3. New concepts must be structured relevant to an existing cognitive structure.

To transition the learner from novice to expert we must have methods of externalizing students' declarative knowledge and the associated knowledge structure We see how the concept map enables this transition.

7.3 Concept map creation

Concept maps are composed of a hierarchy of *nodes* and *links* that represent propositions of knowledge. Nodes represent concepts. Links portray relationships. Novak and Gowin [15] suggest that "concepts and propositions should be structured in a hierarchical format, with the more general positioned at the top of the map" (page 7). Further, they argue the context of the map should be clarified through the use of a *focus question.*

The recent book by Moon, Hoffman, Novck, and Cañas[16], *Applied Concept Mapping*, provides excellent guidance for the creation of concept maps; when the book is combined with the monograph by Novak and Cañas[17] the reader has the preferred reference for practitioners (over many academic papers) as the authors are the original leaders in the development and application of concept mapping.

The emergence of the internet and rules of the road for concept mapping led Alberto J. Cañas to development of the CmapTool[18] that has become popular because of the way it simplifies applied concept mapping for the average student. We encourage exploring the references available on-line and free IHMC Cmap Tools at `http://cmap.ihmc.us/conceptmap.html`. Maas and Leavby [9] outline at least eight other websites offering downloadable mapping tools, and for completeness the student can explore those options, too. A delightfully simple check-list for the creation of concept maps is provided by Moon et al.[16] (chapter 2, p. 24). We see little way to improve the clarity of what is offered, so the following procedure is based heavily on their book. Let's briefly move on to describing the five basic steps for creating a concept map.

1. Define a focus question

It is best to have a context for the concept map centered around a specific issue or question. This is not normally a trivial question and would be expected to involve a number of concepts that are related to the question. For instance, a focus question like, "Is $NPV > 0$ desirable?" has a trivial answer. On the other hand, "What is the role of intellectual property in technology projects?" would be a challenging question involving a multitude of concepts.

2. **Identify key concepts**
   Though a key concept is defined as a "perceived regularity of events ... designated by a label"[15][16], the "concept" for a concept map is a word or short phrase that is typically enclosed by a circle or box when placed on a map. For instance, the concepts of "technology projects," "financial accounting," "intellectual property," "patents," and "cash flow" are concepts that could be related in some way.
3. **Arrange the concepts by some notion of priority**
   Essentially, concepts should be arranged in some form of hierarchy, with the most general (inclusive) concepts presented at the top and subordinate, specific, or less general concepts presented below. For instance, the concept "intellectual property" is more general than a specific type of IP such as "patent" or "trademark" and those categories more general than "provisional" or application." Concepts arranged this way form "levels" of hierarchy; the IP example has three levels.
4. **Create links**
   Concepts linked to other concepts form a prepositional statement, a "meaningful statement" that is the foundation of a concept map. When a series of concepts are linked in a hierarchical fashion, we have constructed a representation of a domain of knowledge. The domain of knowledge should describe, offer insight, or represent a set of concepts related to the focus question. The links eventually provide clarity to the relationship between concepts, with specific wording emerging over time. The four steps discussed thus far are illustrated with a very simple map in Figure 7.2.
5. **Create cross-links**
   As the concept map is built out, there emerge different "domains" on the map. The ability to provide cross-links between domains reflect the "richness" of a map and the potential for new learning can occur. It is often exciting (and memorable) when a learner makes a cross-link between two domains that they had not previously considered interrelated. As a simple example, the concept map of Figure 7.3 seeks to define the role of IP in technology development - a combination of strategic priorities and free cash are responsible for the development of a patent portfolio.

Item 3 (prioritize concepts hierarchically) and Item 5 (domain cross-linking) are two aspects of concept mapping exercises that foster critical and creative thinking leading to memorable learning experiences.

The work of Maas and Leauby in developing concept maps for accounting education[9] has an excellent example illustrating the use of concept maps for GAAP (Maas[9], page 78) and a ready-to-use map for the statement of cash flows (Maas and Leauby[9], page 86); both are worth the time to seek out. The engineering manager is not expected to be an expert in all topics primary and secondary to IP, taxes, depreciation, and other financial decision-making topics. *But*, the ability to see the interrelationships between financial analysis and IP management escapes more managers than you might think. If you are able to sketch Figure 7.3 from memory and use this in team leadership

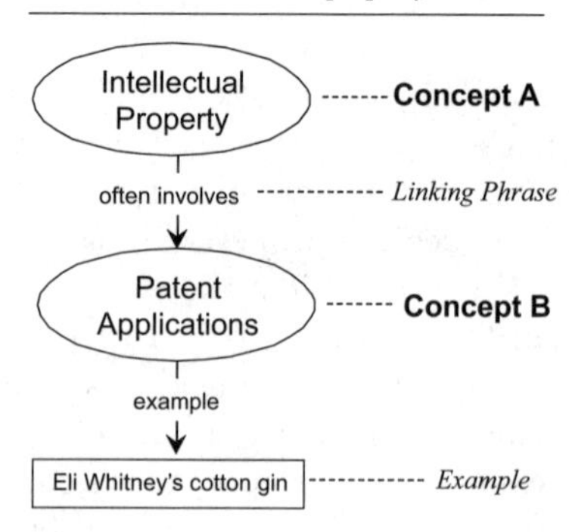

Figure 7.2: A very simple example to illustrate the main components of a concept map.

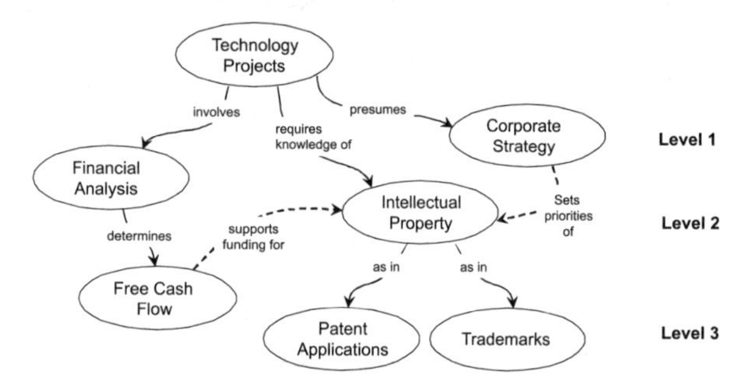

Figure 7.3: A partial concept map to illustrate cross-linking of knowledge domains.

discussions, it will reflect a touch of wisdom that comes from genuine insight on issues of the day. You will find a growing ability to look beyond the obvious, a real mark of the type of decision-maker needed in today's dynamic economy.

7.3.1 Three elements of a good map

Once the focus question has been established, there are three things to look for in a good concept map:

1. There are distinct domains of knowledge characterized by a hierarchy of concepts within that domain. This suggests the learner has a grasp of the key concepts and the hierarchical framework of knowledge within that domain.
2. There is progressive differentiation of the maps, reflecting expansion of knowledge within a domain. It may take several iterations for differentiation to emerge.
3. Cross-linking of concepts across knowledge domains, to show how do-

mains are linked in ways that may not have been originally considered. If care is not taken to be somewhat selective about cross-links, then a spaghetti" map ensues. Strategic cross-links provide what Ausubel[12] calls integrative reconciliation that is the foundation of self-directed learning (Moon et al.[16], p. 73).

Figure 7.4 illustrates the three characteristics of concept maps that can facilitate *meaningful learning* and enable knowledge to be built on a solid foundation of prior knowledge.

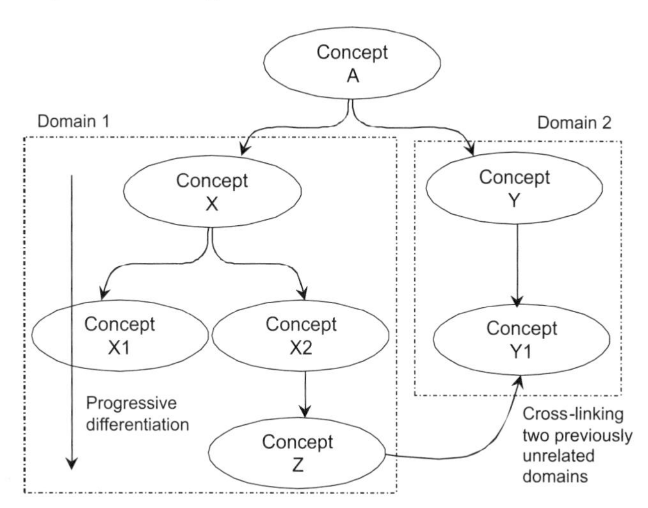

Figure 7.4: Concept map framework to illustrate the desirable progressive differentiation and integrative reconciliation that are sought after in good concept mapping exercises.

7.3.2 Sample concept map on entrepreneurship

The value of a concept map is in the possible links the user may not fully appreciate. In the case below, the student group eventually arrived at the conclusion than the credibility of an entrepreneur might be more important that the presence of a patent.

Examine Figure 7.5 with the following questions in mind:

1. What two (or three) domains are evident? How might they be labeled?
2. For the left-hand domain, how do finance and financial decision-making touch each bubble? Which chapters might be the primary source of insight on the topic in the bubble?
3. Do the chapters of the text flow in the same way the concept map flows? Which chapters are most important and why?
4. Which aspects of the concept map for an entrepreneur are most different than for an on-going concern? Large company versus small?
5. What one or two concepts discussed in the first 6 chapters touch at least 2/3 of the concept map?

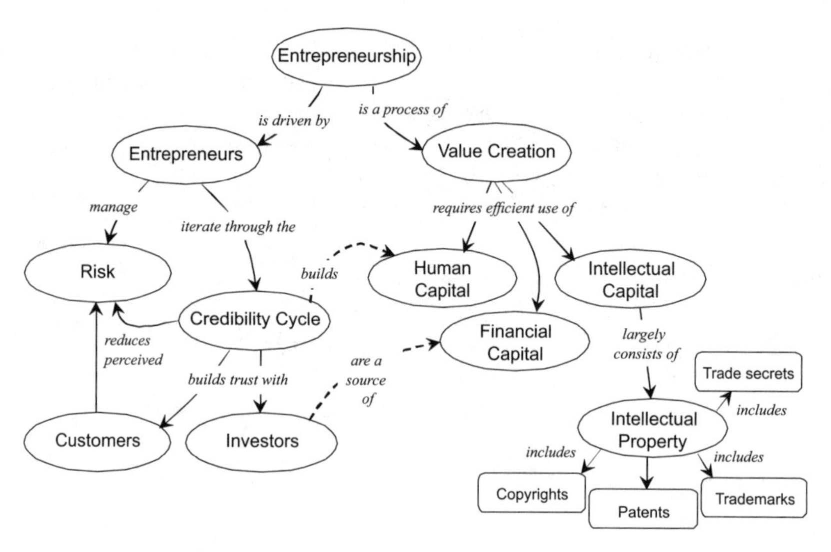

Figure 7.5: A partial concept map to address the focus question "what is entrepreneurship?"

When people claim they are not "creative," this can often be interpreted as an inability to create questions to ask about a given situation. Concept maps enable a wide variety of questions to be asked and ultimately assist in helping refine a focus question for what might otherwise seem to be an intractable and highly ambiguous situation. Section 5.4 on the selection of alternatives – and in particular the Section 5.5 discussion on before-tax and after-tax influences on computed outcomes – illustrated that many of the choices teams make have as much to do with how we articulate the problem nuances and peripheral concerns as the problem in front of us. For instance, the lease-versus-buy decisions in Section 5.8 demonstrate that a choice based on a single financial computation can be questioned when the greater context of finance and operations is taken into account. Leasing *may seem* to be the best option, but the ability to posit an alternative position requires "credibility."

Building credibility does not have to be accidental; Figure 7.5 suggests relationships that can be acquired and used to enable the engineering manager to not just "think outside the box," but to "work in another box" – say, how investors view capital assets on the balance sheet or what "human capital" might be needed to ensure proper utilization of existing capital. Does outsourcing make financial sense but lead to morale problems? It may not take a concept map to understand this is important but a CM can certainly enrich the choices involved. It is worth asking how a CM can help with "why decisions fail" (Section 10.2.1) and why more than finance is at play in Section 10.6 on Die-Cast Testing's production challenges. Remember, one of the important aspects of CM is the ability for the map to help the analyst refine the focus question. When the complexities of engineering management move beyond textbook sample problems, we need to know that the issues to be studied are relevant to the decision that needs to be made, or the decision that needs refinement.

Examine Figure 7.5 and ask how this CM could be used to refine a lease-versus-buy pitch being made to an investor. The exercise always surprises the student and validates what seems academic at first. If you remain unconvinced, ask why most medical-surgical nursing textbooks train on the use of concept maps for diagnostics.

We have argued that the competent engineering manager must have an acute sense of *social capital* in addition to the human, financial, and intellectual capital bubbles in Figure 7.5. In which domain would you put social capital? (Hint: think "value creation.") And, is this idea of social capital of real value – and linked to financial decision-making – or is it just rubbish? The answer relates to the "how" more than "how much." It is one thing to compute the favorable NPW of an investment (the "how much") and quite another to ask how we will *get* the required funds. No question whatsoever that we need capital, but Figure 7.5 should make clear that absent all four types of capital, your project will not create value and is uninvestable from a senior-level managerial standpoint.

7.3.3 Sample concept map on cash flow

Consider the cash-flow concept map shown in Figure 7.6. This is a good example of the way in which broad-scope financial statements can be distilled into a narrower, more actionable set of concepts. In many companies, engineers might find the linkage between investing and operations to be quite vague. A concept map can show that executive outreach to investors and "show and tell" in the board room can have an impact on the ability to procure new equipment for product development.

The beauty of a concept map is that many concepts can be folded into a single diagram. It can be used by engineering managers as a framework for "storytelling" to staff to promote, for instance, more system-wide thinking about the impact of day-to-day operations. In some cases the message is self-evident, but in others there is a linkage that might otherwise escape the attention of staff.

Jayne D. Maas and Bruce A. Leauby [9] use concept maps to present a series of accounting concepts very clearly, and they touch on *many* ways in which concept maps can be used in accounting education. We highly encourage the reader to obtain the Maas and Leauby [9] paper as supplementary reading for this chapter.

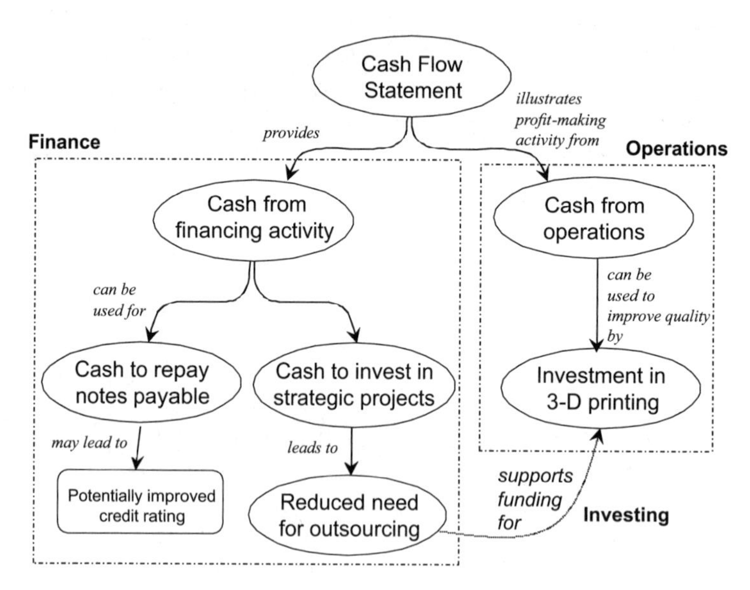

Figure 7.6: Sample cash-flow concept map illustrating the way in which strategic investing activity can enable new quality production improvements for operational competitiveness.

7.4 Systems thinking and causal maps

Concept maps enable the manager to identify pathways to understanding complex processes and to make conceptual links that may not otherwise be visible. Systems thinking takes this awareness to the next level and enables the manager to *understand the impact of taking action* on this new knowledge for a specific subset of organizational activity. Often the refinement and practice of systems thinking is of great benefit to the financial analyst. For instance, the concept of double-loop learning can be of great value in trying to flush out patterns of organizational inefficiency, particularly as it relates to cash-flow problems, ethical lapses, and unrealized benefit of capital equipment investments. Thinking back to Figure 1.8 (p. 11) on day-to-day operations, or Figure 2.3 on the general cash-flow cycle for a business, these diagrams illustrate so-called "single-loop" learning. The idea that "double-loop" learning is central to the analysis of hidden assumptions is discussed extensively by Senge [19]. Figure 7.7 illustrates how this type of thinking evolves.

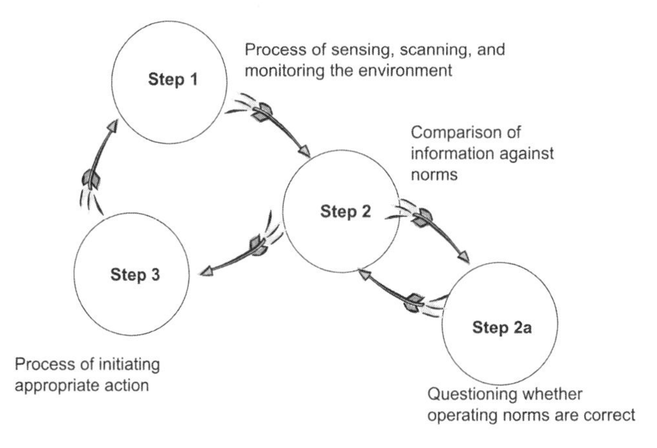

Figure 7.7: Double-loop learning framework, adapted from Senge [19].

While Figure 7.7 seeks to prod the financial analyst to challenge basic assumptions, another – more specific – view of double-loop learning is illustrated in Figure 7.8. In practice this asks for more information about judgement on whether an issue is a symptom or a root cause and whether our "solution" addresses the fundamental problem, or if we are only treating the symptoms and therefore inadvertently "shifting the burden." This overall topic is somewhat challenging, but it typically surprises individuals in class, brainstorming sessions, or workshops how investment needs change when the overall system and attributes are more clearly and explicitly called out and questioned as a team. Treating the symptom shifts the burden of the solution, and the long-run impact on the organization is not immediately obvious, but will be significant over time.

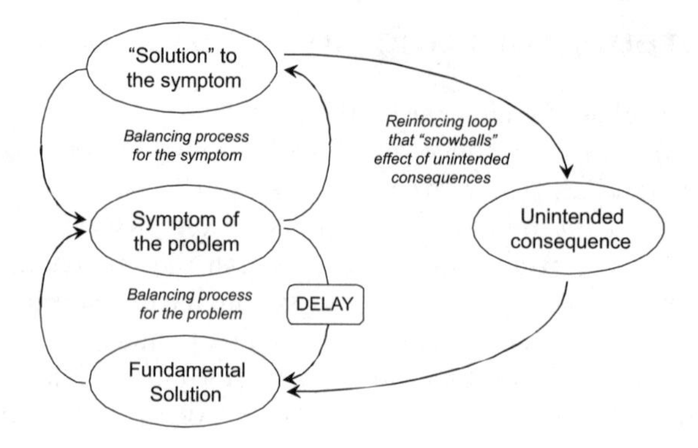

Figure 7.8: Symptoms and root cause diagram for the shifting-the-burden archetype, based on the work of Senge [19], pp. 112-113.

As an example of the application of Figure 7.8, consider the challenge of trying to decide if a lack of revenue is the result of a small sales force. With a single-loop mindset, the challenge of missing revenue targets might be attributed to having too small a sales force. A persuasive argument to hire more sales people might actually cause the problem to get worse, not better, if there is a secondary issue preventing sales from occurring. This form of systems thinking can be very powerful, but considered academic and futile if knowledge about the organization and relationships between key system elements is missing.

Concept maps can help the analyst understand and expand thinking about all the components that need to be considered in addressing a focus question such as "Why is revenue lagging?" but other tools such as double-loop learning help focus on a plan of action. Figure 7.9 highlights the application of systems thinking to the problem of entering the so-called "bidding war" for business when your structural costs are fixed. This may happen when sales is being pushed to make the numbers at the end of a quarter, in which case quality of service is reduced in the long run. The impact of the trade-off between cost and quality will eventually compromise the ability for the entity to compete at all.

Concept maps are sometimes confused with strategy maps [21]. While strategy maps are an excellent way to translate strategic thought into tactical action, they address a different aspect of financial analysis and systems thinking than concept maps. This is not to discount the value of understanding and using these tools in practice, but carefully understanding specifically *how* each of the tools plays into practical work of financial decision-making takes some practice and adaptation to the environment within which you work.

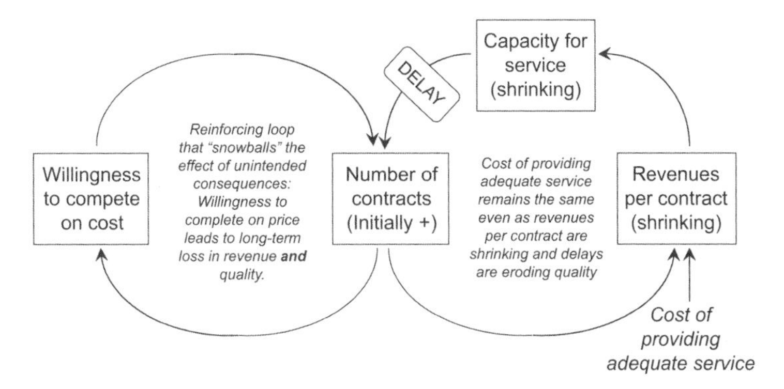

Figure 7.9: Challenge of competing on price in the long run when your fixed costs remain the same, adapted from Senge [20], pp. 151-152.

7.5 Summary

A frequent complaint with the pace of work and assignments is that there is little time to reflect on subject matter to gain a more integrated view of the course material. Working harder is not necessarily the answer, as expanding the need for memorization of facts takes the place of understanding with the consequence that retention drops off once semester exams are over. Concept maps are an active learning tool that can help since the map development process organizes information in a visual way with greater impact than text information; users often find this also helps provide insight on the big picture (Henige[3]). Concept maps are well-known to facilitate self-directed learning and to form a basis on which flaws in the understanding of concepts and relationships can be identified.

1. Rote memorization (arbitrary storage) is not a substitute for meaningful learning.
2. Meaningful learning occurs when new information is learned in relation to existing information.
3. Knowledge *and* knowledge *structure* are critical to assimilation of new concepts.
4. New concept assimilation depends on the "adequacy" (correctness) of existing concepts.
5. Concepts organized according to a hierarchy enable cross-linking of concepts.
6. Existing concepts must be placed in context or framed by a *focus question.*
7. Revise concepts, links, and crosslinks as the concept map evolves.
8. Iterate, iterate, iterate.

There are limitations to simply working harder, and sometimes you "have to do something different to get different results." That is one of the key points of this chapter.

7.6 Problems to explore

A good measure of your development in financial decision-making is the ability to exhibit self-directed learning and thus concept maps should come naturally and not be difficult to construct, despite the fact that there are several topics yet to be covered in this monograph. There is significant educational value to the exercises to follow when taken seriously. The map and associated relationships provide a unique opportunity to understand this course from a different (and memorable) perspective and test your ability to deal with ambiguity.

Problem 7.1
Concept maps can be fun to create, and there is some excellent software available to assist in the creation of your own maps. With the cash-flow concept map of Figure 7.6 in mind, explore the following links and describe which software you might use to create a cash-flow map; explain why.

See: IHMC Cmap Tools, `http://cmap.ihmc.us/conceptmap.html`, and "A survey of concept mapping," `http://datalab.cs.pdx.edu/sidewalk/pub/survey.of.concept.maps/`.

Problem 7.2
The work of Jayne D. Maas and Bruce A. Leauby [9] provides an illustration of the underlying theory of a concept map as it applies to GAAP. Create your own version of a GAAP concept map, based on (a) your individual interpretation of the work of Maas and Leauby and then (b) a map derived by group exercise; include linkages to other courses you are now taking. Reflect on the differences between your individual perspective and that of the group on the role and importance of GAAP in finance today.

Problem 7.3
Maas and Leauby [9] describe concept maps as a "planning tool" (p.84). Explain how this can be applied to your classroom learning of finance and accounting.

Problem 7.4
Figure 7.4 provides a concept map framework illustrating progressive differentiation and integrative reconciliation that are sought after in good concept mapping exercises. Explain the terms "progressive differentiation" and "integrative reconciliation," and use the concept of "depreciation" as a focal point of your discussion.

Problem 7.5
Create a concept map around the focus question "How do I advocate a new capital project within a large, conservative company?"

References

[1] Joseph D. Novak. *A theory of education.* Cornell University Press, 1977.

[2] Daniel C. West, J. Richard Pomeroy, Jeanny Park, Elise Gerstenberger, and Jonathan Sandoval. Critical thinking in graduate medical education: A role for concept mapping assessment? *JAMA*, 284(9):1105–1110, September 2000.

[3] Kim Henige. Use of concept mapping in an undergraduate introductory exercise physiology course. *Advances in Physiology Education*, 36:197–206, 2012.

[4] D. Torre, B. Daley, Tracy Stark-Schweitzer, Singh Siddartha, Jenny Petkova, and Monica Ziebert. A qualitative evaluation of medical student learning with concept maps. *Medical teacher*, 29:949–955, 2007.

[5] Douglas Luckie, Scott H. Harrison, and Diane Ebert-May. Model-based reasoning: using visual tools to reveal student learning. *Advances in Physiology Education*, 35:59–67, 2011.

[6] Pamela M. Schuster. *Concept mapping: A critical thinking approach to care planning.* F. A. Davis Company, 2012.

[7] Suzanne Smeltzer, Brenda Bare, Janice Hinkle, and Kerry Cheever. *Brunner and Suddarth's Textbook of medical-surgical nursing*, volume 1. Lippincott Williams and Wilkins, 11th edition, 2008.

[8] Helen Irvine, Kathie Cooper, and Greg Jones. Concept mapping to enhance student learning in a financial accounting subject. *Proceedings of the Accounting Educators Forum*, pages 1–19, November 2005.

[9] Jayne D. Maas and Bruce A. Leauby. *Concept mapping - Exploring its value as a meaningful learning tool in accounting education.* Global perspectives in Accounting Education, volume 2, 75-98 edition, 2005.

[10] ABET. *Criteria for accrediting engineering programs.* ABET, 2010.

[11] Michelle M. Buehl and Helenrose Fives. Best practices in educational psychology using evolving concept maps as instructional and assessment tools. *Teaching Educational Psychology*, 7(1):62–87, 2011.

[12] David Ausubel. *The psychology of meaningful verbal learning.* New York: Grune and Stratton, 1963.

[13] David Ausubel Joseph Novak and H. Hanesian. *Educational psychology: A cognitive view.* New York: Holt, Rinehart and Winston, 2 edition, 1978.

[14] John Dewey. Science as subject-matter and as method. *Science*, 31(787):121–127, 1910. www.sciencemag.org/content/31/787/121.

[15] Joseph D. Novak and D.B. Gowin. *Learning how to learn.* Cambridge University Press, 1984.

[16] Brian M. Moon, Robert R. Hoffman, Joseph D. Novak, and Alberto J. Canas. *Applied concept mapping.* CRC Press, 2011.

[17] Joseph D. Novak and Alberto J. Canas. *The theory underlying concept maps and how to construct and use them.* Florida Institute for Human and Machine Cognition, 2008. Technical Report IHMC 2006-001 Rev 01-2008.

[18] Alberto Canas, G. Hill, R. Carff, N. Suri, J. Lott, and T. Eskridge. *CmapTools: A knowledge modeling and sharing environment.* Pamplona, Spain: Universidad Publica de Navarra, proceedings of the first international conference on concept mapping, vol. i, pp. 125-133 edition, 2004.

[19] Peter M. Senge. *The fifth discipline: The art and practice of the learning organization.* Doubleday Press, 1990.

[20] Peter M. Senge. *The fifth discipline fieldbook.* Doubleday Press, 1994.

[21] Robert Kaplan and David Norton. *Strategy maps: Converting intangible assets into tangible outcomes.* Harvard Business School Press, 2004.

Chapter 8

Risk and Uncertainty

8.1 Learning objectives

In much of our prior work we have placed great confidence in our ability to forecast revenue streams, estimate interest rates, and generally being able to predict the future! In reality, there is a great deal of uncertainty in our ability to estimate future values of parameters. There are many sources of uncertainty, and future values have a degree of randomness that should be accounted for in our analysis efforts. This chapter will expand on earlier discussions about risk by introducing common methods of analysis. In this chapter we discuss discrete random variables, explore probability trees, and provide several worked problems for decision trees. We also briefly look at the mathematical conveniences associated with variables distributed according to a so-called "normal" Gaussian distribution.

After reading and discussion sessions, the student should be able to:

1. Describe the difference between risk and uncertainty in decision-making.
2. Describe the four common sources of uncertainty:
   (a) Inaccuracy of cash-flow estimates.
   (b) Type of business involved relative to the future health of the economy.
   (c) Type of asset involved.
   (d) Length of the study period.
3. Understand how the various analysis methods (NPV, etc.) are impacted by uncertainty.
4. Perform a project analysis using a probability tree.
5. Understand and create decision trees.

8.2 Introduction

It is easy to forget that many of the calculation procedures outlined in previous chapters have *assumed* the events we have forecast will actually occur as planned with *some degree of certainty.* Parameters like cash flow, interest rate, and service life were routinely part of calculations that sought to support or assist with decisions about future project options and possibilities. In Section A.2 we will discuss how reality intrudes on the confidence with which we can make such predictions, and we introduced the two terms *risk* and *uncertainty* as they relate to decision-making and in the characterization of project outcome predictability. The purpose of this chapter is to provide some tools and tips to enable the analyst to accommodate risk and uncertainty in decision-support calculations.

Although definitions vary, Michael Mauboussin[1] has provided some definitions of risk and uncertainty that are frequently recited since they are quite simple and easy to remember:

1. Risk: We don't know what is going to happen next, but we *do know* what the possible distribution looks like.
2. Uncertainty: We don't know what is going to happen next, and we *do not know* what the possible distribution looks like.

You master complexity when you understand the different sources of uncertainty and the different risk characteristics of each uncertainty.

We will find later (Section A, p. 347) that in lay usage, "risk" often seems to imply that something "bad" or negative will happen with some degree of probability. In contrast, "uncertainty" is a bit more innocuous (that is, harboring less of a value judgment) and implies a simple indefinite or incalculable nature of events. Fundamentally, though, in common usage, both terms refer to a similar situation, in which some aspect of the future cannot be foreseen.

By accounting for risk and uncertainty a good manager is simply providing a type of insurance in forecasting situations which seem the most likely to happen. Mathematically, we are talking about how to deal with variances in expected outcomes.

To some extent the difference between risk and uncertainty is academic, and you can even view the two terms as endpoints on a continuum of our confidence and ability to assign numbers to possible distributions in outcomes. This fits nicely with our discussion in Section A.2 (p. 349) when discussing *deterministic* and *stochastic* problems. That discussion highlights that engineers would be more familiar with deterministic problems (all the information needed to solve the problem is known with a great deal of certainty) and the solution of stochastic problems more difficult (information behaves in a probabilistic fashion). Essentially, this chapter works through ways of treating stochastic problems; like most things in life, the passage of time creates instability and the engineering manager must know how to anticipate change.

8.3 Frame of reference

Accountants "live in the past," stockholders "live for today," and engineering economists "live for the future." As mentioned in the previous section, a great deal of our work thus far has involved estimates about future activity – relevant, of course, to building leadership and decision-making skills – but now it is time to get specific about how to account for errors in our estimates. It is not a question "Am I wrong?" but moreso What is the impact of how wrong I am?" There are actually quite a few ways to estimate how wrong we are and an analysis challenge is narrowing the field of methods to treat the topic of uncertainty and risk. Fortunately the topic of uncertainty and risk is not new to the world of engineering management, and this chapter will simply look at the following three topics:

1. Sensitivity analysis
2. Discrete frequency probabilities
3. Frequency distribution functions

We'll begin assessing the impact of uncertainty with sensitivity analysis, the simplest of the three calculation methods listed above.

8.4 Sensitivity analysis

Sensitivity analysis can take on a number of forms:

1. In the **characteristic scenario** analysis, we recognize that a combination of parameter changes produces characteristic *scenarios* for which project performance criteria are computed and compared against each other. Projects may have a required $NPV > X$ or $IRR > Y$, and we seek to establish which set of characteristic scenarios meet the performance criteria. Section 5.2 (p. 138) provided guidance and examples on the essential aspects of this method.
2. A **characteristic range** analysis has some of the attributes of the characteristic scenario method, but the difference is that rather than have combinations of variables defining scenarios, we use *characteristic ranges* of specific parameters that we believe will have a significant impact on performance criteria. Here, we establish a set of realistic ranges of *optimistic* and *pessimistic* estimates of a parameter that provide the upper limit and lower limit surround a baseline *most likely* estimate. This approach will look familiar to anyone who has experience in six-sigma process control management.
3. The **parametric range** analysis is an extension of the characteristic range method but is actually closer to what is traditionally associated with sensitivity analysis. Here, all other parameters assume a fixed mean value while we incrementally vary a specific parameter of interest around its mean value. We will elaborate on this further in this chapter, but the output of the analysis is a plot of performance impact (dependent variable plotted on the ordinate axis) against the entire range of parameter

variations (independent variable on the abscissa). When more than one parameter is modified, then this is sometimes called the *spider plot.*

4. Generally, **probabilistic sensitivity** analysis is much less common that the other methods because of the difficulty many engineers and managers have in interpreting the results. In this method, a set of, say, five variables is allowed to vary according to a probability distribution function, and a set of outcomes for the dependent variable, say, PW or IRR, is expressed in the form of a probability distribution for that dependent variable. We then use the cumulative probability distribution to perform statistical probability calculations that suggest a specific outcome will occur.

In this section of the chapter we restrict our attention to the more commonly used *characteristic range* and *parametric range* approaches. This is not to discount the value of the characteristic scenario method, only that the principle is already covered in Section 5.2 and the outcomes are less specific to observation of the sensitivity of a dependent variable due to variation in a specific independent variable.

8.4.1 Characteristic range method

As mentioned above, we use *characteristic ranges* of specific parameters that we believe will have a significant impact on performance criteria. You may recall we briefly touched on this method when discussing `IRR` in Section 4.2 (p. 101), and so the calculation sequence should be rather familiar. Table 8.1 illustrates a simple example where we establish a set of realistic ranges of an *optimistic* and *pessimistic* estimate of the initial project investment, expected duration of use, and estimated annual net revenue.

Project Parameter (8% MARR)	Pessimistic	Baseline	Optimistic
Initial project investment, P	$22,000	$20,000	$18,000
Expected duration of use, N	5 years	6 years	7 years
Estimated annual net revenue	$4,500	$5,150	$7,500

Table 8.1: Characteristic ranges for Option B in the capital investment project featured in the example in Table 5.1, p. 140

The NPV of each option is easily computed using:

$$\begin{aligned} NPV_{Baseline} &= -\$20,000 + \$5,150(P/A, 8\%, 6) \\ &= -\$20,000 + \$5,150(4.6229) \\ &= -\$20,000 + \$23,808 = \$3,808 \end{aligned}$$

The IRR is computed using the spreadsheet solution outlined in Table 4.1, Section 4.2, p.101. Results are summarized below, along with a computation of the IRR in each case.

Note how quickly the feasibility of the project declines in the pessimistic scenario. Our pessimistic lower limit reflects that a 10% increase in investment, 1-year decrease in life, and 12% loss in annual revenue all combine to take an otherwise healthy IRR > 10% and diminish it to under 1%. We leave it to the reader to show that if the pessimistic scenario were to *only* reflect the 12% loss in revenue (keeping $P = \$20,000$ and $N = 6$), the NPW rises to $802 and the IRR rises to 9.3%, barely exceeding the 8% MARR floor.

Project Parameter (8% MARR)	Pessimistic	Baseline	Optimistic
Net Present Value, NPV	$(4,033)	$3,808	$21,048
Internal Rate of Return	0.8%	14.0%	37.1%

Table 8.2: NPV and IRR results for the characteristic ranges from Table 8.1.

Overall, the characteristic range method is simple and readily amenable to a spreadsheet solution, and is thus quite popular. What is lacking however, is clarity on how *rapidly* dependent variables change with small changes in the independent variable; the *parametric range method* to follow addresses this issue.

8.4.2 Parametric range method

The *parametric range* method expands our application of sensitivity analysis to do more than just examine dependent variable responses to independent variable changes. Earlier, we saw what a difference ±10% in purchase price would do to IRR, but (by definition) results simply described overall differences from the baseline case. We now want to refine the prior method and use an entire range of values in calculations that illuminate trends and the *robustness* of dependent variables relative to the problem independent variable. The idea is to understand the influence that variations in key problem variables can have on project success factors and thus help in decision-making.

There are four basic steps involved in this sensitivity analysis:

1. First, select those variables you believe will most likely change from the baseline estimates. Interest rate, period of use, and cash-flow streams are common.
2. Second, establish the *range* of the variable (say, ±10%) and the increment of change to explore, for instance, $\delta i = 0.2\%$.
3. Recompute the dependent variable (PW, EUAC, ...), varying one independent variable (MARR, i, ...) at a time, assuming for the purpose of analysis that the variables are *independent of each other.*
4. Finally, chart or plot the results.

An example of the result of this process is shown in Figure 8.1, based on Table 8.1.

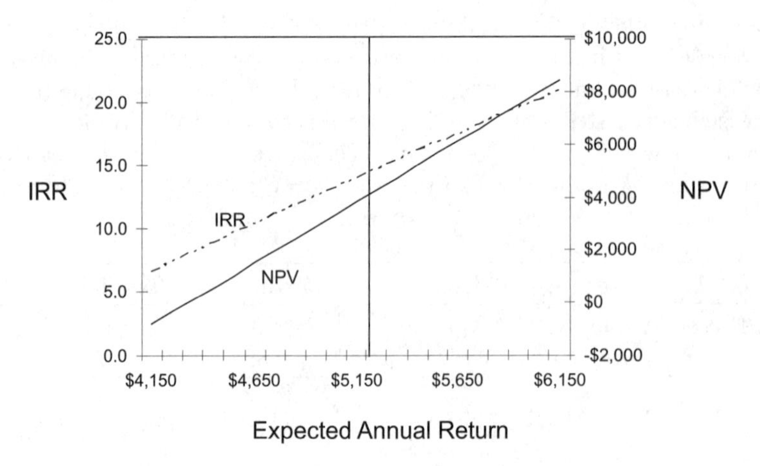

Figure 8.1: Option B sensitivity of IRR and NPV to changes in expected annual revenue.

Figure 8.1 illustrates how a project NPV or IRR will vary with changes in the expected annual return for the project. A few items to note:

1. It is clear that there is a difference in slope for each of the lines, but we should not be misled into thinking the sensitivities are the same, since the scaling (left side for IRR and right side for NPV) for the plots are not normalized. We really can't say much about the slopes except in a general way that are not too surprising.
2. The abscissa for expected annual returns ranges from \$4,150 to \$6,150 and is centered at the baseline value of \$5,150, but it is hard to extract whether a drop in return of \$1,000 is significant. If the horizontal axis were marked $\pm 20\%$, then it would be easier to judge, for instance, that a 10% drop in revenue leads to a 60% (\$2,311) drop in NPV that would command a *lot* of management attention! Similarly, the drop in IRR from 14% to 10.8% would also be of concern.
3. The positive slope of the NPV and the use of dual ordinates (one for IRR and the other for NPV) seems to detract from the fact that an approximate 10% drop in annual return places NPV < 0, whereby the IRR is also less than the MARR. These critical facts do no jump out at the reader and do not communicate the real risk of revenue departures from the baseline value.

Consider Figure 8.1 modified as shown in Figure 8.2, to improve communication of sensitivity.

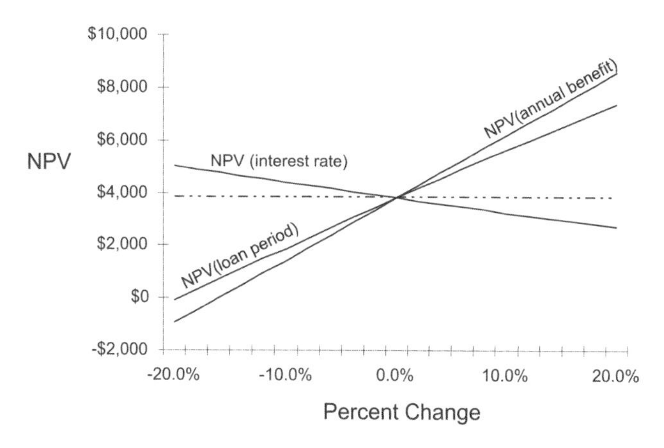

Figure 8.2: Example from Figure 8.1 reformatted with a single ordinate for NPV that varies with percentage change along the abscissa for any of the arguments ϕ for NPV(ϕ).

8.4.3 Combined parameter changes

Up to this point we have computed dependent variable sensitivities to changes in a single parameter. In the sample problem of the prior section, we explored NPV varied with ±20% changes in period, interest rate, and annual revenue with all parameters held at the baseline value *except* the single variable for which sensitivity was to be examined. We now wish to expand on the idea of the prior section and observe how the sensitivity of, say, NPV depends on changes in *multiple* parameters against a baseline. The idea is to take two or more parameters and vary them to exacerbate what might be observed with single parameter changes. Implementing the logic of multi-parameter changes does not require any new theory, so a simple example can be used to make salient points clear.

Consider the baseline case of Table 8.1, and how ±20% changes in period *and* annual revenue affect NPV outcomes. This situation is illustrated in Figure 8.3.

Several features of this plot are worth noting:

1. The combined effect of period and return is much greater than each individual change.
2. There are a multitude of permutations of independent variables, so unless there is some strategic focus to the activity, it may be difficult to extract useful information about sensitivity.
3. Multi-parameter plots provide trends that would link the extremes of the scenarios, the outcomes of which were already conveyed in Table 8.1.

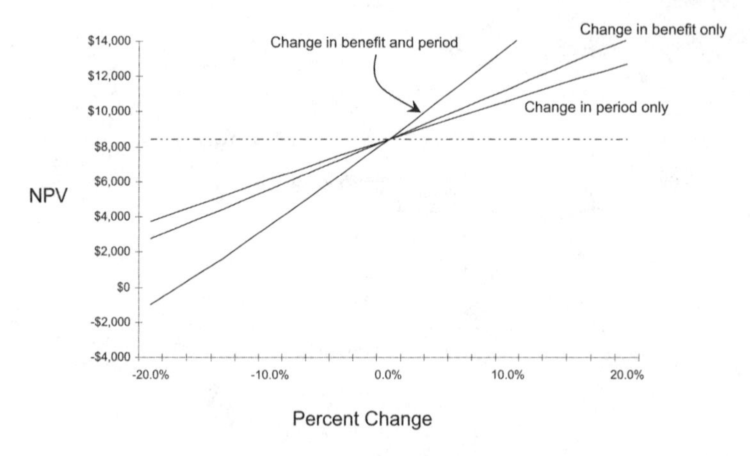

Figure 8.3: Typical spider plot.

8.4.4 Indifference point for staged investments

Earlier, in Section 4.5 (p. 116), a simple, typical *payback period* problem was presented for Electronic Health Records, Inc. (EHR) related to the upgrade of a server farm. Like most payback problems the idea was to find the point in time that the accrued revenue of additional production capacity equaled the project investment itself. As is most commonly the case, the question was whether the payback period was quick enough to meet the corporate criteria (for EHR, it did). Simple payback problems can be solved either graphically or analytically with a "one-equation-one-unknown" formulation. The choice is usually based on personal preference. In corporate presentation situations, a graph might be more appealing to the audience. However, one special situation in which a graph seems to lend clarity to decision-making is when seeking a *point of indifference* for single or multiple *staged investments.* No new theory or equations are required here, so we'll demonstrate the procedure by expanding on the EHR example.

> Electronic Health Records, Inc. would like to bid on a large hospital network services contract, but the uptime and expanded personnel "help line" requirements demand a slightly larger operations space. They can expand and renovate the entire second floor of their existing technology office space for about \$117,000. Business has been pretty good for the last 8 years, but management feels they have been inefficient in capital expenditures with constant incremental facility improvements. They are presented the option to *also* remodel the third floor (now vacant) at the bargain price of \$58,000 *today*, but the option to remodel will expire within a year and will require a less advantageous (but delayed) investment of \$75,000 within the next five years ($N \leq 5$) years and an investment of \$105,000 afterward ($N \geq 6$ years). The cost of capital is still assumed to hover around 8%.

If we look at the problem from an NPV perspective, investing a lump sum into the upgrade of the facility for floors 2 and 3 *today*, the total one-time investment is simply

$$\begin{aligned}
\texttt{NPV} &= \sum_{n=1}^{N} F(P/F, i\%, n)_{Floor[k]} \\
&= \texttt{NPV}_{Floor2} + \texttt{NPV}_{Floor3} \\
&= \$117,000 + \$58,000 = \$175,000
\end{aligned}$$

Alternatively, if the investment were to be considered in stages, then we compute the NPV as the sum of two parts: the baseline investment for renovation of floor 2 (needed today) and the NPV for each of the investment options.

$$\begin{aligned}
\texttt{NPV} &= \texttt{NPV}_{Floor2} + \texttt{NPV}_{Floor3} \\
&= \$117,000 + \texttt{NPV}_{Options}
\end{aligned}$$

For the Floor 3 renovation, we have two future options to consider, in which the NPV is computed from the familiar equivalence formulas:

$$NPV_X = \$117,000 + \$75,000(P/F, 8\%, n), \quad n = 1, 2, 3, 4, 5$$

$$NPV_Y = \$117,000 + \$105,000(P/F, 8\%, n), \quad n = 6, 7, 8, ...$$

In this case we work through each of the calculations using a computational factor table, calculator, or spreadsheet to obtain the NPV plot shown in Figure 8.4.

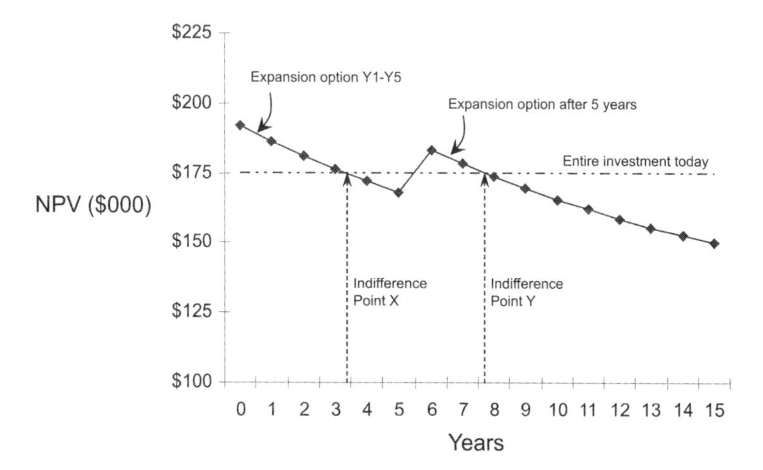

Figure 8.4: Indifference point for two staged investment options for EHR.

Two points of *indifference* can be observed in Figure 8.4. The first occurs at a point very close to but slightly greater than 3 years, and the second occurs between years 7 and 8. One could find the exact value in each case by solving the equations

$$\$175,000 = \$117,000 + \$75,000(P/F, 8\%, n), \quad 3 < n < 4$$

or

$$\$175,000 = \$117,000 + \$105,000(P/F, 8\%, n), \quad 7 < n < 8$$

For the first indifference point,

$$\$175,000 = \$117,000 + \frac{\$105,000}{(1 + 0.08)^n} \tag{8.1}$$

from which the analytic answer extends from very basic algebraic manipulation to find n:

$$\begin{aligned} 0.773 &= 1/(1.08)^n \\ (1.08)^n &= 1.29366 \\ n \ln(1.08) &= \ln(1.29366) \\ n &= 0.2575/0.07696 = 3.34 \quad \texttt{years} \end{aligned}$$

In a similar way we find that for Option Y the break-even point is $n = 7.71$ years. The algebraic approach, with the use of the *ln* function, seems a bit outdated, and despite the beauty of logarithms, spreadsheet solutions via the Excel `NPER` algorithm are popular. As well, we leave it to the reader to explore the workbook that accompanies this chapter for the related Excel calculations used to produce Figure 8.4. The format will simplify calculations for problems of a similar nature.

The presence of two indifference points might seem a bit odd but is logical in this case because of the out-year opportunity to renovate in year 6, and the fact that – all other variables remaining the same – the time value of money works in favor of the client.

Other aspects of this simple problem are unrealistic or incomplete:

1. The problem data suggest that the $105,000 renovation price would remain constant and valid in the out years, though were we to account for expected rates of inflation in the MARR, the values might be more reasonable.
2. We assume that if the extra space for Option Y were to be constructed, the company would be able to put the space to good use so that the return at least offsets the asset depreciation.
3. We assume here that the majority of the risk is in the "stand-alone" project risk that the company assumes as corporate risk, but there is significant market risk as well. If the company is leasing the space, then there needs to be considered the lease-hold improvement they are probably making to a facility they do not own; also a majority of the improvement costs are lost if business does not continue to grow as expected.

4. We have not looked at other options, like potentially outsourcing non-core competencies and thus keeping operations contained within Floor 2. While seemingly a highly conservative approach, the computer services industry is quite volatile, and options like the "cloud" have become surprisingly reliable.
5. There may be tax credits for capital investments that could help offset the investment *today* that lower the effective price of the renovation.

8.5 Probabilistic methods

We have alluded to the importance of undertaking only those investments where the anticipated rate of return is adequate to compensate the investor for the risk involved. Section 8.4 highlighted several categories of risk and related analysis techniques. In the previous section, our scenario and sensitivity analyses implied that we had little knowledge or a minimal basis for the *likelihood of probabilities* that could influence knowledge of outcomes. In this Section, we do not propose that the future is so vague; through the assumption of discrete and continuous probability distributions for independent variables, we are able to develop a firmer basis on which to estimate risk and uncertainty for the investor. Discrete and continuous probability distributions are both discussed within this section under the general heading of *probabilistic methods.*

Readers may already be aware the that topic of probability and statistics is a fundamental part of many undergraduate courses of study. We do not attempt to reproduce or provide tutorials for the required probabilistic formula and their derivations here since so much information can be found elsewhere in math handbooks [2]; management science textbooks [3]; financial management textbooks [4], [5]; and, of course, many textbooks on engineering economics [6], [7], [8]. Citations here are not intended to be comprehensive; there are *so* many online references, bookstore websites, or online courses (`www.edx.org`; `www.coursera.org/course/introstats`, etc.) to help with any curiosity or knowledge gaps arising from the terse probabilistic discussion to follow.

A distinction to understand clearly is that we either have *some firm ideas* about the *probability* of an event occurring or really *no idea* at all. This is one way to distinguish between risk and uncertainty:

1. Risk: We *can make a reasonable estimate* about what the form of the underlying distribution looks like.
2. Uncertainty: We *do not know and cannot estimate* what the possible distribution looks like.

Probabilistic methods help us conveniently characterize *risk* since we can use statistical distributions for estimating the *probability* of an event occurring. There are two basic types of random variables we use in the analysis of risk: discrete and continuous. Knowing the difference is important since the type of variable relates to the underlying distribution we use in problem analysis.

1. Discrete variables can only take *one of a specific set* of a finite number of values within a specified range of numbers.
2. Continuous variables can take *any* number of an infinite number of values within a specified range.

The choice of variable dictates the form of the underlying probability distribution. The probability of an event occurring can range from a value of "0" which means the event does not occur, to a value of "1" whereby we are assured the event *will* happen. Our understanding of the underlying distributions might come from historical data, experiments, design experience, or just our intuition and expert judgment. It is quite common for the characteristics of problem data to immediately identify whether our probability calculations are based on discrete or continuous random variables. The classic case is the flipping of a coin. There is a 50% probability that the likelihood of a coin flip will be "heads." Indeed, for any series of four tosses of the coin, the probability of getting a "heads" is shown in Table 8.3.

X	P(X)
0	0.0625
1	0.2500
2	0.3750
3	0.2500
4	0.0625

Table 8.3: Discrete probability distributions for the flip of a coin.

So, the flip of a coin can only yield a *discrete* (finite set) of probable outcomes. For a set of four flips of the coin, Table 8.3 tells us there is a 25% probability you will get 3 heads. The finite set of results is that you either get 0, 1, 2, 3, or 4 heads, with the probabilities shown. Probabilities cannot be negative nor can the sum of all values exceed a value of 1. Figure 8.5 is a plot of the data from Table 8.3.

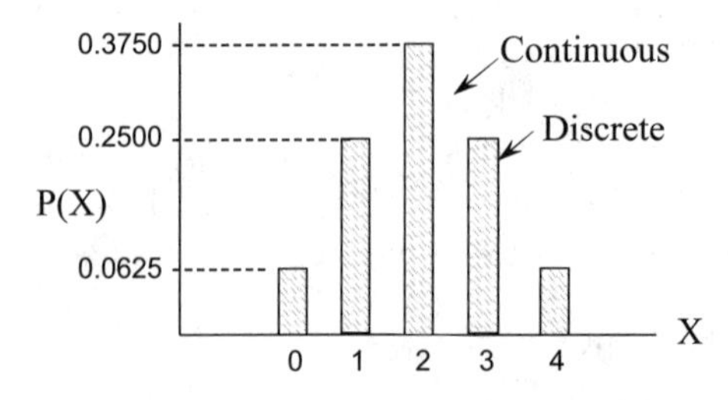

Figure 8.5: Probability distribution for the flip of a coin

Discrete variables are shown in Figure 8.5 as having 1 of 5 possible values, with the sum of all values equal to 1.0.

$$\sum_{n=0}^{4} P(X_i) = 1.0 \tag{8.2}$$

The flipping of a coin is a clear example of a problem involving discrete random variables.

8.5.1 Shape of probability curves

Figure 8.5 also shows a curve that provides a view of the "skyline" we would expect if we had a different problem involving many more variables. In fact, as the number of variables approaches ∞, the curve represents the distribution of *continuous random variables* and is called the *probability density function* $f(x_i)$. And, rather than the convenience of summing a few discrete variables, we have to treat the sum of the probabilities as an integral quantity.

$$\int_{-\infty}^{+\infty} f(x_i)dx \;=\; 1.0 \tag{8.3}$$

The treatment of continuous probability distributions is different than for discrete probability distributions. We will see later that the idea of a "single" value for a continuous probability distribution does not make sense, since the probability for a single value is equal to 0. Instead, we compute the probability that the independent variable will fall between a *range* of values. The types of problems and calculations are very different indeed. These are, of course, *objective* probabilities that can be independently verifies they are not *subjective* probabilities that are simply personal opinions or beliefs.

The shape of the probability curves can reveal a lot about processes and process control. For instance, the normalization of discrete probabilities might reveal the "diffusion" of the normalized curve over time. (Some engineers will recognize this as similar to the jet-spreading of turbulent flow.) Figure 8.6 illustrates this trend.

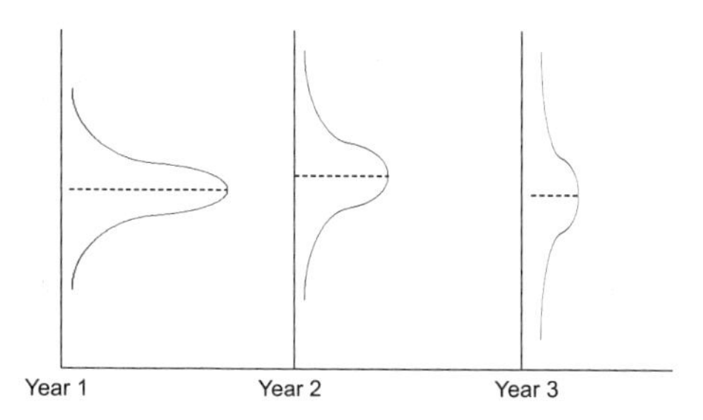

Figure 8.6: Diminished peak outcomes for a process or phenomenon as a function of time.

Although it may not be clear from Figure 8.6, the area under the curve in each case is unity (= 1) as indicated in Equation 8.3. This "conservation of the probability density" occurs by definition. The difference in the span of the curve is marked by differences in the value of the standard deviation.

Different projects will exhibit different characteristic curve shapes, as shown in Figure 8.7; only through the product specification projects can the "preferred" or "better" curve be established. That curves "A" and "B" may have different peak values and standard deviations may not be out of specification, and the shapes of the curves are adequate for the characterization of risk.

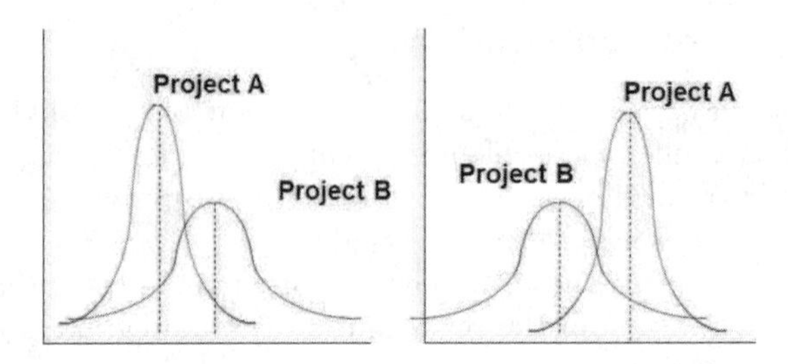

Figure 8.7: Difference between mean and standard deviation shifts.

We primarily direct our attention to the simple probability $P(X)$, though as shown in Equation 5.1, p. 139, joint probabilities are often of interest when trying to establish independence or mutual exclusivity of events (in essence, the probability that such events will occur together). There is also a *conditional* probability $P(X|Y)$ that examines the occurrence of A if B has occurred. As a practical matter, the discrete random variable approach has widespread use and is not difficult to understand, but having several worked problems is critical to showing the value of the formula that will follow on page 236. There are many uses for this technique, and the bulk of Section 8.5.2 is consumed with two worked problems.

After the discussion of the discrete random variable approach, we move immediately to the topic of *decision trees* in Section 8.5.3. Some would argue it is better to examine continuous probability distribution frequency methods, but in a condensed course the proximity of discrete variables and decision trees actually makes good use of study time and linkage of examples. Were this a more rigorous and comprehensive book, the more traditional approach might be better.

Section 8.5.4 on frequency distribution methods of analysis is somewhat more challenging, especially if the students have not had a course in statistics, or maybe did not take the course seriously (which could easily be the case if there was a lack of relevance). We work through a few examples of how the frequency methods really *can* be of value in problem analysis if a Gaussian probability distribution can be assumed for the random variables.

8.5.2 Discrete random variables

Application of discrete random variable theory to the estimation of economic uncertainty is actually quite easy. Most problems arise from a lack of familiarity with the calculation steps, not any real mathematical challenge. To facilitate the application of this technique, we will first outline (without derivation or proof) the essential equations needed. This will then be followed by two classic problems quite often solved with this method.

Probability variables and relations

We have previously discussed in the coin flip example that a discrete random variable X is limited to a *finite* number of values. For a wide variety of engineering situations the number of variables is actually quite small, and the number of discrete variables is often less than 9. This statement might seem odd at first, but in the sample problems we see that, for instance, the number of variables is closely linked to a specific set of probabilities we establish for a situation. The balancing act is that we want to have enough variables to test excursions from a baseline value, but not so many that it is hard to make decisions. We'll come back to this idea in a bit after summarizing some key equations. Given the discrete random variable X, we typically want to know four different things:

1. The probability `P` of a value `X` occurring
2. The expected value `E` of `X`
3. The variance `V` of the expected value
4. The standard deviation `S`

Let's begin with the random variable X, which has a finite number N values

$$X = x_1, x_2, x_3, ...x_N \tag{8.4}$$

From this range of values, the probability that a variable has a specific value $X = x_i$ is given by

$$P[X = x_i] = p(x_i), i = 1, 2, 3, ...N \tag{8.5}$$

However, the formula has a slightly different form if we want to determine the probability that X is in the *range* $a \leq X \leq b$:

$$P[a \leq X \leq b] = \sum_{i=a}^{b} p(x_i) \tag{8.6}$$

Equation 8.6 will be useful in the discussion of Sample Problem 2, to follow. Beyond the probability of occurrence of X, we often need to find its expected value, E, too. That calculation simply requires us to sum all the products of the values of the variable X and its probability:

$$E[X] = \sum_{n=1}^{N} x_i p(x_i) \tag{8.7}$$

We can think of the expected value of a probability distribution as the mean of the distribution, but it should be clear that the expected value is *not* the peak value, but again, the weighted sum of all values.

The behavior of a random (variable) event is expressed in the form of the variables probability distribution. Thinking of a mean value (mean behavior) described by a probability distribution should also bring to mind the *variance* of the distribution, which tells us how "close" the average value is to the mean. No mystery that variance V about the mean of the variable X is found through the weighted average of the difference between the mean and a specific value:

$$V[X] = \sum_{n=1}^{N} (x_i)^2 p(x_i) - (E[X])^2 \tag{8.8}$$

Now, from V[X], we can find the standard deviation of the random discrete variable X quite simply from a square root calculation:

$$SD[X] = \sqrt{V[X]} \tag{8.9}$$

At this point, Equations 8.5 to 8.9 are essentially all we need to work through some sample problems.

For a very select number of problems we may need to compute the product of discrete random variables. In the joint probability distribution $Z = XY$, the expected value is given by

$$E(Z) = E(X)E(Y) \tag{8.10}$$

and the variance is simply

$$V(Z) = E(X^2)E(Y^2) - [E(X)E(Y)]^2 \tag{8.11}$$

The work of Sullivan [9](p. 500) expands further on the topic of joint probabilities and provides example calculations.

Some areas where joint probability distribution could be useful are in trying to estimate if a drop in the value of the dollar correlates to a drop in the S&P 500, if two stocks will rise on the same day, or whether a union will strike at the same time steel imports rise. We really don't look at those problems here, actually, and our worked problems are of a much simpler type.

Worked problems

There are two classic problems to work in this area that illustrate typical usage. The first problem involves the expected monthly revenue from an equipment investment, and the second problem is an expected PW problem.

Problem 1: Expected monthly revenue from an equipment investment

Sam is the vice-president of sales for EHR and was excited that senior management approved the purchase of the computer hosting server upgrades discussed back on p. 116. Sam is eager to push engineering to get the equipment installed and on-line ASAP. There are certain realities to bring the server farm on-line, but Jason is the engineering manager who has "been there, done

i	Capacity	$p(x_i)$	x_i
0	0.50	0.10	\$3,000
1	0.60	0.25	\$4,500
2	0.70	0.35	\$6,000
3	0.80	0.25	\$7,500
4	0.90	0.05	\$9,000

Table 8.4: Probability of capacity utilization and revenue for EHR production server ramp-up.

that" and he estimates that within 45 days they will be on-line with the production capacity estimates listed in Table 8.4.

Sam likes the idea of reaching 90% capacity within the next two months, since the \$9,000 revenue would help his bonus check. Still, he decides he needs to be a bit conservative and is pleased when Jason suggests that within the first year it is not uncommon for systems to operate at or near 70% capacity. With this information, what is the expected annual revenue and its variance?

Solution. From Equation 8.7:

$$\begin{aligned} E[X] = & \sum_{n=1}^{N} x_i p(x_i) \\ = & (.10)(\$3,000) + (.25)(\$4,500) + (.35)(\$6,000) \\ & \qquad + (.25)(\$7,500) + (.05)(\$9,000) \\ = & \$5,850 \end{aligned}$$

from which the variance is

$$\begin{aligned} V[X] = & \sum_{n=1}^{N} (x_i)^2 p(x_i) - (E[X])^2 \\ = & (.10)(\$3,000)^2 + (.25)(\$4,500)^2 + (.35)(\$6,000)^2 + (.25)(\$7,500) \\ & \quad + (.05)(\$9,000)^2 - (\$5,850)^2 \\ = & \$36,675,000 - \$34,222,500 = \$2,452,500 \end{aligned}$$

and the standard deviation is

$$SD[X] = \sqrt{V[X]} = \sqrt{2,452,500} = \$1,566.04 \qquad (8.12)$$

It should be alarming, of course, that the standard deviation is so very large!

Problem 2: Expected present worth

A small internet application service provider (ASP) is looking at the purchase of a high-speed server that will cost about \$6,000 and will last no more than 3 years (before it becomes obsolete and is of no value). The idea behind the server is to enable the promotion of cloud computing and draw in a major new EHR client. The company is excited about the revenue possibilities, but is also realistic that they could get caught with a bunch of "computing cloud" hype and do not realize any value from the purchase. Management estimates the sales revenue as shown in the table below, also assuming that the probability of losing the contract is 25%, "wait and see" is 50%, and the probability of winning is 25%. Assume that i = MARR = 18% per year and determine if the equipment should be purchased.

Scenario	A	B	C
	No contract	Wait and see	Win contract
Year	p = 0.25	p = 0.50	p = 0.25
0	-$6,000	-$6,000	-$6,000
1	+1,500	+2,000	+4,000
2	+1,750	+2,500	+4,500
3	+2,000	+3,000	+5,000

Solving this problem can be performed in three easy steps:

1. For each scenario (lose, wait, win), convert the revenue streams into an equivalent form by calculating the PW for each year.
2. Sum the equipment investment and the PW of the revenue streams in each case.
3. Compute the expected worth of the investment using Equation 8.7 (p. 237).

	A	B	C	D	E	F
1	Expense		PW of Annual Expense		Equivalent Uniform Annual Cost	
2	Year	Expense	Value	Equation for "C"	Value	Equation for "E"
3	0					
4	1	$ 2,000	$ 1,818	= B4/(1+$B$10)∧A4	$ 2,000	= PMT($B$10, A4, C4)
5	2	$ 2,250	$ 1,860	= B5/(1+$B$10)∧A5	$ 2,119	= PMT($B$10, A5, C4:C5)
6	3	$ 2,500	$ 1,878	= B6/(1+$B$10)∧A6	$ 2,234	= PMT($B$10, A6, C4:C6)
7	4	$ 2,750	$ 1,878	= B7/(1+$B$10)∧A7	$ 2,345	= PMT($B$10, A7, C4:C7)
8	5	$ 3,000	$ 1,863	= B8/(1+$B$10)∧A8	$ 2,453	= PMT($B$10, A8, C4:C8)
9	6	$ 3,250	$ 1,835	= B9/(1+$B$10)∧A9	$ 2,556	= PMT($B$10, A9, C4:C9)
10	MARR = 10%					

Table 8.5: Excel calculation of PW for potential sales revenue stream.

We can calculate the PW of each scenario by hand or use a spreadsheet as in Table 8.5:

$$
\begin{aligned}
PW[A] &= -\$6,000 + \$3,745 &= -\$2,254.73 \\
PW[B] &= -\$6,000 + \$5,316 &= -\$683.73 \\
PW[C] &= -\$6,000 + \$9,665 &= +\$3,664.81
\end{aligned}
$$

Now that we have the revenue streams in equivalent form, the expected value is

$$
\begin{aligned}
E[X] &= \sum_{n=1}^{N} x_i p(x_i) \\
&= (.25)(-\$2,254.73) + (.50)(-\$683.73) + (.25)(+\$3,664.81) \\
&= -\$563.68 - \$341.87 + \$916.20 \quad = \quad \$10.66
\end{aligned}
$$

Given that E[X] is essentially 0, it is questionable whether to proceed.

We leave the details to the reader to show that if MARR = 12%, the situation is much more favorable, with E[X] = $675.93.

8.5.3 Decision trees

In the prior discussion of discrete random variable analysis, we demonstrated that the inclusion of *probability* can diversify the set of problem outcomes. Our understanding of risk evolved in a more structured way than, say, the simple characteristic scenario analysis methods discussed back in Section 5.2 (p. 138). Each of the worked problems in the prior section was intended to enable a more realistic problem framework to be addressed. The current section on *decision trees* continues that trend and incorporates increased problem complexity in two ways:

1. A decision tree allows refined branching of project pathways *and* enables outcomes from one decision to have an impact on decisions further down the path.
2. Event probability is closely linked to specific project decisions and events *within* the project and not just at the outset (as in scenarios).

We are now able to compute a much wider distribution of, say, project `NPV` outcomes, with the range of `NPVs` based on branch probabilities. The solution set that arises is a form of a discrete probability distribution for a problem (you may have to read this sentence twice to understand more clearly!). As well, capital budget projects frequently involve decisions that can only be made once a certain amount of work has been accomplished; although we may not be able to make certain decisions until we reach that decision point, we *do* know the choices and their probabilities of success once we get there. Clearly, the tree problem formulation has intrinsic randomness embedded in the structure!

An interesting variety of complex problems can be analyzed with decision trees, but tracking all the branch options can seem cumbersome, especially when working though decision trees by hand. The popularity of software might seem to make working out branch diagrams antiquated, but this is one of the few topics where understanding the underlying logic of decision trees is essential. The content of this section assumes the reader has fully grasped the worked problems of the prior section (p. 238), and they are close by if you need to review them again. Calculation patterns are similar. Decision-tree logic marks a critical point in understanding the ways that financial decision-making can be achieved with a mosaic of methods. Just to keep the discussion as simple as possible, we limit the scope of analysis in this section to the evaluation of the NPV of a project. Trees have a broad role in managerial decision-making[3] that might be suggested here. The chapter proceeds by providing a quick summary of common decision-tree architectural components in the next section, and then working through three sample problems:

1. A single-level tree to understand a preference between project options.
2. A multi-level tree related to new product development decisions.
3. A multi-level tree illustrating a multiple probability node problem.

Decision-tree solution strategy overview

Decision trees involve some symbols and logic that have evolved over time and vary only slightly between textbooks. With reference to Figure 8.8, we see the following:

1. The tree structure begins with a fundamental decision that should be made. The decision node in Figure 8.8 might be asking if we should redesign a product.
2. From the basic design decision node, there could be additional possible outcomes from a decision. Each of the decisions becomes a *branch* on the tree. Decision nodes are often denoted by a square symbol. Often the decision involves just a few options ("yes or no", "low, medium, or high", "success or failure"), but in principle there is no restriction.
3. Nodes can also be probability nodes, shown as circles in Figure 8.8, that represent probabilistic events leading to different outcomes. Each branch of the tree will have the probability of an outcome noted. The sum of the probabilities coming from a node must equal 1.0.
4. We continue to build out the tree with as many decision and probabilistic (chances) nodes as necessary to represent the problem. Figure 8.8 illustrates one decision branch and one probability branch.
5. Each branch of the tree will note some associated economic value (PW, FW, UAW) final. Final outcomes are listed at the end of each branch (on the right-hand side). In the examples to follow we will demonstrate how we support the original decision by working back from right to left.

Decision trees can help with forecasting how project outcomes can be affected and identify decision entry-points for the engineering manager once a project has been launched. In this textbook we construct decision trees based on an implied belief about the existence of probabilities (to drive calculations), though an equally valid perspective is to generate the probabilities algorithmically based on a set of specified outcomes (for an optimized error function).

Worked example: Rethink an investment?

To illustrate computational procedure, let's consider a very simple example. Suppose a proposed production system's project plan estimates NPV = -$3

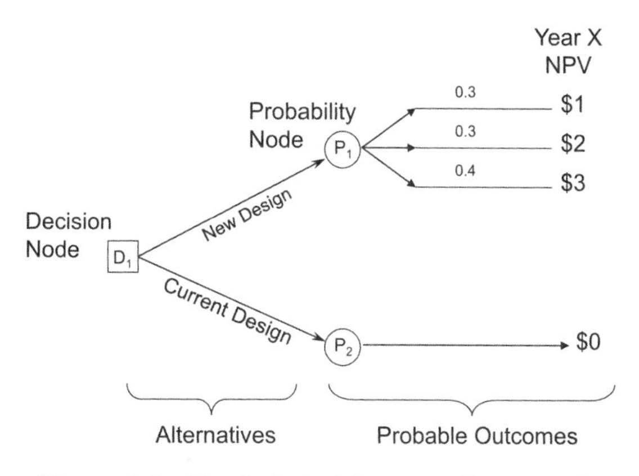

Figure 8.8: Typical decision-tree framework.

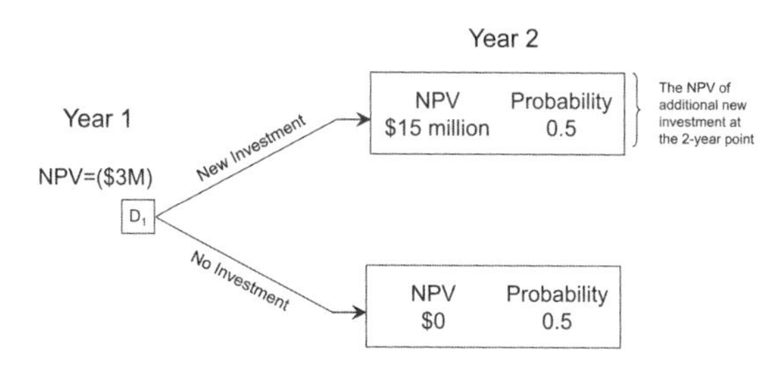

Figure 8.9: Impact of decision trees on decision-making

million for Year 1 which means that the project is currently a "no go." Now, new marketing information is available suggesting that in Year 2, the product might be a "blockbuster," and there is a 50/50 chance of producing $\texttt{NPV} = \$15$ million in that year. The situation is depicted in Figure 8.9. Does the new marketing information change the original decision?

First, let's convert the \$15 million into an equivalent form for Year 1, assuming an 18% MARR.

$$\texttt{NPV} = \frac{\$15\texttt{M}}{(1+18\%)^2} = \$10.77\texttt{M}$$

The project worth is now the sum of the product of the NPV of each branch and its respective probability of occurrence:

$$\texttt{NPV} = -\$3\texttt{M} + \{0.5(\$10.77\texttt{M}) + 0.5(\$10.77\texttt{M})\} = \$2.38\texttt{M}$$

Quite evidently, the prospect of a second-year success story makes the investment more attractive – understanding, of course, there is risk involved!

Worked example: Self-insure a project or buy insurance?

A slightly more complex example helps understand calculations a bit more. In Figure 8.10, we have purchased a high-technology blood centrifuge, and the dealer asks if we would like a warranty. The decision DO is simply whether to self-insure and accept the risk of loss or mitigate the downside by picking up the warranty. The warranty costs \$1,200 per year and has a \$500 deductible. It seems so unlikely that there would be a total loss of the centrifuge we assign a 2% probability to this branch. Other data are shown in Figure 8.10.

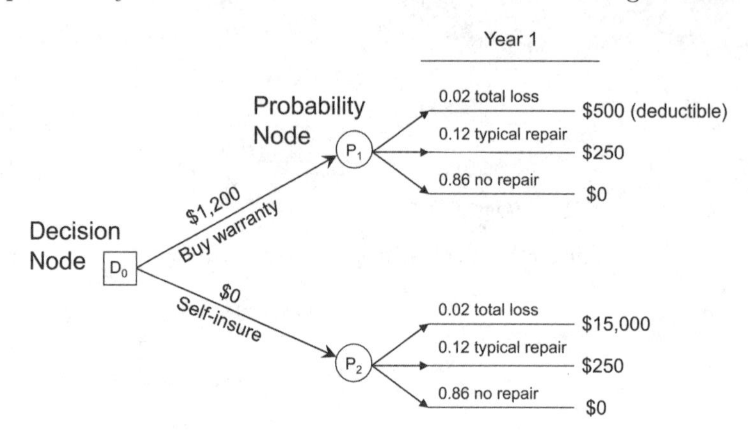

Figure 8.10: Simple model for choice of warranty or self insurance.

Since all costs are already annualized, let's use the EUAC approach to the problem.

$$\begin{aligned}\texttt{EUAC}_{P1} &= 0.02(\$500) + 0.12(\$250) + 0.86(\$0) = \$40\\ \texttt{EUAC}_{P2} &= 0.02(\$15)K + 0.12(\$250) + 0.86(\$0) = \$330\end{aligned}$$

So, in the case of an accident, the expected cost is \$40 if we *do have* insurance, and \$330 if we *do not have* insurance; thus, purchasing the \$1,200 insurance policy lowers the expected cost of an accident by \$290. To be fair, we have to include the cost of insurance in our decision, and we find that

1. Without insurance the total cost of an accident is \$0 + \$330 = \$330.
2. With insurance the total cost of an accident is \$1,200 + \$40 = \$1,240.

From these statements, we find that the *total* cost to self-insure is simply the difference: \$1,240 - \$330 = \$910. Of course while it may seem like a better use of funds to self-insure, the maximum corporate exposure in the event of total loss is \$15,000 versus the \$500 deductible, and most people will logically purchase insurance because of the benefit of pooled risk.

Several aspects of this problem are unrealistic, primarily the timing of cash flows. Insurance premiums are typically paid at the start of the period that would cover the potential loss. Were this problem to be expanded over the course of several years, the time value of money would also enter into the mix.

Worked example: License IP or market a new product?

Probability trees can be built out using a combination of decision nodes and probability nodes, with multiple branch outcomes to assess, as shown in Figure 8.11. Consider the case of a new sensor manufacturing technology developed in R&D and management must consider the decision to (A) act on the opportunity internally, (B) license the technology to someone else, or (C) "wait and see" and to simply "put the technology on the shelf."

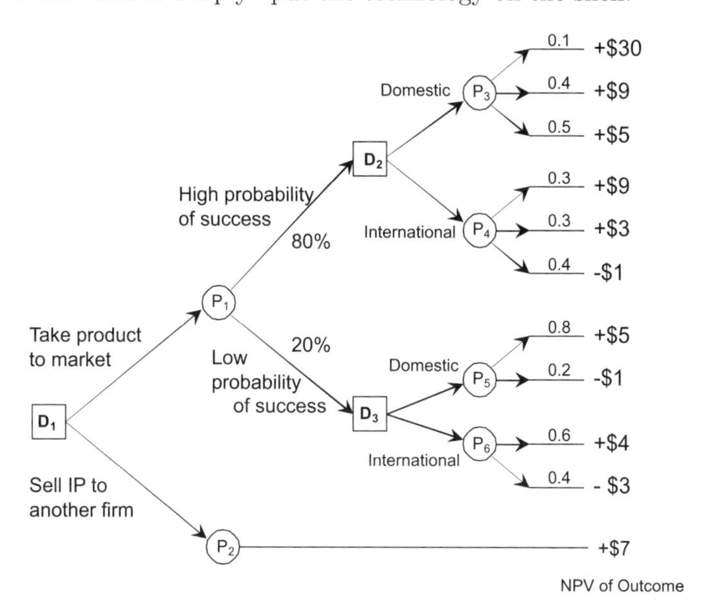

Figure 8.11: Decision tree to assess whether a company should sell or market a technology.

It is reasonable that the "go forward" strategy takes into account other corporate considerations, as shown in Figure 8.12, such as the strategic fit of the technology to the current project objectives, and the possibility of abandoning the technology altogether.

Let's work through the details of this decision tree to illustrate problem-solving logic, as the "backward tracing" method can be a little confusing at first.

Figure 8.11 tells a story that begins with our first decision point, D_1, in which the option of selling or licensing the IP must be weighted against the company creating a product they will develop and take to market themselves. This license/sell option has the trivial probability of $P_2 = 1.0$, and the NPV of that option is estimated to be approximately \$7 million. For simplicity, assume that NPVs expressed for any branch of the tree have been converted to equivalent form at $t = 0$.

Most of the work in this problem has to do with converting the NPV for each of the branches, with branch options pruned along the way by summative values that are unfavorable. The internal go-to-market strategy of path D_1 to P_1 presents a high (80%) probability of a successful internal effort, and only a 20% probability for low success. Assuming success, management estimates they have a 50/50 chance of being successful in either an international or

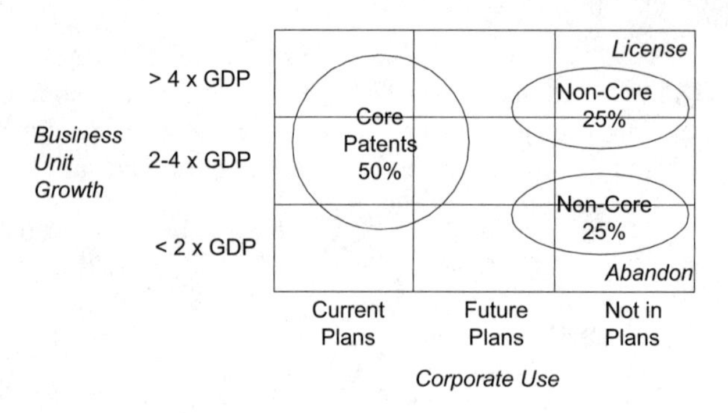

Figure 8.12: Strategic fit of technology IP to company portfolio.

domestic market.

For this company, the domestic branch D_2 to P_3 has an overall expected NPV much higher than for the international market, possibly due to the relative inexperience and additional costs of moving into unknown territory. If the company is successful domestically, marketing has produced estimates suggesting a 10% probability of NPV = \$30 million, but alternatives of 40% that NPV = \$9 million and 50% that NPV = \$5 million. International NPV estimates have the probabilities shown along the lower branch emanating from D_2 to P_4. Let's zoom in on the calculations related to D_2. First, the NPV at P_3 and P_4 is readily found as

$$
\begin{aligned}
E[X] &= \sum_{n=1}^{N} x_i p(x_i) \\
&= 0.1(\$30.0) + 0.4(\$9.0) + 0.5(\$5.0) \quad = \quad \$3.0 + \$3.6 + \$2.5 \quad = \quad \$9.1 \quad \texttt{P3} \\
&= 0.3(\$9.0) + 0.3(\$3.0) + 0.4(-\$1.0) \quad = \quad \$2.7 + \$0.9 - \$0.4 \quad = \quad \$3.2 \quad \texttt{P4}
\end{aligned}
$$

Clearly the NPV of the domestic option has a much higher value than the international option; notwithstanding any other criteria we assign $\texttt{NPV}_{P3} =$ \$9.1, as shown in Figure 8.13. The international tree branch from D_2 to P_4 is thus pruned, and the healthier branch from D_2 to P_3 survives.

Computing NPV for D_3 from P_5 and P_6 quite simply follows the pattern from the prior discussion:

$$
\begin{aligned}
E[P_5] &= 0.8(\$5.0) + 0.2(-\$1.0) \quad = \quad \$4.0 - \$0.2 \quad = \quad \$3.8 \\
E[P_6] &= 0.6(\$4.0) + 0.4(-\$4.0) \quad = \quad \$2.4 - \$1.6 \quad = \quad \$0.8
\end{aligned}
$$

Now, of these two branches, P_5 has the higher estimated NPV, so D_3 is assigned the \$3.8 value from P_5. As illustrated in Figure 8.14, we can now "roll-back" to examining P_1, as the sum of the 80% high probability of success for D_2

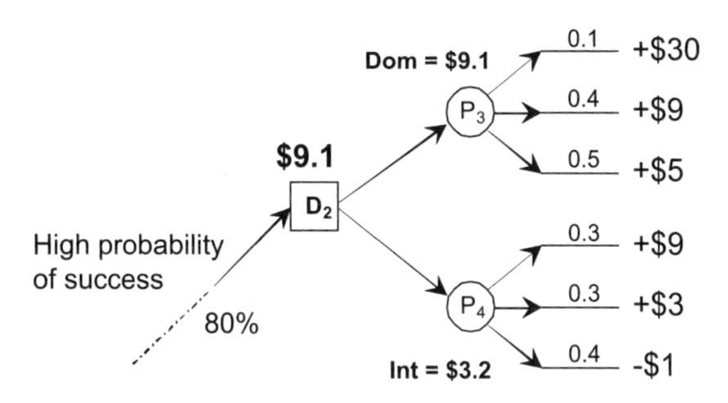

Figure 8.13: Equivalent NPV = \$9.1M at D_2 from options P_3 and P_4.

value, \$9.1, and the 20% low probability of the D_3 value, \$3.8.

$$E[P_1] = 0.8(\$9.1) + 0.2(\$3.8) = \$7.28 + \$0.76 = \$8.04$$

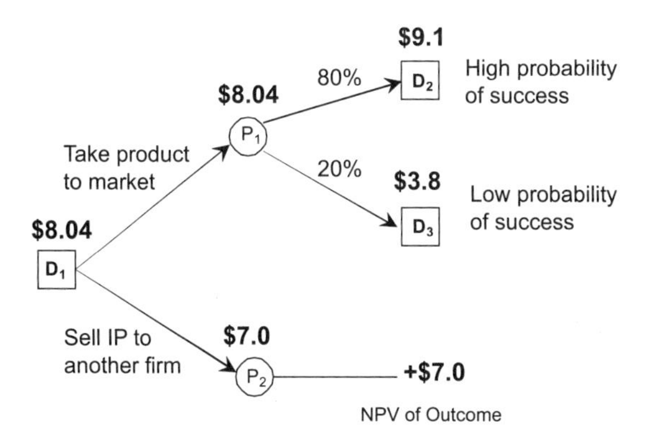

Figure 8.14: Equivalent NPV = \$8.04M at P_1 from decisions at D_2 and D_3.

Worked example: Enter a new market?

A concept for a new, low-cost, simple-to-use malaria detection system for patients in remote and underserved regions has been prototyped, and preliminary clinical results are encouraging. Although the target market for the detection system requires that the per-unit cost be under \$500, many of the system parts are off-the-shelf. The inventors decide to start a company and become systems integrators marketing the product to global health charitable organizations. Figure 8.15 illustrates the key product development pathways. The inventors feel there is a 75% chance the product will be highly successful (given weak competing products), and estimate generating \$100K in net revenue in the first year. Further, an \$87K equipment investment is expected to yield higher production efficiencies in years 2-5. Investment leads to the dual benefit of lowering unit cost and raising unit volume; in this scenario, new revenue of \$175K is projected for each of the following four years. Absent the investment, the plan is to simply "keep doing what we're doing" and accept a lower annual revenue of \$150K in the out years. Since the project is the personal passion of the inventor, a home-equity loan is being used to fund the project, and the cost of capital can be assumed at (a very low rate of) 8%.

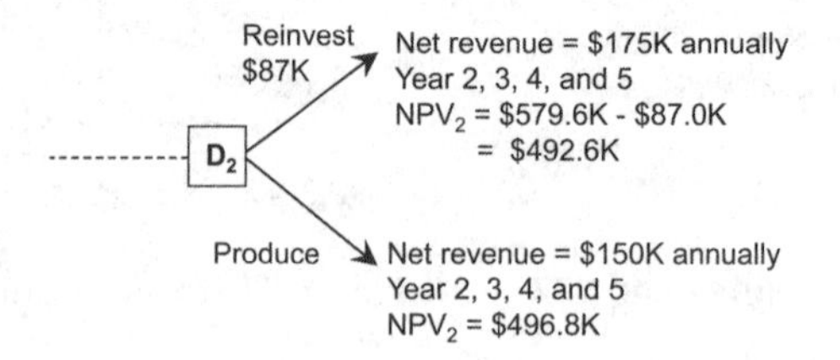

Figure 8.15: Malaria detection device.

The pattern for the solution procedure is as follows:

1. Compute revenue stream NPV for each tree branch emanating from D_2 and D_3.
2. Identify the favorable NPV for D_2 and D_3 to use for Year 1 calculations.
3. Use Equation 8.7 to find the probable NPV at P_1.
4. Compare P_1 and P_2 at D_1 and select the best option.

Net revenue at node D_2 is *not* given in equivalent form, so we must account for the TVM, as illustrated in Chapter 4. Now, NPV for the reinvestment path D_2 is easily computed from Equation 4.1 (p. 98):

$$NPV = -\ \$87,000 \quad \text{Present value of the reinvestment}$$
$$+\ \$175,000(P/A, i\%, N) \quad \text{Present value of the annual revenue}$$

$$NPV = -\ \$87,000$$
$$+\ \$175,000(P/A, 8\%, 4) \quad \text{n= 2, 3, 4, and 5}$$

This leads to

$$\begin{aligned} NPV &= -\ \$87,000 + \$175,000(3.312) \\ &= -\ \$87,000 + \$579,600 = \$492,600 \end{aligned}$$

Similarly, the alternative path from D_2 is the NPV for simply continuing production with the system as is:

$$NPV = \$150,000(P/A, 8\%, 4) = \$150,000(3.312) = \$496,800$$

The results for D_2 are illustrated in Figure 8.16.

Basing the decision at D_2 on strictly the NPV, then avoiding the reinvestment and simply continuing with the production system as is, has a \$4,200 advantage. As a practical matter, the advantage is less than 1% of the revenue, and such a small amount might suggest other out-year benefits. But if

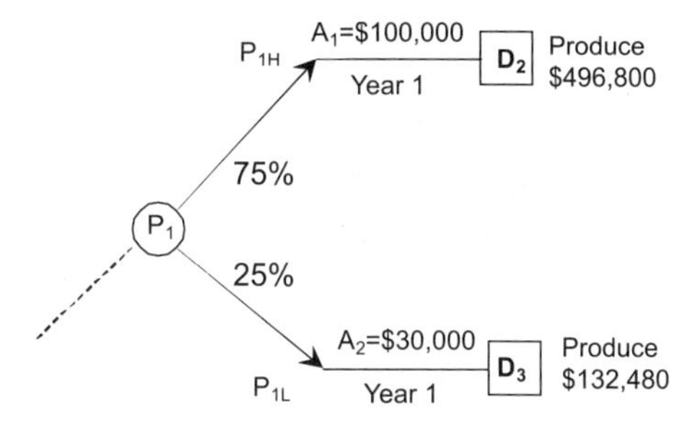

Figure 8.16: Malaria detection device, high probability of success after Year 1.

the company is using a home-equity loan to fund the investment, there may be a high opportunity cost relative to other needs of the business at this time.

Continuing on, the results for D_3 follow the method outlined for D_2. We have the "produce" pathway yielding

$$NPV = \$40,000(P/A, 8\%, 4) = \$40,000(3.312) = \$132,480$$

This amount is actually slightly more than the "terminate" salvage value of $125,000.

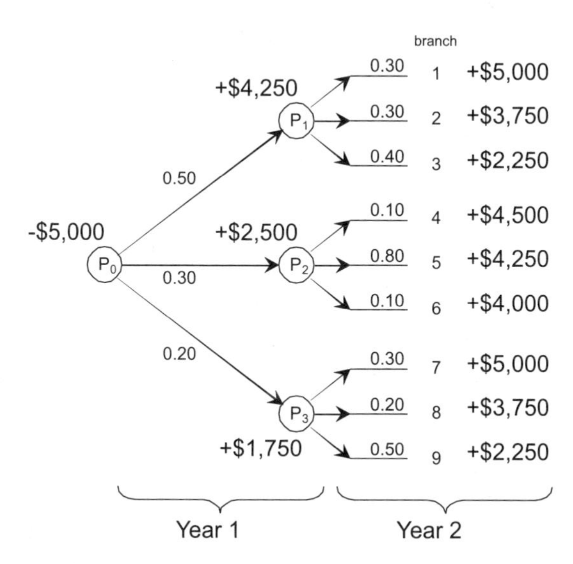

Figure 8.17: Node P_1 summary of NPV with outcomes for D_2 and D_3.

The expected values of each branch P_{1H} and P_{1L} are as follows:

$$NPV[P_1H] = \$100,000 + \$496,800(P/A, 8\%, 1)$$
$$= \$100,000 + \$496,800(0.926) = \$559,296$$
$$NPV[P_1L] = \$30,000 + \$132,480(P/A, 8\%, 1) = \$152,646$$

Thus the expected value at node P_1 is

$$E[P_1] = 0.75(\$559,296) + 0.25(\$152,646) = \$419,472 + \$38,161 = \$457,633$$

Given that $E[P_1] = 0$, the decision to make is quite obvious, and we choose to develop a new product.

The multiple steps involved are summarized as follows:

1. Compute favorable revenue stream for branch D_2 = \$496,800.
2. Compute favorable revenue stream for branch D_3 = \$132,480.
3. Use Equation 8.7 to find the probable NPV at P_1 is \$457,633.
4. Compare P_1 = \$475,633 and P_2 = \$0 at D_1, from which the best option is to develop a new product.

What would have made the problem more interesting is to have had a more sophisticated set of branch options, such as those shown in Figure 8.12, where the possibility of loss of revenue is apparent. In the malaria detection tree in Figure 8.15, no losses seem to be projected, so it is challenging to think you would *not* proceed with the decision to develop the new product!

Worked example: What is the probability of losing money on new equipment?

Max runs a moving service and would like to expand into the transport of sensitive electronic instruments; as a small owner-operator, the OEM's promise of a maximum of \$5,000 annual revenue is appealing. Discussions with an OEM have gone well, but to limit liability and keep insurance rates low, Max has to purchase a special trailer for moving electronic equipment that sells for \$5,499 (if he buys within the next 7 days, he can get it for \$5,000, taxes and title included!). Max likes the idea that he can get the \$5,000 price, since even if the contract is voided after one year, Max figures he'll break even in the worst-case scenario. As a very small owner-operator Max has negotiated a deal with the OEM that the contract will have a "floor" of 35% of the expected maximum annual moving revenue, about \$1,750. Several of Max's colleagues think he is crazy risking a lot with little promise of revenue just to secure a big-name client. Max says, "Even if I lose money, I think of this as a marketing effort to open up a new line of business."

Max's friend at the racing casino has produced the revenue and probabilities shown in Figure 8.18 to help bracket revenue streams. The analysis period is two years, and MARR is 19% per year. Based on this information, *what is the probability Max will lose money on this deal?* Thinking back to

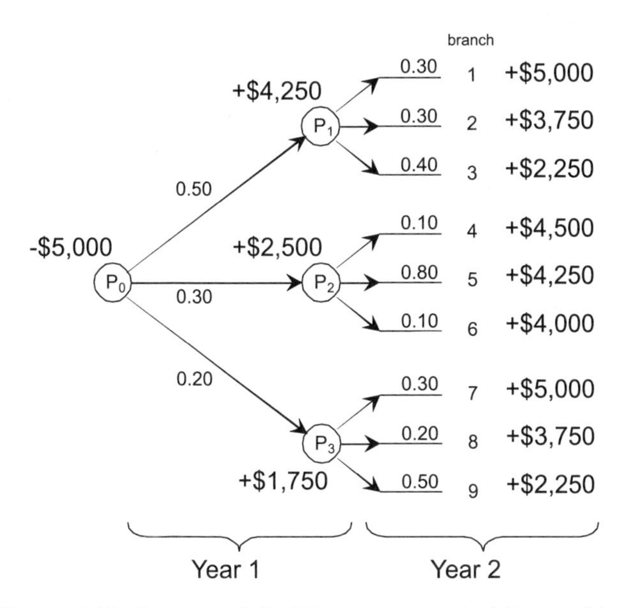

Figure 8.18: Impact of decision trees on decision-making.

Section 8.3 (p. 225), we begin to see the reasoning behind the comment that six to nine branches seems to be the norm for many problems. Not too few, not too many! Figure 8.18 provides enough branches to cover a full landscape of win/lose options, but the set of calculations is not trivial either. We have "optimistic," "typical," and "pessimistic" scenarios, each with three probability branches, to essentially give us nine scenarios to examine. Nine scenarios can be computed by hand easily enough, though a spreadsheet can be really helpful, too. (Suppose, for instance we want to vary MARR; see, for instance, Table 8.6.)

Branch of Tree, j	T0 Cash Flow	T1 Cash Flow	T2 Cash Flow	PW(j) T0+T1+T2
1	-$5,000	$4,250	$5,000	$2,102.25
2	-$5,000	$4,250	$3,750	$1,219.55
3	-$5,000	$4,250	$2,250	$160.30
4	-$5,000	$2,500	$4,500	$278.58
5	-$5,000	$2,500	$4,250	$102.04
6	-$5,000	$2,500	$4,000	-$74.50
7	-$5,000	$1,750	$5,000	$1.41
8	-$5,000	$1,750	$3,750	-$881.29
9	-$5,000	$1,750	$2,250	-$1,940.54

Table 8.6: Impact of decision trees on decision-making: Part B.

Let's examine a few branches of the tree in Figure 8.18 to illustrate where the numbers in Table 8.6 come from. For the case $j = 1$, we have:

$$\begin{aligned} \texttt{NPV}_1 &= -\$5,000 + \$4,250(P/F, 19\%, 1) + \$5,000(P/F, 19\%, 2) \\ &= -\$5,000 + \$4,250(0.840) + \$5,000(0.706) \\ &= +\$2,102 \end{aligned}$$

Similarly, for $j = 6$,

$$\begin{aligned} \texttt{NPV}_1 &= -\$5,000 + \$2,500(P/F, 19\%, 1) + \$4,000(P/F, 19\%, 2) \\ &= -\$5,000 + \$2,500(0.840) + \$4,000(0.706) \\ &= -\$74.50 \end{aligned}$$

Clearly, the PV year 1 and 2 revenue streams create a scenario where the purchase of the trailer does not lead to a break-even condition. To be conclusive, however, we need to build out the weighted probability of the branches, again using Equation 8.7 (p. 237) and referring to Table 8.7:

$$E[X] = \sum_{n=1}^{N} x_i p(x_i) = \$331.72$$

Branch of Tree, j	PW(j) Amount	p(j) Revenue	A PW(j)*p(j)	B PW(j)2̂	C p(j)*PW(j)2̂
1	$2,102.25	0.15	$315.34	$4,419,466	$662,920
2	$1,219.55	0.15	$182.93	$1,487,294	$223,094
3	$160.30	0.20	$32.06	$25,696	$5,139
4	$278.58	0.03	$8.36	$77,608	$2,328
5	$102.04	0.24	$24.49	$10,412	$2,499
6	-$74.50	0.03	-$2.24	$5,550	$167
7	$1.41	0.06	$0.08	$2	$0
8	-$881.29	0.04	-$35.25	$776,679	$31,067
9	-$1,940.54	0.10	-$194.05	$3,765,699	$376,570
			$331	**$10,568,406**	**$1,303,784**

Table 8.7: Impact of decision trees on decision-making: part C.

One may wonder where the $p(j)$ values came from; these are simply the product of probabilities along any one branch.

The value of $E[X]$ is actually very small relative to the initial investment, and we can easily compute the probability that PV < 0 by summing the probability values associated with each of the negative PV values in Table 8.7:

$$p(X) = p(x_6) + p(x_8)p(x_9) = 0.030 + 0.040 + 0.100 = 0.17 = 17\%$$

From this, we see a 17% probability that PV < 0.

The standard deviation of E[X] is computed in two steps from Equations 8.8 and 8.9:

$$V[X] = \sum_{n=1}^{N} (x_i)^2 p(x_i) - (E[X])^2 = \$1,303,784 - (\$331.72)^2 = \$1,193,746$$

From V[X] we can find the standard deviation for the random discrete variable X quite simply from a square root calculation:

$$SD[X] = \sqrt{V[X]} = \sqrt{\$1,193,746} = \$1,092.58$$

The standard deviation of the expected value should be a red flag that this situation is very risky and should be viewed by Max as quite unfavorable. In this case, his friends are right!

Present worth problems involving branch probabilities can be a bit daunting to consider at the outset, but the use of graph paper or workbooks can make the process quite easy. Generally the calculations themselves are simple to execute, once we know what to do! As the example problem has demonstrated, there are three major steps in the solution process:

- **Step 1.** Organize the branches of the tree in a chart and ensure the cash flows are on the correct lines for each branch. A nine-branch, three-level problem has 27 entries to place correctly! Make life easy by computing the PW for each branch on this chart.
- **Step 2.** Organize the combined probabilities along branch lines, too, and compute the p(j) for each branch.
- **Step 3.** Despite the appearance of redundancy, format the final table with all the necessary data visible to compute expected revenue, variance, and standard deviation.

Table 8.8 illustrates the three-step process as has been implemented in an Excel worksheet and can be accessed in the supplementary material for this textbook.

8.5.4 Frequency distribution functions

Probability distributions are mathematical models in which the value of a random variable is related to the probability of occurrence. In the previous section we featured *discrete variables* – the random variables can only take on *specific* values – which had the probability mass function $p(x)$ and the cumulative distribution function $P(x)$. Now we consider the situation where the random variable is based on a *continuous scale*, and the *probability density function* $f(x)$ is related to the *cumulative distribution function* $F(x)$ through, for example, the function

$$Px \leq a = F(a) = \int_{-\infty}^{a} f(x)dx \qquad (8.13)$$

Equation 8.13 tells us that the probability that x is less than a specified value a is given by the value $F(a)$. Earlier, in Figure 8.5, p. 234, we illustrated the shape of the curve $F(x)$. Many engineers have heard of the "Gaussian" (or, *normal*) distribution represented by

$$f(x, \mu, \sigma^2) = \frac{1}{\sigma\sqrt{2\pi}} \exp -\frac{1}{2}(\frac{x-\mu}{\sigma}^2) \tag{8.14}$$

through the change of variable

$$z = \frac{x-\mu}{\sigma} \tag{8.15}$$

Then the integral in Equation 8.13 can be performed as a function of z, and Equation 8.13 reduces to

$$P(x \leq a) = P\left\{z \leq \frac{x-\mu}{\sigma}\right\} = \Phi\left(\frac{x-\mu}{\sigma}\right) \tag{8.16}$$

Here the *cumulative distribution function* Φ has values tabulated as a function of z in Appendix C. Some interesting features of the probability distribution function are illustrated in Figure 8.19.

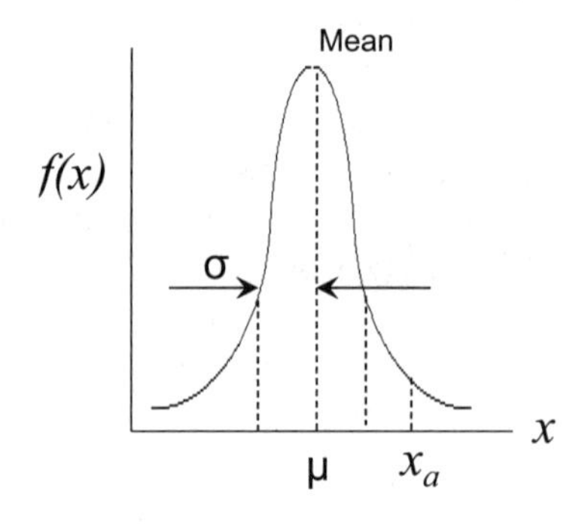

Figure 8.19: Key features of the probability density function.

We have one last comment on the shape of the probability distribution curves and understanding "drift" in performance as measured by the *mean* versus *standard deviation.* This was touched on briefly when discussing Figures 8.6 and 8.7. In Figure 8.20 we illustrate how the curves can reflect differences in system performance.

Normalized probability distributions

We saw in the previous section how the use of the normalization parameter z defined by Equation 8.15 assisted in performing the integral of Equation 8.13 to produce the general form of the cumulative probability distribution function, Equation 8.16. Φ has values tabulated as a function of z in Appendix C.

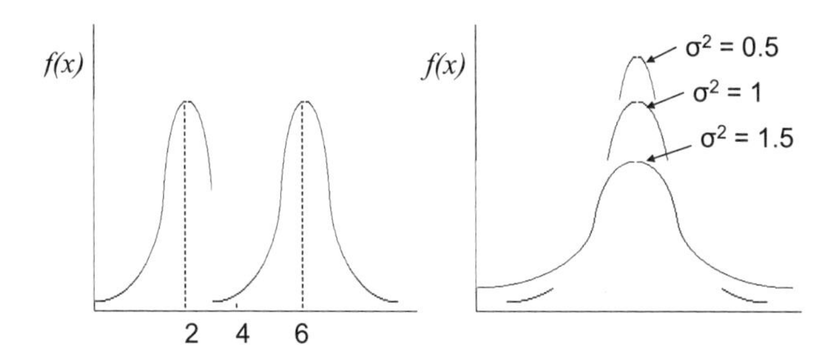

Figure 8.20: Difference between mean and standard deviation shifts.

As with most things, several interpretations and usages of z are possible, so the best way to illustrate essential ideas may be to work some examples. Some individuals have gotten quite creative about how to best understand probability density functions, and a quick visit to sites like `www.mathsisfun.com/data/standard-normal-distribution-table.html` can be as informative as they are fun!

Before working through examples, it helps to have an understanding of the tables in Appendix C. There are two sets of tables. Tables C.1 and C.2 assume that the mean of z = 0 and the z-value in the chart are the values *from the mean.* Tables C.1 provides values of $z < 0$, and Table C.2 provides values for $z > 0$. The second set is for the *cumulative* value of the probability distribution function. Table C.3 provides values of $z < 0$, and Table C.4 provides values for $z > 0$. The only difference between the tables is that the values in Tables C.1 and C.2 are the difference between the mean of 0.5 and the values in Tables C.3 and C.4. For instance,

$$\begin{aligned} \texttt{Table C.2: } P(z = 1.25) &= 0.3944 \\ \texttt{Table C.4: } P(z = 1.25) &= 0.8944 - 0.5000 = 0.3944 \end{aligned}$$

What about negative values of z? From Table 8.10, we see that P(z=-1.82) = 0.0344.

Example A: Events within a specified range

Assume that the length of time required to replace a computer kiosk on the factory floor is "normally distributed." Over the past few months, the mean time required to replace the kiosk is 60 minutes with a standard deviation of about 8 minutes. For any given service technician, what is the probability the technician will take between 60 and 70 minutes to replace a kiosk? The first step is to compute the normalized z function

$$z = \frac{x - \mu}{\sigma} = \frac{70 - 60}{8} = 1.25$$

Thus, $0 < z < 1.25$. Looking up the corresponding values of P(z) from Figure 8.9

$$P(0 < z < 1.25) = 0.8955 - 0.5000 = 0.3944$$

So there is a 40% probability the technician will take between 60 and 70 minutes to replace the kiosk.

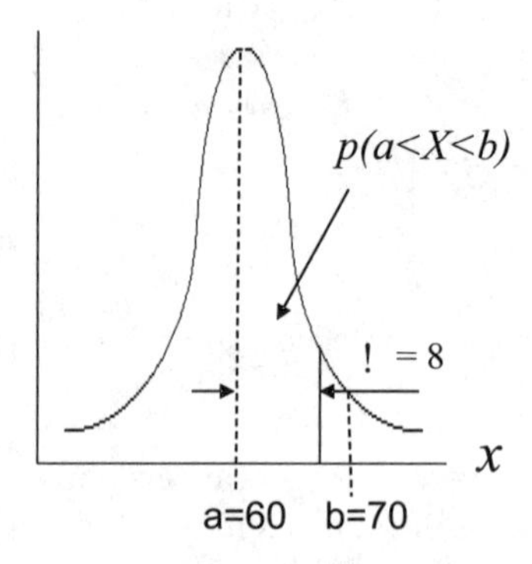

Figure 8.21: Probability distribution function.

Example B: Probability of NPV < 0

A public utility project has an NPV of $2,052 and a standard deviation of $1,140. What is the probability NPV could be 0 or less?

We first compute the normalized z function:

$$z = \frac{x - \mu}{\sigma} = \frac{0 - \$2,052}{\$1,140} = -1.8$$

The NPV of 0 lies 1.8 standard deviations to the left of the mean. Looking up the corresponding values of P(z) from Appendix C,

$$P(z < -1.8) = 0.0359$$

Thus, there is a 3.6% probability the project will lose money.

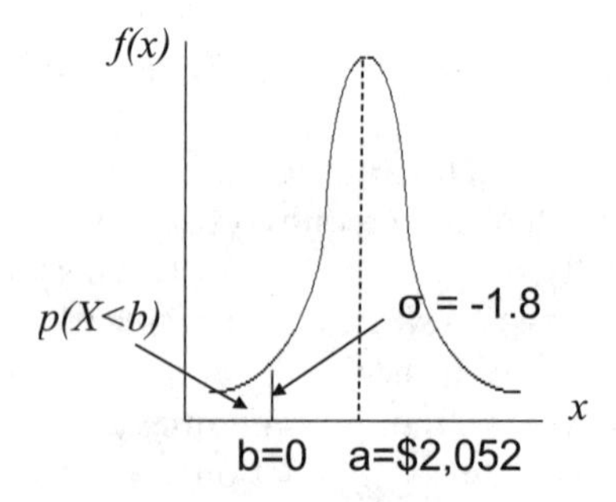

Figure 8.22: Probability for σ = -1.8.

Example C: Audit variance changes

An audit of the energy project has revealed the NPV was overstated, and a new estimate shows an NPV of \$1,234 with a standard deviation of \$1,700. What is the probability NPV could be 0 or less?

As before, we first compute the normalized z function:

$$z = \frac{x - \mu}{\sigma} = \frac{0 - \$2,052}{\$1,140} = -0.726$$

The NPV of 0 lies 0.726 standard deviations to the *left* of the mean. Looking up the corresponding values of P(z) from Appendix C,

$$P(z < -0.726) = 0.23$$

Thus, there is a 23% probability the project will lose money.

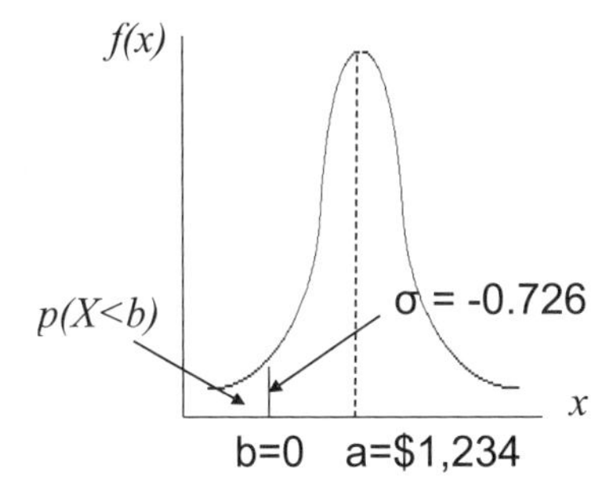

Figure 8.23: Probability for σ = -0.726.

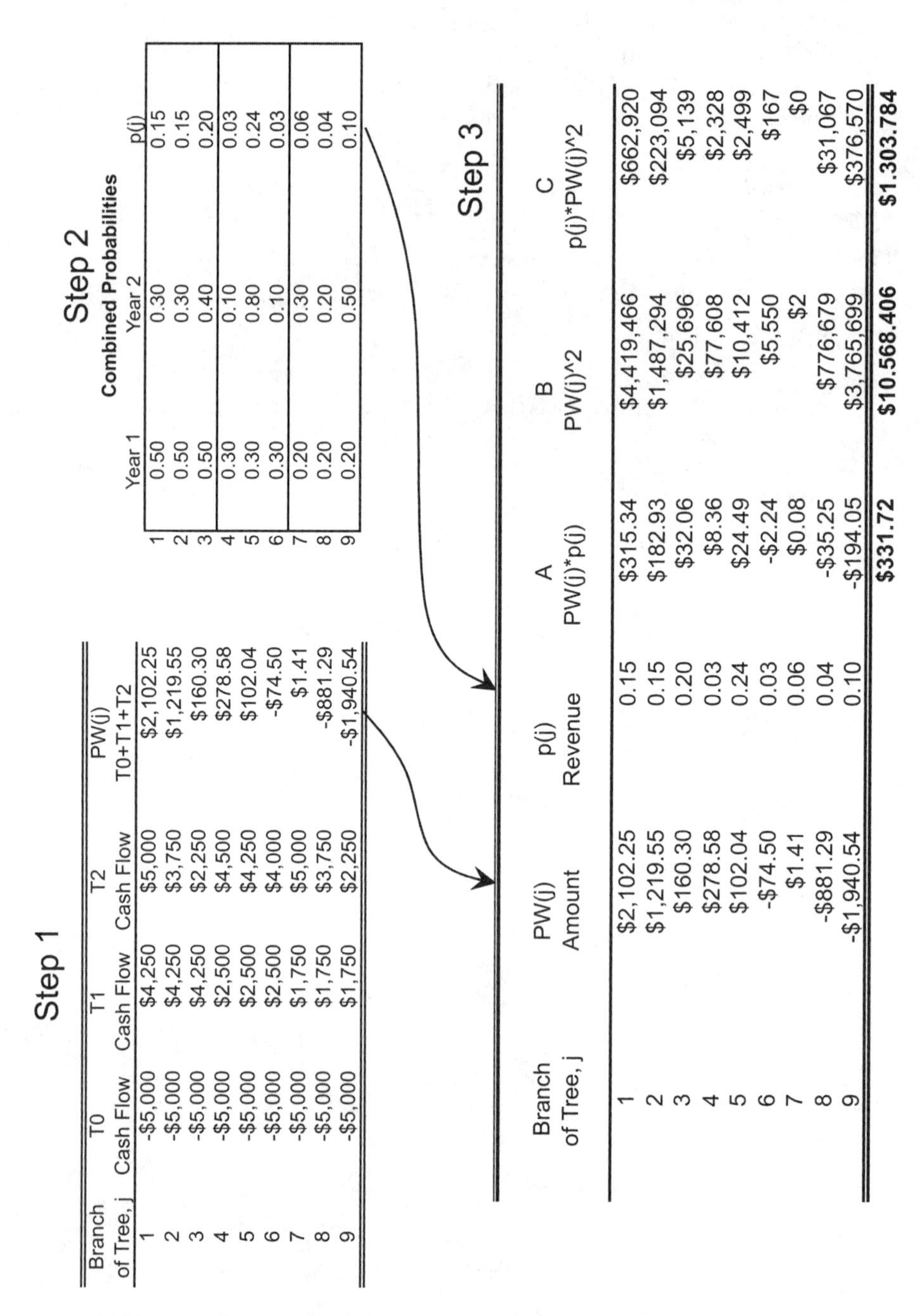

Step 1

Branch of Tree, j	T0 Cash Flow	T1 Cash Flow	T2 Cash Flow	PW(j) T0+T1+T2
1	-$5,000	$4,250	$5,000	$2,102.25
2	-$5,000	$4,250	$3,750	$1,219.55
3	-$5,000	$4,250	$2,250	$160.30
4	-$5,000	$2,500	$4,500	$278.58
5	-$5,000	$2,500	$4,250	$102.04
6	-$5,000	$2,500	$4,000	-$74.50
7	-$5,000	$1,750	$5,000	$1.41
8	-$5,000	$1,750	$3,750	-$881.29
9	-$5,000	$1,750	$2,250	-$1,940.54

Step 2

Combined Probabilities

	Year 1	Year 2	p(j)
1	0.50	0.30	0.15
2	0.50	0.30	0.15
3	0.50	0.40	0.20
4	0.30	0.10	0.03
5	0.30	0.80	0.24
6	0.30	0.10	0.03
7	0.20	0.30	0.06
8	0.20	0.20	0.04
9	0.20	0.50	0.10

Step 3

Branch of Tree, j	PW(j) Amount	p(j) Revenue	A PW(j)*p(j)	B PW(j)^2	C p(j)*PW(j)^2
1	$2,102.25	0.15	$315.34	$4,419,466	$662,920
2	$1,219.55	0.15	$182.93	$1,487,294	$223,094
3	$160.30	0.20	$32.06	$25,696	$5,139
4	$278.58	0.03	$8.36	$77,608	$2,328
5	$102.04	0.24	$24.49	$10,412	$2,499
6	-$74.50	0.03	-$2.24	$5,550	$167
7	$1.41	0.06	$0.08	$2	$0
8	-$881.29	0.04	-$35.25	$776,679	$31,067
9	-$1,940.54	0.10	-$194.05	$3,765,699	$376,570
			$331.72	**$10.568.406**	**$1.303.784**

Table 8.8: Overview of the three steps of branch tree calculations.

z	0.00	0.01	0.02	0.03	0.04	0.05	0.06
0.0	0.5000	0.5040	0.5080	0.5120	0.5160	0.5199	0.5239
0.1	0.5398	0.5438	0.5478	0.5517	0.5557	0.5596	0.5636
0.2	0.5793	0.5832	0.5871	0.5910	0.5948	0.5987	0.6026
0.3	0.6179	0.6217	0.6255	0.6293	0.6331	0.6368	0.6406
0.4	0.6554	0.6591	0.6628	0.6664	0.6700	0.6736	0.6772
0.5	0.6915	0.6950	0.6985	0.7019	0.7054	0.7088	0.7123
0.6	0.7257	0.7291	0.7324	0.7357	0.7389	0.7422	0.7454
0.7	0.7580	0.7611	0.7642	0.7673	0.7704	0.7734	0.7764
0.8	0.7881	0.7910	0.7939	0.7967	0.7995	0.8023	0.8051
0.9	0.8159	0.8186	0.8212	0.8238	0.8264	0.8289	0.8315
1.0	0.8413	0.8438	0.8461	0.8485	0.8508	0.8531	0.8554
1.1	0.8643	0.8665	0.8686	0.8708	0.8729	0.8749	0.8770
1.2	0.8849	0.8869	0.8888	0.8907	0.8925	0.8944	0.8962
1.3	0.9032	0.9049	0.9066	0.9082	0.9099	0.9115	0.9131
1.4	0.9192	0.9207	0.9222	0.9236	0.9251	0.9265	0.9279
1.5	0.9332	0.9345	0.9357	0.9370	0.9382	0.9394	0.9406
1.6	0.9452	0.9463	0.9474	0.9484	0.9495	0.9505	0.9515
1.7	0.9554	0.9564	0.9573	0.9582	0.9591	0.9599	0.9608
1.8	0.9641	0.9649	0.9656	0.9664	0.9671	0.9678	0.9686
1.9	0.9713	0.9719	0.9726	0.9732	0.9738	0.9744	0.9750
2.0	0.9772	0.9778	0.9783	0.9788	0.9793	0.9798	0.9803

Table 8.9: Probability distribution values from Appendix C, Table C.4.

z	**0.00**	**0.01**	**0.02**	**0.03**
-3.4	0.0003	0.0003	0.0003	0.0003
-3.3	0.0005	0.0005	0.0005	0.0004
-3.2	0.0007	0.0007	0.0006	0.0006
-3.1	0.0010	0.0009	0.0009	0.0009
-3.0	0.0013	0.0013	0.0013	0.0012
-2.9	0.0019	0.0018	0.0018	0.0017
-2.8	0.0026	0.0025	0.0024	0.0023
-2.7	0.0035	0.0034	0.0033	0.0032
-2.6	0.0047	0.0045	0.0044	0.0043
-2.5	0.0062	0.0060	0.0059	0.0057
-2.4	0.0082	0.0080	0.0078	0.0075
-2.3	0.0107	0.0104	0.0102	0.0099
-2.2	0.0139	0.0136	0.0132	0.0129
-2.1	0.0179	0.0174	0.0170	0.0166
-2.0	0.0228	0.0222	0.0217	0.0212
-1.9	0.0287	0.0281	0.0274	0.0268
-1.8	0.0359	0.0351	0.0344	0.0336
-1.7	0.0446	0.0436	0.0427	0.0418
-1.6	0.0548	0.0537	0.0526	0.0516
-1.5	0.0668	0.0655	0.0643	0.0630

Table 8.10: Probability distribution values from Appendix C, Table C.3.

8.6 Summary

Quite a few topics on uncertainty, probability, and risk have been covered in this chapter. We have highlighted some of the most critical items about which the engineering manager should have at least some cursory working knowledge. All of this is to emphasize and dispel the notion that many of the calculations in the prior chapters are as precise as we may want them to be! Most real problems involve at least a few estimates (and a few guesses), and it is not a situation to be timid about. Rather, we should feel good that we have the comfort level with a variety of tools that can take vague or ambiguous situations and shape concrete options and recommendations.

It would be wrong to not recognize and recommend the reader that seek out other references to supplement the current chapter. There are *soooo* many books, websites, and media on statistics, probability, and risk that it is somewhat overwhelming. The fact that we've condensed so many topics into several dozen pages should underscore the role of this book as a "field guide," not as a definitive document.

A few things should be remembered:

- *Sensitivity analyses* should be done much more frequently than seems to occur in practice. Unfortunately, sensitivity is viewed as a hassle to some analysts, but it is a very powerful tool to understand the *robustness* of a financial model.
- *Indifference points* are very approximate and often involve a lot of simplifying assumptions, but they can help *frame* whether the project makes sense to the corporation on the most fundamental level. If the corporate mantra is that "every project has to have a 24-month payback period," you can save a lot of energy if this basic criteria is not met.
- *Probabilistic methods* are no substitute for knowing the real data, but if you had the real data, you'd already know the risk you had taken! The discrete and continuous tool-sets are wonderful ways to characterize uncertainty and provide tractable pathways of analysis to help the engineering manager predict the future.
- *Decision trees* give us the chance to explore specific scenarios, including the *option to explore how future investments might salvage a project* deemed unworthy today. Options management is what you do in reality: if you are right half the time, then you'll always review all the capital investments annually anyway, and decision trees let you mix decisions and probability.

Recall that each chapter is designed to be covered in *one* or maybe *two* 3-hour class sessions in a condensed course of study. If the chapter provided a basic understanding of the four points above and helped you appreciate that probability statistics can be an asset to financial analysis, then that is a success in our book!

8.7 Problems to work

Problem 8.1
Provide a description of "risk" and the idea of a "risk-taker" versus someone who "manages risk." What relationship exists between *risk* and *return* (if any)?

Problem 8.2
Frequently people will claim that they are "risk-averse." But it seems that many of these same people like to take a holiday in Las Vegas, and many even play the Lotto regularly. It seems like this behavior is irrational. What basis exists for being risk-averse, but still engaging in gambling?

Problem 8.3
Some people claim that putting money in the stock market is a gamble and that the return on investment can be modeled as a random variable. Is the stock market all that random? How direct is the correlation between stock price and company performance?

Problem 8.4
It seems that the idea of "putting money at risk" would apply to *both* the concepts of investing and gambling. Why is there such a difference in the public perception of the value of the two ideas? After all, many people manage their retirement accounts by investing, not buying secure bonds.

Problem 8.5
In July 2013, heavy rains in Cleveland caused a retaining wall inside the Terminal Tower for the Regional Transit Authority (RTA) train to collapse. It was estimated that between repairs and lost ticket sales, the occurrence of this highly unlikely event was going to cost the city about $350,000. It had been about 15 years since a similar accident occurred, and the statistics jocks estimate that something like this should only occur once every 25 years. The city has proposed rebuilding the wall with an additional special storm sewer line that would prevent the retaining wall from *ever* collapsing again. Of course, the problem is that the storm sewer upgrade would cost an additional $35,000, and would also require annual maintenance of about $1,000 per year. Given that the storm sewer upgrade could reasonably be expected to have a life of 15 years before it has to be replaced, should the storm sewer upgrade be implemented? Assume the repairs are funded with bond money paying 5%.

Problem 8.6
For the probability distribution shown below, calculate the expected return and standard deviation.

Return	Probability
6%	0.05
8%	0.20
10%	0.50
12%	0.20
14%	0.05

Problem 8.7
During Hurricane Sandy in October 2012, there was structural damage done to the Ninth Street Marina (NSM) and this is the third time since 1992 this has happened. Gen-

erally the contract engineers for the insurance company claim that the design safety margin is so large that, "Don't worry, the pier is built like a rock." The city disagrees and has decided to help tourism by totally rebuilding the marina to "really" keep the damage from occurring. The local port authority has proposed the probability of Sandy (in some form of hurricane level) coming back as shown below, along with the investment required to build a pier to withstand the hurricane level cited.

Hurricane Category	Probability of Occurrence	Preventative Investment
4	0.025	$40,000,000
3	0.050	$30,000,000
2	0.075	$20,000,000
1	0.100	$10,000,000

As in the case of the Terminal Tower incident of Problem 8.5, there is bond money to finance the project earning 5% annually. If the average damage to the pier during a hurricane is $172,000, which hurricane protection level should the city advocate?

Problem 8.8

Weekly profit on pennant flags at the baseball stadium is simply the unit profit, F, multiplied by the number of units sold, N. The product line is not really doing that well, and the concessions manager would like to discontinue the pennant flags and sell more candy and diet soda. Create a probability distribution plot for the pennant flag data in the table below.

Unit Profit		Weekly Sales	
Value	Prob	Value	Prob
$0	0.25	100	0.5
$10	0.25	500	0.4
$20	0.45	999	0.1
$30	0.05		

1. Compute the anticipated mean and standard deviation for the weekly profit.
2. The concessions manager thinks that sales might increase if the pennant flags had a "Made in Cleveland" tag attached, though that would increase the price by 10% for each unit. Is this still profitable?
3. The local newspaper has caught wind of the idea that the pennants might be discontinued, generating some bad press that the pennants have been part of the ball club for 60 years, and fans should boycott the concession stands. What would you do?

Problem 8.9

The local ice hockey club would like to retrofit all the Zamboni machines with digital displays that fans can Tweet and text messages to in support of various popularity polls. It will take about $68,000 to get the LED displays custom built and mounted with all the wireless support, but marketing is confident they can net $24,000 in sponsorship revenue the first year (12 games at $2,000 per major sponsor), and they think they could build revenue 9% year over year. A problem, though, is that due to cold temperatures, a wet environment, and a generally harsh environment, student intern announcers have guestimated the probability the Zamboni displays will last as follows:

Useful Life (Seasons)	Probability
1	0.05
2	0.05
3	0.10
4	0.25
5	0.50

Does this project have any hope of being economically acceptable? Assume that MARR = 12%.

Problem 8.10
An electronic health records (EHR) system costs $100,000 to implement. Relative to other physician office practices, there are projected cost saving in the first year of operation in addition to improved workflow proficiency. Is estimated that approximately 5% cost improvements will occur year-over-year. Although the installation of the system is becoming the industry standard in patient record-keeping, the managers of the practice want to know if this will have been "worth the effort five years from now." Assume a hurdle rate of 15% per year.

Performance Results	Probability	Cost Savings in First Year
Optimistic	0.30	$60,000
Most Likely	0.55	$40,000
Pessimistic	0.15	$18,000

Problem 8.11
The film thickness for a conformal coating has a mean value of 40 μ and a standard deviation of $\sigma = 2$. What is the probability that a particular point on a sample will have a thickness of at least 35 μ?

Problem 8.12
The market value of a Venturi pump averages $195.50. There are a large number of suppliers of the pump, and prices vary only slightly from the mean, but are still normally distributed. If the variance in market price is 10 (units = $\$^2$), what is the probability the market price at any point is at least $190.00?

Problem 8.13
In a high-speed coating system, the mass flowrate of polyester powder has an average value of 120 gm/sec. There is a 75% probability the flowrate is between 110 gm/sec. and 130 gm/sec. What is the variance of the power flowrate? Assume the flowrate is normally distributed.

Problem 8.14
A project has an NPV of $3,444 and a standard deviation of $1,780. What is the probability that NPV could be 0 or less?

Problem 8.15
In Problem 8.14 it turns out that the project manager overestimated the project NPV, and the best estimate is that now NPV = $2,444 with a standard deviation of $1,500. What is the probability that NPV could be 0 or less?

References

[1] Michael Mauboussin. *More than you know: Finding financial wisdom in unconventional places.* Columbia University Press, revised and expanded edition, 2008.

[2] Samuel M. Selby. *Standard mathematical tables.* CRC Press, 1975.

[3] Kamlesh Mathur and Daniel Solow. *Management science: The art of decision making.* Prentice Hall, 1994.

[4] William R. Lasher. *Practical financial management.* South-Western CENGAGE Learning, 2011.

[5] Eugene F. Brigham and Michael C. Ehrhardt. *Financial management.* South-Western CENGAGE Learning, 2008.

[6] W.G. Sullivan, E.M. Wicks, and C.P. Koelling. *Engineering economy.* Prentice Hall, 15th edition, 2011.

[7] Leland T. Blank and Anthony J. Tarquin. *Engineering economy.* McGraw Hill, 1989.

[8] Donald G. Newnan, Jerome P. Lavelle, and Ted G. Eschenback. *Engineering economic analysis.* Oxford, 2009.

[9] John V. Sullivan. *How our laws are made.* United States Congress, 2007. `http://thomas.loc.gov/home/lawsmade.toc.html`.

Chapter 9

Capital Budgeting and Replacement Analysis

9.1 Learning objectives

Capital financing and *capital allocation among projects* are the two fundamental facets of capital budgeting an engineering manager must understand. Managing the sources and uses of capital funds is a highly strategic activity by corporate-level decision-makers; the ability of an engineering manager to make meaningful contributions to the capital budgeting process adds value to the corporation (you differentiate yourself in the job market when these skills have been proven). Companies are willing to seek outside sources of capital to fund those projects under the assumption the opportunity cost fits strategic goals and objectives. The ratio of debt to equity for project financing and other capital financing concepts discussed earlier are now expanded and applied. The second half of this chapter is centered on *replacement analysis*, a special form of decision-making that is aligned with capital budgeting but also draws on the project selection techniques developed in Chapter 5. After reading this chapter you should be able to

1. Describe the capital financing and allocation functions.
2. Explain the five basic steps in typical capital budgeting and in what ways the process is interdisciplinary.
3. Distinguish between debt and equity capital as well as between WACC and the MARR and the impact of taxes.
4. Understand the weighted criteria and ROI approaches to project investment selection.
5. Work through conventional replacement problems and recite the three unique aspects of replacement problems relative to other project selection methods.

9.2 Capital budgeting

Investments in long-term assets often involve sizable outlays of funds that can command the attention of company leadership because of the strategic nature of the decisions. Entirely new courses of action can result from investment decisions; a careful analysis of each situation is essential to minimize decision-making risk. The process of *capital budgeting* involves the identification, evaluation, and selection of long-term asset investments that decision-makers believe will create or maximize value to the company. Unlike many of the topics covered in previous chapters, the process of capital budgeting is highly interdisciplinary, and within this chapter we attend to many non-monetary issues that legitimately are a part of the capital budgeting process. Sometimes people initially feel capital budgeting is more of an "art" than a "science" but this extends from thinking about budgeting as exclusively based on financial calculations and not being fully aware of how long-term asset decisions require an understanding of (and the context for) corporate strategy, too.

9.2.1 Focus on long-term strategic investments

Managers spend a lot of time allocating, juggling, and finding the resources needed for the development and production of goods and services. Earlier, in Section 1.3, we noted that management activities can be delineated in three ways:

1. **Operating** activities related to transactions impacting net income.
2. **Investing** activities related to the purchase and sale of long-term assets.
3. **Financing** activities related to issuing and repayment of debt and equity.

These activities are themselves multi-faceted; different concerns exist at different management levels (president, vice president, division director, director, engineering manager, engineer, etc.):

1. **Operational** activities are centered around implementing and executing policies and procedures on the "front lines" of management. Operational projects tend to be "stand-alone" in nature.
2. **Tactical** activities originate with strategic decisions. Tactical-level decisions are used to prescribe or frame operational activities. Projects in this domain are loosely coupled to other projects.
3. **Strategic** decision-making cascades from the top management of an organization, setting company-wide goals and objectives. Projects here might be much broader in scope and seek to integrate functions across organizational departments of supply chain elements.

Although these lists encompass a wide variety of activities, it is convenient to realize that most decisions are associated with **either** *day-to-day operating expenses* **or** *long-term strategic project expenses.* The latter (strategic expenses) are also known as *capital* expenses, and *capital budgeting* is a decision-making process by which projects are evaluated for suitability for funding and alignment with corporate strategy. Consider the example of a corporate-level vice-president (VP) *strategic decision* to enter a new industrial electronics market; the opportunity is financially quite attractive, but to gain this business will require the company to undergo supplier certification though the implementation of a new ISO-compatible production system. Under the direction of a VP, the strategic business unit (SBU) manager may ask engineering managers to identify new process changes or specific equipment upgrade scenarios that could meet OEM specifications. In the process of envisioning *tactical and operational* changes, the engineering manager might be asked to present and defend capital expense options to upper-level managers. It is often a good sign and a compliment to the engineering manager to be included in such decision-making activities, as it is rewarding to have the chance to see things at a strategic level. You might even have the chance to influence the course of the company! The engineering manager earns a place at the table with well-prepared financial interpretation of technical matters, enjoying unique insight into company affairs at the same time. Career growth often follows demonstration of strategic competencies in that setting.

9.2.2 Typical capital budget scenarios

Three general categories of capital budget scenarios are the source of many of the capital budgeting activities the engineering manager might encounter and undertake:

1. Replacement (or renewal) of equipment that is wearing out or is obsolete.
2. Expansion to handle infrastructure or sales growth.
3. Investment in a "NewCo" to grow into new product or service areas.

Replacement. In Section 5.2 we introduced a framework for analysis for comparing alternatives. We found that four basic decision-making *contexts* (see p. 136) would influence analysis technique; we will not repeat all the details here, except we will expand on presentation of the theory behind *replacement analysis*. Our discussion on replacement analysis draws heavily on sample problems since the organization of data is more challenging than the enabling theory (already covered previously). Relative to other investments a corporation might make, replacement analysis involves minimal capital risk since a company is essentially doing what it has always done and assumes satisfaction with the status quo. Section 9.3 explores replacement problems in more detail. The closely related scenario of *renewal* is sometimes undertaken.
Expansion. When the decision involves *expansion of production*, this typically means doing the same things but on a larger scale. There is, of course, more risk relative to a replacement problem since we are doing something similar but *not exactly* as in the past. As expected, such uncertainty carries greater risk, and the way to handle the uncertainty in an expansion project is to draw upon the probabilistic methods presented in Section 8.5.3; specifically, consider the worked problem on the investment in new equipment on page 250. Some of the risk inherent in an expansion project is that the forecast revenue may be just a guess and it is entirely probable no benefit will materials at all.

New venture. Investment in a NewCo (short for "New Company") brings with it probabilistic *and* decision risk, as outlined in Section 8.5.3. In some cases the company may be venturing into entirely new business lines, and it can help to visualize or estimate overall venture risk (in comparison to other alternatives) with so-called rules of thumb as shown in Figure 9.1. In a more rigorous assessment of risk, a NewCo investment analysis is based on anticipated revenue and expenses, as discussed in Appendix B, p.357, leading to, say, the calculation of MARR (Section 4.4).

As we progress from replacement to expansion and then to new venture analysis, the level of corporate exposure (risk) increases along with the degree of complexity of assessment, the number of assumptions that must be made, and the uncertainty of the investment (or market) opportunity. Whether an internal investment (i.e., equipment replacement within the company) is perceived to have more or less certainty than an investment outside the company (business line diversification through a NewCo) is often a matter of opinion. In either event we have the basic tools to proceed with the capital budgeting process. In the next section we look at five key steps in capital budgeting.

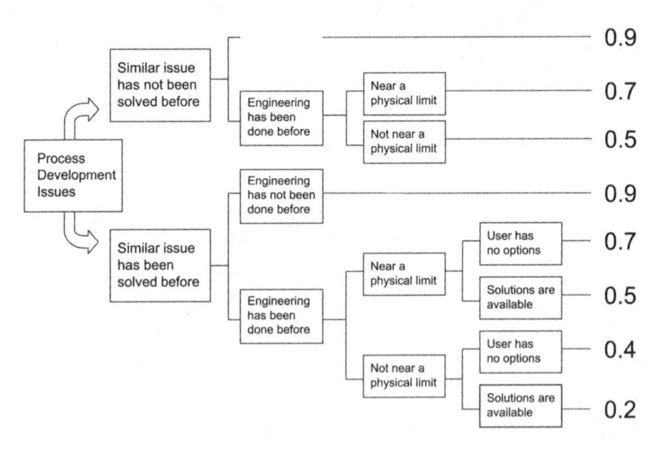

Figure 9.1: Simple method to estimate the risk threshold for new venture activity

9.2.3 Key steps in the capital budgeting process

There seem to be as many prescriptive techniques for capital budgeting as there are textbooks, but a fairly common five-step process can be outlined. The process has many of the attributes suggested in Appendix B. **Step 1 - Generate investment proposals**

The strategic and operational impact of the capital budgeting process often reaches across many disciplines and departments within a company and frequently is timed to support the annual corporate strategy and budgeting process. Often, there is a call for investment proposals from many levels within the organization, to be submitted to the corporate investment team (or committee of similar charter). To harmonize input and simplify subsequent committee effort it is not unusual to require the investment proposals to be in a standard format with relevant financial assessment completed and business purpose narrative included.

Step 2 - Review and analysis of proposals

The investment committee will spend several weeks reviewing proposals. Often there is an *administrative review* to check if the proposal meets the basic financial metrics established in the call for proposals or even if the proposed investment qualifies as a long-term investment. For instance, the investment in a major marketing plan might require capital funds to launch, but may not meet the requirement of a long-term asset suitable for the balance sheet. The committee may ask clarifying questions of the proposal teams, explore tactical and strategic impact, conduct due diligence, and convert the financials into a standard form or integrate them into a master budget decision-making workbook. The committee will prepare a summary report that is submitted to upper-level management for consideration. We might think of this step as involving the *screening* of proposals for suitability in meeting the minimum standards of investability.

Step 3 - Evaluation and investment recommendations for proposals

While there is a lot of background work and financial analysis activity during Step 2, the evaluation and proposal investment recommendation is a much-anticipated crescendo of the capital budgeting process. Much validation of corporate strategy ("Strategy is what you do, not what you say") unfolds from the very visible decisions about capital investments. This is particularly true since many decisions are made at the corporate board level, given that the size of the investments requires that level of fiduciary accountability and authorization for commitment of funds (above a prescribed threshold, of course). Another way to look at such decisions is that the board had the chance to select from several acceptable proposals and thus made corporate *preference decisions.* Knowing what was *not* funded – but what *did* pass an administrative screening test – will indicate a lot about corporate strategy.

Step 4 - Authorization of recommended investments

Once the board or investment committee has completed proposal evaluation and discussed recommendations for investments, there is typically a vote to approve certain projects and authorization to company managers to proceed with a project. Given the over-arching strategic perspective of board-level decisions it is not unusual for the proposal evaluation process to result in a request for a modified proposal. Depending on the significance and extent of the modifications, there may need to be follow-up to the president or CEO, who will be empowered to act on behalf of the board to bring closure to this step in the process.

Step 5 - Follow-through and follow-up

Project managers are fond of the phrase "Plan the work - work the plan," and this is so very true for small and large projects alike! For very large projects it is common for the project to be funded in phases or tranches based on the ability of the implementation team to meet certain performance milestones. The tracking of project progress is frequently a recurring agenda item at monthly and quarterly operations reviews in which "actuals" are discussed from both a financial and strategic perspective. For large projects the benefits trail the investment by months (if not years), and this is where the original metrics of performance and methods for collecting relevant data are critical to success in ensuring that the original investment objectives have been met.

While these five basic steps are common to many organizations (profit or non-profit) some steps are more visible than others and have different activity time-scales. For instance, Step 1 might require 1-2 months of pre-board-meeting effort by engineering teams in the development of a proposal. The board-level process of evaluating the reports from the investment committee and subsequent authorization of funds could even occur over the course of just a few days.

The more extensive diligence of tracking metrics and benefits related to Step 5 may receive less attention (or might be largely ignored), though for those companies with integrated enterprise resource systems the tracking of metrics through performance dashboards or other IT tools simplifies and makes more accessible data for managerial review (thus improving monitoring and compliance with plans). This chapter tends to focus more on analysis (Step 2) and evaluation (Step 3) of proposals as these steps are closely linked with engineering manager decision-making interests and activities. Here, we consider "analysis" to mean that set of activities required to understand, infer, or establish the meaning or purpose of the capital budget proposal. *Analysis* renders an interpretation of the capital budget item in a form suitable for management consideration. In comparison, the *evaluation* of a proposal involves a judgment about the quality of the proposal that leads to a decision about whether to move forward and authorize funds for the project or to decline any further action.

Little has been mentioned about the nature of capital budgeting in the international context, but it would be incorrect to trivialize the differences as merely accounting for exchange rates. To be sure, *currency fluctuations* can impact a project, but this variance can be mitigated by incorporating offshore and conducting all affairs in the local currency. This way – at least over the long haul – the performance of the asset will self-adjust and be de-coupled from exchange rate risk. Another potentially more significant risk is *political.* If, for instance, the company invests in assets that are part of a joint venture, then at least some operational control or leverage is provided to the company; you are not completely insulated from political issues (e.g., property seizure), but if you have the right joint venture partner, then at least you are in a stronger negotiating position. As well, the company can act like a bank and loan the subsidiary the funds to purchase the asset, and in this way hold *debt*, not *equity.* Overall, debt tends to have more universal support as allowable transfer payments (from the subsidiary back to headquarters) instead of equity (cash) payments that are more easily blocked or held by the foreign country.

It is worthwhile to quote the ABET [1] definitions for assessment and evaluation since they illustrate the widespread utility of quality improvement processes, whether we are talking about capital budgeting or educational programs:

> **Assessment**: Assessment is one or more processes that identify, collect, and prepare data used to evaluate the attainment of student outcomes. Effective assessment uses relevant direct, indirect, quantitative and qualitative measures as appropriate to the objective or outcome being measured. Appropriate sampling methods may be used as part of an assessment process.
>
> **Evaluation**: Evaluation is one or more processes for interpreting the data and evidence accumulated through assessment practices. Evaluation determines the extent to which student outcomes are being attained. Evaluation results in decisions and actions regarding program improvement.

The steps in the capital budgeting process are fairly general; what makes the conversation lively are details of analysis that vary depending on whether the scenario of interest is production expansion, replacement equipment, rent, renewal, or a new venture. Within these scenarios are issues and concerns that test the reader's judgment, which makes things fun!

In the second half of this chapter we focus exclusively on *replacement problems* due to many special considerations that differentiate analysis; Table 9.1 illustrates where other problem types are treated.

Scenario	Asset Nature	Analysis data	
Expansion	Both assets are new	Data often available	Sect. 5.2
Replacement	One new / one old	Old asset data can be speculative	Sect. 9.3
Rent/Lease	Avoid asset ownership	"Lease versus buy" work-up	Sect. 5.8
Renewal	Refurbish old asset	Required data highly speculative	Sect. 9.5.3
New Venture	New asset / new market	Market growth data is speculative	Appendix B

Table 9.1: Characterization of key capital investment decisions.

9.2.4 Weighted average cost of capital, (WACC)

Investment projects are frequently funded through some combination of equity (retained earnings, stock) and debt (bonds, credit lines). Although government policy can affect MARR through the cost of capital charged to banks, it is safe to say that federal monetary policy is generally of secondary concern to most engineering managers. If all the capital needed by a company were derived from stockholders, then the required *rate of return on equity* would dominate the MARR decision, but that is not typical. However, with capital markets still quite conservative post-2008, capital projects with an investment as small as $P = \$1$ million might be financed from a combination of debt, equity, and economic development tax incentives. The *weighted average cost of capital* (WACC) is obtained by "weighting" the various components of capital. In the most basic form, we could calculate a weighted average as simply:

	Fraction	Risk	
Debt	0.60	8%	4.8%
Equity	0.40	15%	6.2%
		WACC =	11.0%

Weighting the cost of capital is a concept that simply says that available funding comes from different sources, and we wish to establish a single "blended" rate to simplify calculations in support of managerial decision-making. As an example, suppose we estimate roughly a 12% IRR on a project and then discover that our credit rating is so poor that banks are demanding 17% interest; we might eagerly re-think that this "color of money" is simply not for us! As a general rule we only want to invest in those projects whose rate of return is higher than the cost of capital available to us.

A more rigorous approach would include the impact of taxes, producing (Marshall et al. [2]):

$$\texttt{WACC} = W_e\ r_e + W_d\ r_d(1 - T) \tag{9.1}$$

Where

$$\begin{aligned} W_e &= \text{fraction of project funding through equity} \\ W_d &= \text{fraction of project funding through debt} \\ T &= \text{combined federal and state tax rate} \\ r_e &= \text{prevailing equity returns to stock investments} \\ r_d &= \text{before-tax cost of debt} \end{aligned}$$

If we use the data from the table above and assume a tax rate of 40%, then

$$\texttt{WACC} = 0.40(.15) + 0.60(0.08)(1 - 0.4) = 0.1095 = 9.08\%$$

This example clearly illustrates the favorable impact of including tax in `WACC` calculations; taxes reduce the effective level of debt involved in the project since the expense interest paid on debt is a deductible expense. So, some projects that might initially have a low `NPV` and thus be viewed as unfavorable due to an elevated pre-tax WACC, might subsequently be considered acceptable if tax-adjusted rates were in play.

In some cases there may be a significant difference between *preferred* and *common* stock – particularly if a start-up company is going through several rounds of equity financing – and the `WACC` equation can be modified (again, Marshall et al. [2] or Brigham and Ehrhardt [3]):

$$\texttt{WACC} = W_{pe}\ r_{pe} + W_{ce}\ r_{ce} + W_d\ r_d(1 - T) \tag{9.2}$$

Where

$$\begin{aligned} pe &= \text{preferred equity} \\ ce &= \text{common equity} \end{aligned}$$

Some care should be taken with some types of refinements since they may imply work at a greater level of accuracy than other more general assumptions admit. For instance, use of a firm's desired rate may not be achievable and prevailing market rates might be different; it may not matter what the best price is if your credit standing has eroded. It may be helpful to pause for a minute and revisit the MARR and WACC discussions from Chapter 4, p. 112. There are a variety of factors that the firm cannot control that will affect the cost of capital available to them. It is important to review those factors to see a broader managerial context of capital investments (it is easy to get caught up in the numbers).

Most often we view the WACC as the so-called "floor" of the minimal acceptable rate of return (MARR), even though the WACC refers to project financing and MARR is the minimum expected return of anticipated capital allocation. Generally we seek a situation where NPV > 0 prevails for MARR > WACC, though certainly there are investments where MARR > WACC; typical examples would be in investments for environmental compliance, plant safety, employee morale and welfare (recreation), and other non-monetary reasons[4].

Usage of WACC is not without controversy. Some believe it is an overly simplified view of the true cost of capital, particularly in a post-Enron era where we wish to do everything possible to maintain investor confidence. For instance, the over-simplification of tax treatment can be viewed as a source of uncertainty and thus mislead investors; that is, the real financial structure is much more complex than can be characterized with a single number, especially in an era when spreadsheets can accommodate financial details with great ease. Among other concerns might be that equity is highly volatile and will fluctuate over time, which the classic WACC equation cannot easily accommodate. Similarly, it is argued that tax rates can fluctuate with government policy. Most engineering managers overlook the nuances of WACC precision in favor of the generally acceptable trends offered.

9.2.5 Assigning MARR

Investment uncertainty can be quantified in one of several ways, and it can be challenging to know what is the "best" method. Sometimes we have access to risk-adjusted data through a decision tree as in Section 8.5.3. Other times, a rule-of-thumb given by Figure 9.1 or Table 9.2 can provide adequate insight. Figure 9.1 implies that some analysts may emphasize historical results that approximate the overall investment risk, versus, say, a more conservative approach that may focus on short-run performance through high MARR values. As discussed in Chapter 4, setting the MARR is not an "exact science" and MARR can be expected to vary as a result of any combination of factors outlined in Section 4.4.1.

It is prudent to explore *several* perspectives to bound the assignment of MARR; despite an engineering preference for a formula or technique, the reality is this this often distills down to a core team of managers discussing company strategy and how the prevailing year's key performance objectives

Project Type	MARR
Replacement scenario	8%
Product line expansion	12%
New venture in current market	18%
New venture in new market	25%

Table 9.2: Comparison of risk-adjusted MARR.

translate into tolerance for risk. Reference data like that provided in Table 9.2 can help set the stage for the discussion.

9.2.6 Sale of assets: A bit more taxing

We briefly mentioned the impact of taxes on the WACC, and another brief note on the taxes on sale of assets is important, too. Managing disposal, replacement, and obsolescence projects quite obviously deals with possible purchase or sale of assets. Taxes must be considered whether new *or* old assets are involved. If the asset has been used in service, then its values have been apportioned on the income statement, and in this way a portion of the asset becomes the source of taxable income. The tax calculation of interest to us here is the tax *gain or loss* when the asset is sold above or below book value. We noted earlier that market value plays virtually no role in depreciation deductions, but at the time the asset is sold in the market a tax adjustment to the projected depreciation schedule must be accounted for through a tax on gain or loss on book value. Just to be clear, when the asset is sold for *less* than the current book value, the prior tax paid as part of depreciated expense within the income statement means a *tax rebate* is allowed. If the asset is sold *at book value*, then no tax is due. If the asset is sold above book value, then a *tax on the capital gain* is owed.

9.2.7 Project portfolio: Weighted criteria ranking

Up to this point the discussion of project merit has been established by conditions where the estimated NPV > 0, where NPV is based on a selected MARR hurdle rate corresponding with the nature of project risk. While this may define an investable project, it does not mean capital will automatically be allocated to proceed with the investment. A project might be an attractive investment but still unaffordable. Most firms have fixed capital (equity and debt limits) to work with, and this constraint dictates that only a subset of investable projects can be acted upon. Project portfolio ranking helps with the constrained capital budget decision-making process. Project feasibility certainly counts in decision-making, but corporate strategy, the investment team, and the current business environment also influence the final decision. It can be difficult for an analyst working in relative isolation to know what matters. The reality tends to be that the engineering manager should already at least understand a few of the issues on the minds of the investment committee. Such knowledge will exist for the engineering manager where the numbers have been worked in parallel with managerial feedback and other information obtained by networking with key organizational stakeholders.

A search of the Internet will yield a very large number of tips, analytic methods, and software programs that can assist with project portfolio ranking. But *assist* is the operative word. At the root of ranking are a lot of subjective criteria that must be factored into an investment decision. Certainly, numerical ranking systems can help *start* the process and set the stage for the discussion and often a system that is appropriately adapted to external industry and market conditions will contribute to investment team discussions. In the end, it is the *team, not the technology* that will select a project. There are three general steps to follow:

1. Establish the *criteria* for assessing project importance and a *rating scale* for the extent to which a project meets the rating criteria. Criteria can be on a scale of 1 to 10.
2. Arrange the criteria in (descending) order of importance and establish the *weighting* of each criteria. Weighting might be such that the sum of all weights is 100.
3. Projects are scored by the product of the weighting of the criteria and the degree to which the project meets the criteria. This produces a prioritization score.

The mechanics of implementation vary widely, and this is where software is claimed to be helpful, though in many situations a simple spreadsheet is adequate for the task. For instance, we may want to rank projects on four major criteria:

1. Does the project meet the established financial criteria for the organization?
2. Is the project aligned with the current organizational development and strategic goals?
3. Does the project create value to the customer or other key stakeholders?
4. Is the project required to meet legal or regulatory demands of our business?

A sample rating scale for the degree to which a project meets each of the criterion for assessing project importance is illustrated in Table 9.3 below.

The degree to which a project meets specific criteria involves a different value judgment than for setting the relative weights of each criteria. For the example above we might have the following project weighting criteria: Certainly, the scale used in weighting says quite a bit about what is important to executive decision-makers. It is prudent for the engineering manager to have a good grasp on the "what and why" of all criteria and weighting (actually, it is important to know what the boss cares about for *anyone* at *all* levels of the company!). Avoid making assumptions; ask the questions to avoid embarrassing misunderstandings.

It should not be lost that *at a minimum*, the starting point for any discussion often begins with a defensible estimate for (a) the total investment in

Project criteria		Scoring of criteria
Financial criteria met?	10	Financial goals exceeded
	1	Meets minimum financial goals
Aligned with strategy?	10	Aligned with all strategies
	5	Aligned with some strategies
	1	Not aligned with strategy
Creates customer value?	10	Creates significant customer value
	5	Creates some customer value
	1	Little to no customer value created
Is project mandatory?	10	Essential for compliance
	1	Is not required or not applicable

Table 9.3: Four common project criterion and scoring scales.

Financial criteria met?	20%
Aligned with strategy?	30%
Creates customer value?	40%
Is project mandatory?	10%

the project, (b) the expected return on investment (ROI), and (c) the MARR. Some people suggest going beyond ROI and also giving *detailed product costs* in the investment summary (versus just the ROI), but normally the supporting documents for the proposed project have cost information – or cost can be backed out easily enough from the ROI. Finance is fundamental.

It may be a challenge to know exactly *which* criteria and weighting to use to prioritize and rank opportunities, but therein lies the value of an engineering manager's understanding what the company's strategy is today and what it was historically. Sometimes, though, developing the practical understanding, shrewdness, or savvy about strategy is just a matter of time. Experience counts!

9.2.8 A typical scenario

Exploring a typical scenario can help remove some of the ambiguity related to project portfolio ranking. Let's assume Bethesda Imaging, Inc. has $50 million cash on the balance sheet. Executives at the annual board meeting determine that a moderately conservative business strategy in a dynamic market is to set aside $10 million for future investments, preserving $40 million as a hedge against cash flow and other market issues (they are in a capital equipment market with high sales fluctuations). In response, the chief operating officer (COO) asks her direct reports (a) investment ideas that would meet the minimum MARR set forth in Table 9.2 and (b) that each investment would not exceed a $5 million cap. The four division managers under the COO came up with six projects; managers have tried to be persuasive with the value of their proposals with very long documents and supporting studies, but in the end the COO simply extracted the data shown in Table 9.4.

Project	Project Type	Investment	ROI
A.	Replace coil-winding machine to maintain production reliability	\$1.2M	10%
B.	Relocate assembly of hand-held products to lower-labor costs	\$2.8M	20%
C.	Expand new plant locally (with investment tax breaks)	\$5.1M	12%
D.	New imaging accessory product line in existing market	\$4.5M	18%
E.	Quality system to improve software regulatory compliance	\$2.5M	32%
F.	New wound-healing pulsed magnetic product for consumer market	\$3.9M	25%

Table 9.4: Investment options summarized by Bethesda Imaging's COO.

Project managers have worked to position projects as desirable by proposing either low-capital budget ideas with a high return *or* those that are the most strategic, safest growth bet for the company. The total of all investment options is \$20 million, twice as much money as the board of directors wished to allocate to new projects. Management has decided the floor of the MARR is 10%, but also promoted the idea that higher MARRs will show that the company is not entirely risk-averse.

Table 9.4 illustrates that the managers took the COO's guidelines seriously, as the mean project investment was \$3.3 million (σ = \$1.4 million) and only one project exceeded the maximum of \$5 million (and that was by just 2%). Recalling MARR trends from Table 9.2, the anticipated ROI for the projects had an average of 19.5%, aligning well with the strategic push by corporate.

If every project essentially meets the financial criteria, what now? Arranging projects in order of ROI is shown in Figure 9.2. This is sometimes referred to as a *waterfall* or *opportunity cost diagram.* Essentially, Bethesda Imaging's situation it that \$20 million in qualified projects are on the table and they have a budget of \$10 million. This capital constraint means that some form of prioritization of projects has to be made. Either that, or the budget must be expanded to fit the desired portfolio of projects (to avoid eliminating projects from the investment pool).

All the projects in Table 9.4 have a positive `NPV`, and some have very attractive estimated `ROI` values. In Figure 9.2 we have ranked projects in order of decreasing `ROI`, under the assumption that the higher the (achievable)

ROI, the more value this will create for the company. As well, each project "block" is labeled to correspond with the projects in Table 9.4, with a height corresponding to the ROI and the width representing the capital investment required for that project. The abscissa value represents the cumulative capital investment. For instance, at point "Z" in Figure 9.2, between project block "B" and "D" we see that the cumulative investment in projects "E", "F", and "B" is $9.2 million, which leaves less than 8% of the investment pool available to support other projects.

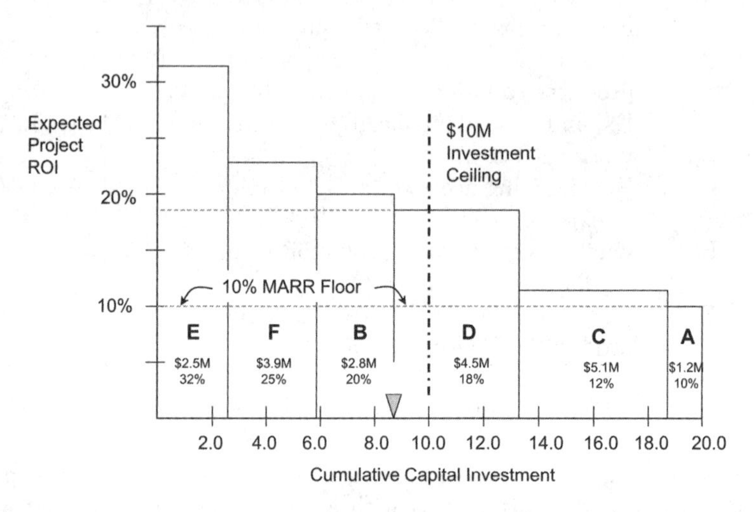

Figure 9.2: Bethesda Imaging, Inc., potential capital investments for annual budgeting process. Projects ranked according to estimated ROI in a cascade.

Under the assumption we would not subdivide projects (not in this discussion, but as a practical matter, partial investments are frequently made), then projects "D", "C", and "A" do not make the pay line and are eliminated from consideration. The next step in this *capital rationing* exercise is to find the *next best* investment opportunity that meets both the ROI criteria and has an investment requirement less than the remaining investment funds. Were we able to renegotiate the investment in project "E" from $1.2M to $0.8M, then we could fit project "E" into the portfolio, but as is our capital budget limitation means we can only accommodate projects "E", "F", and "B."

An underlying yet very important aspect of this process is that we are assuming each of the projects has computed NPV very accurately to have produced the ROI ranking shown. As well, the capital budgets for each project may well need to be re-examined. If warranted, then the project management charts are adjusted to reflect critical path items, as project milestones may be important, but not essential to the project. There is nothing inferior about a staged project. In the investment world, projects are often funded in stages or tranches (French for "slice") that in aggregate sum to the proposed project capital investment. Let's consider the project portfolio in Table 9.4 from the perspective of the weighted criteria ranking approach discussed earlier on page 276. For simplicity, let's use the criteria outlined in Table 9.3; results are shown in Table 9.5.

Project criteria		Project A	Project E
Financial criteria met?	20%	5.0	9.0
Aligned with strategy?	30%	9.0	8.0
Creates customer value?	40%	9.0	4.0
Is project mandatory?	10%	6.0	10.0
Total weighted score:		7.9	6.8

Table 9.5: Bethesda Imaging, Inc., project portfolio weighted criteria ranking.

What a difference in outcome! Project "E" with a projected ROI of 32% and a weighted score of 6.8, is now moderately unfavorable relative to Project "A" with a projected ROI of 10% (the minimum) and a weighted score of 7.9. In this case we would include "A" in the investment portfolio.

Financial calculations are an essential part of the decision-making process, but do not *always* include all factors involved in investments. This is a matter of degree and project scope, too. It may be that selecting a copy machine vendor is viewed as a "commodity decision" and the mutually exclusive decision is amenable to rigorous financial analysis. As projects become more strategic in nature, the role of finance and engineering economics must fit the context of the greater goals of the company. Projects are less likely to be considered mutually exclusive. Value may be intrinsically non-monetary. Nothing is more frustrating to an engineering manager who "followed the book" in generating documentation, spreadsheets, and a persuasive PowerPoint presentation on a project proposal only to have executives delay or decline the request due to "other factors." Naively, the engineering manager concludes, "Executives must not know what they are doing." So, having spent much of the earlier chapters exploring numerous rigorous mathematical and statistical tools in project analysis, this chapter tosses in a few attributes of projects that are bound to occur but are not amenable to financial analysis. If we could distill everything into NPV, then life would be easy, but in reality finance is *necessary but not sufficient* in a wide variety of situations. Multi-attribute decisions are common, involving the choice of attributes, measurement scales (often mismatched), and different pathways for establishing some way of assigning value or comparative indices for those situations where it is essential to force a decision on a quantitative basis.

If we step back and consider the *outside view* discussed back in Section 2.3.1 (p. 23), we noted that frequently the "story is in the numbers" and there was considerable benefit to examining an enterprise through the lens of the financial statements. This remains true since the financial statements are a historical record of the affairs of the company, and when combined with the annual and news reports, a fairly good characterization of the company emerges. Here we would like to bring up the "outside view" on another aspect of capital investment that can help shape our thinking about portfolio management. Consider an investment in 10 projects as shown in Table 9.6.

	Portfolio	Initial Value	Final Value	Return (ROI)	Approx ARR
	Year:	0	5		Year 0-5
1	Product X "home run"	$1,000	$10,000	900%	58.49%
2	Product Y "aggressive"	$1,000	$2,500	150%	20.11%
3	Product A "base hit"	$1,000	$1,500	50%	8.45%
4	Product B "base hit"	$1,000	$1,000	0%	0.00%
5	Product C "base hit"	$1,000	$1,000	0%	0.00%
6	Product D "base hit"	$1,000	$1,000	0%	0.00%
7	Product Q "dog"	$1,000	$0	-100%	-100.00%
8	Product R	$1,000	$0	-100%	-100.00%
9	Product S	$1,000	$0	-100%	-100.00%
10	Product T	$1,000	$0	-100%	-100.00%
		$10,000	$17,000	70%	11.20%

Table 9.6: Portfolio of investments.

This portfolio is a bit academic, but is designed to make a specific point about the need to have a "home run" project within the R&D portfolio. Despite the declaration that most upper-level managers would like to "foster innovation" among projects, studies have shown that when it comes right down to investing in such projects, they tend to be overlooked in favor of more "base hits." This could be for entirely good reasons, as we just saw with the Bethesda project "A" and "E" comparison. At this time we also want to draw on Figure 1.4 (p. 4) where a "cost plus" mentality was discussed relative to product pricing set by *what the market will bear.*

In Table 9.6 the investment committee has decided to take some risk on product "X," which they believe will be a home run. If the product offers a return that is 10 times the original investment, it can make a difference in the profitability of the entire portfolio. Investors in companies see things a similar way. What Table 9.6 underscores is the difficulty in making project selections. The projects for Bethesda outlined in Table 9.4 are simply *estimates and projections* and some may have little or no historical backing. So, it is possible that one of the projects may be a home run and the others entirely unprofitable. Parameters surrounding the capital budgeting decision are changing as the sophistication and production rates of rapid prototyping – in particular, 3D printing – improve in terms of rate of unit output and quality. In highly fragmented markets where customers have very small lot sizes (as in "mass customization") we can now envision production in an entirely new light. Rather than fabricate an expensive product component mold (or *several* molds!) to then be used to make parts through injection molding, the possibility exists now to skip a few steps in the manufacturing process. This type of system clearly changes the capital budgeting process. Indeed, the entire field of replacement analysis is undergoing change as these new types of production move to the factory floor. The distinction between capital equipment needs for traditional design of high-volume production for commodity products will emerge separate from the low-volume needs of niche markets whereby flexibility and changeover are figures of merit. Said another way, the engineering manager must be as adept at the economics of a machine with a 20-year machine life as one that now will be depreciated in 3 years, being replaced when it might seem it was just installed!

It is delightful when the engineering manager has the opportunity to participate in executive-level discussion of capital investments. Many engineers are trained throughout their college years to quickly solve well-framed problems with great precision. Financial decision-making offers a blurred view of the world in which aesthetics, politics, regulatory mandates, employee morale, and unconventional methods of manufacture will have a direct bearing on decisions. The work of Patterson[5] would be a great read in preparation for an engineer's first round of negotiations.

It is prudent for the engineering manager to continually apprise the methods of replacement analysis and question relevance of techniques. Section 9.3.4 on "objectivity in analysis" will play an increasingly important role in analysis as the project horizons shorten, the economics of outsourcing becomes more complex and permanent to production, and sustainability and responsible use of material grow in importance. Decisions about carbon-footprint, use of energy, and re-use and recycling of products will broaden the scope of concern. All of this must be translated into financial terms for decision-making and accounting for assets. This might include the ability to develop entirely new techniques of analysis that might emerge from concept mapping (Chapter 7) or where new leadership styles (Chapter 10) will be needed to change the "status quo" of problem definition.

9.3 Fundamentals of replacement analysis

At first it might seem odd for *replacement analysis* to be placed in the same chapter as capital budgeting; an initial reaction might be, "Didn't we already cover something like this during the project selection technique discussions of Chapter 5?" Well, yes, many of the basic mathematical techniques have application in replacement problems, but the context of replacement analysis draws on many more similarities with long-term asset management techniques than with the conventional project selection of Section 5.4. This issue was touched on briefly when discussing Table 9.1.

It helps to understand the distinctive nature of *replacement problems* (relative to the project selection framework of Chapter 5) by elaborating on several features:

- **Investment context**. The typical example problems of Chapter 5 tend to be *forward-looking* in the sense that a choice is being made about a *new* project investment. In replacement analysis the project is *already under way*, and the debate is whether an old asset needs to be replaced with a new asset.

- **Economic life**. The old asset in a replacement problem may have additional physical life even though its economic life has expired; the physical or economic life may be under 1 year. The new asset is chosen specifically because it has a much longer (unequal) life. In conventional problems we anticipate equal lives of the asset or work to make that intentionally the case. We have to remember that in a replacement problem *we already own the old asset* but are concerned that it is near the end of its life. The new asset can have whatever life we wish to have (for the right price). By definition, asset lives are *different.*
- **Principal investment relevance**. New investments involve an initial principal cash P outlay for which there are long-term revenue A benefits; in the examples of Chapter 5.4, both options involve an initial investment P in analysis. In a replacement analysis we have a *forward-looking* situation involving an aging asset. The prior investment P_X in the existing system is considered a "sunk cost" that is *not relevant to the replacement analysis*, but the new replacement system has an investment cost P_Y that *is* relevant.
- **Capital gain or loss tax**. In the event the system to be replaced is sold at a *market price* that is above or below the current depreciated book value of the system, there will be tax implications whether the market price is higher (capital gain) or lower (capital loss) than book value. There is no equivalent gain/loss for the replacement system. In conventional comparison problems from Chapter 5, taxes *may* have an impact, but *both* systems are under consideration so the net impact on decision-making is neutralized.
- **Dissimilar economic characteristics**. Conventional project selection involves a forward-looking problem with projected expenses that can be used throughout the *expected useful life* of the asset. In replacement analysis, we may not know or have good projections for the expected costs of keeping the asset being considered for replacement; economic life does not align with physical life. This data will be *known* for the new system, and so generally the dissimilarity of new-versus-old cash streams requires special attention.
- **Relevance of need**. In conventional project choice situations the decision has already been made to act on the purchase of a system. In replacement problems there may be the attitude that as long as the current system is "working" then "Why spend the money? If it is not broken, don't fix it."

The very last point underscores some of the special engineering management problems one must contend with, in particular the idea of "professional" versus "bureaucratic" accountability in organizations [6].

Replacement analysis gained traction in the late 1940s as the productivity and efficiency of the factories of the day started exhausting the surplus of "modern" inventory left over from the scaling of production capability during the Second World War. Many of the analytic techniques and methodologies we use today were outlined in a series of works by the Machinery and Allied Products Institute (MAPI), then under the direction of George W. Terborgh, beginning with *Dynamic Equipment Policy* first printed in 1949 [7]. With a bit of imagination we can even remark on how similar the analysis table looks to those we create today (Terborgh [7], p.31). It was Terborgh who said that "the obsolescence of an asset must be defined in terms of its relation to its job, not its age, and [the asset] must defend its tenure." Terborgh's work was the origin of the term "defender" and later "challenger" that we still use today. The idea of replacement analysis extended from concerns that the practice of renting and leasing might not be the best way to augment factory production (Terborgh[7], p. 29 "Replacement of Rented Equipment"). For a long time the MAPI method was synonymous with replacement analysis. Over time, as analysis methods were refined and expanded to suit many more complex situations, what eventually emerged was the "art of replacement analysis" we know today.

Replacement analysis is quite relevant for companies seeking to optimize capital expenses whenever possible. Many maintenance folks joke that they are required to "do more with less," but what has emerged after the 2008 financial crisis is that maintenance and replacements are being deferred – sometime with disastrous results – as people are being asked to "do *even more* with *even less*!" Even some more historical applications of replacement analysis have not gone away. The recent collapse of the Skagit River Bridge in the state of Oregon is bound to spark dialogue a bit, as terms like "functionally obsolete" take on a new, more urgent meaning for maintenance, repair, and replacement.

Replacement analysis builds directly on prior decision-making competencies identified in Chapter 5. In the current chapter, the basic question about asset replacement analysis is really quite simple: we would like to know if an existing asset (the "defender") should be retired, kept in service, or replaced with a new asset ("the challenger"). In this chapter we explore a variety of reasons that replacement analysis is conducted and work through some typical problems. The concepts involved tend to be cumbersome more than difficult, meaning that in practice, solutions are obtained with the use of spreadsheets.

9.3.1 Defender and challenger revisited

In our earlier discussion of alternatives (Section 5.4, p. 140), we presented the idea of a "defender" and a "challenger" for multiple project selection. Here, we present slightly different descriptions, specific to replacement problems:

> **Defender** - The defender is an existing asset we *own that is currently in service.* The defender itself may have been a challenger years ago, so the idea that it is being considered for replacement means it is near the end of its useful life. We may not know much about the exact length of remaining life or the cost to reach that

point. As each year of services passes by, three characteristics of the defender are that (a) the EUAC ownership cost has been decreasing, (b) the salvage value is decreasing, and (c) often the cost of maintenance is increasing.

Challenger - The challenger is an asset under consideration for *purchase as a replacement for the defender.* The challenger itself may have been identified as the best possible alternative to the defender as the result of a comparison of two or more **new** investment options using the techniques of Section 5.4. The challenger (a) will have a *longer* life than the defender, (b) typically have much *lower and more predictable maintenance costs* than the defender, and (c) is assumed to provide the same level of revenue as is currently provided by the defender.

A key assumption is that the challenger has been maintained so that its production and quality of output are at a satisfactory level, and that the challenger will meet the same criteria. As such, the discussion of replacement is one about *cost comparisons*, often using EUAC since the lives of the two assets are different. The contest between the defender and the challenger is quite simple:

1. Has the economic life of the defender been reached?
2. Is there a satisfactory alternative to the defender?

In the end, we are simply saying that as long as the defender is the best economic alternative, we will keep it in service. At the point in time when the challenger becomes more attractive, the defender will be replaced.

The "life" of an asset can take on several meanings. We need to be clear about what specific interpretation of "life" is intended when working through replacement analyses. Although we briefly mentioned economic life on p.283, it helps to provide an engineering economics interpretation of "life" and to underscore the differences between *physical life*, *economic life*, and *technological life*.

1. *Physical life* is the elapsed time between the moment the asset in put into service and that time at which the asset is removed from service. Age or decrepit appearance define life less than productivity and the ability of the asset to meet production and performance criteria. A critical assumption is that we've made the necessary investment to maintain the asset and that at any given point the asset could *technically* be refurbished and restored to requirements. Physical life does not begin when the asset is manufactured, but when it is placed into service.
2. *Economic life* relates to profitability. EUAC is an important tool in determining economic life. Economic life begins when the asset is placed into service and sometimes ends when the EUAC is at a minimum. Economic life generally takes into account the cost (or current market value) of the asset, operating costs, and estimated salvage value. At the end of life an asset may still be *functional*, but might not be *useful* for the intended purpose, and thus would still have *salvage value* since someone else has a useful purpose that the asset satisfies. This effort is sometimes

called the *minimum cost life analysis.*

3. *Technological life* is the period of time after the start of service that the asset is still technically relevant to the system requirements. The asset becomes *obsolete* for the job at hand if its service value is no longer aligned with the production performance requirements. Electronic devices are a classic example, in which innovations are generally believed to last no more than 18 months to 2 years. Think about the life of the Intel 8086 and the manufacture of products based on that platform; the 5-year period of relevance was made obsolete by the Intel 486 and subsequent products. The chips themselves had a physical life that was much longer than their technical life (by premature termination of use, not shelf life by initial date of manufacture!).

Other engineering economies reference books may use more or fewer terms for "life" and wording is always slightly different, but in all cases the authors are simply trying to underscore the difference between physical use, economic value, and technical utility.

It will emerge that there are three different comparison methods for "challenger versus defender" decision-making, and the *preference of method depends on the data available.* To begin to understand the differences, let's examine the composition of the EUAC for a typical asset (be it challenger or defender).

9.3.2 EUAC revisited: The "total EUAC curve"

Unless the concept of equivalent uniform annual cost (EUAC) is clear, it is difficult to appreciate the subtle difference in mathematics between the three methods of replacement analysis we need to consider later in this chapter. The *total EUAC curve* is generally composed of three components, and our ability to work replacement problems requires fluency in applying data to calculations. At the risk of some duplication of Section 4.6, a discussion of total EUAC as the sum of *capital recovery*, *operating costs*, and *maintenance costs* follows.

1. Capital recovery. As the time horizon of asset usage expands, the EUAC cost of the asset is lower with each passing year. This is logical. As the original capital cost of an asset is apportioned over a longer period, we expect the average yearly cost to be lower.
2. Operating costs. Normal wear and tear of machinery often results in escalating costs to operate the machine. Such costs might be as mundane as increased use of compressed air to drive raw material though the system or greater electrical consumption as motors age. Equally valid might be the need to have more machine operators attending to a system to manage variance of output.
3. Maintenance and repair. Annual or quarterly maintenance schedules may call for part replacement, cleaning and purging of systems, or replacing lubricants and other consumables. The cost of labor to conduct this maintenance is important.

Let's take a closer look at each of these three costs and clarify calculation procedures through examples.

EUAC of capital recovery

At any given year n within an operational horizon of $n = N$, the EUAC for capital recovery is given by the sum of the capital recovery of the principal (first cost) of the asset investment; this calculation simply involves the capital recovery factor given in Equation 3.11,

$$A = P\left\{\frac{i(1+i)^n}{(1+i)^n - 1}\right\} = P\left(\frac{A}{P}, i\%, n\right)$$

Consider an asset with an acquisition cost of $9,000 and for a company with a pre-tax MARR of 10%. The EUAC for this asset retired at the close at the fifth year of service $n = 5$ would be simply:

$$EUAC_5 = \$9,000\left(\frac{A}{P}, 10\%, 5\right) \;=\; \$9,000(0.2638) \;=\; \$2,374.20$$

For any other year within the project horizon, the EUAC is found by substituting the pertinent value of n.

$$EUAC_8 = \$9,000\left(\frac{A}{P}, 10\%, 8\right) \;=\; \$9,000(0.1874) \;=\; \$1,686.60$$

Calculating EUAC over the period $n = 0, 1, 2, 3, ...N = 10$ produces the values shown in Column A, Table 9.7.

Note that the EUAC for $n = 1$ has a value of $9,900, which is $900 *higher* than the original $9,000 purchase price. This might not seem right at first, but in this case if you retire the asset after the first year of usage, you have the original price of the asset distributed *only* over the first year *and* the cost of the use of the funds for that year (10% of $9,000). Clearly the EUAC drops rapidly with each additional year as we expand the project horizon.

EUAC of operating costs

While the EUAC for capital recovery decreases over time, typically operational expenses increase over time. Of the many different components of operational cost that exist, the most common might be utilities and labor. Utility usage estimates might be provided by the OEM, or could be estimated from comparable systems. Labor is not *always* included, especially if the comparison or selection involves assets with equivalent labor needs. Sometimes, though, a certain type of skilled labor may be needed or the number of workers required to run the machine might vary. It is not unrealistic to assume a fixed percentage annual increase in expense. In that case, the operating cost `OC` for any given year n is quite simply

$$OC_n \;=\; OC_{n-1}(1 + i_O)$$

	A	B	C	D	E	F	G	H
Year Ending	EUAC CRC	Operating Expenses Cost	PV	EUAC	Maintenance Expenses Cost	PV	EUAC	Total EUAC
0	$9,000	12.0%			6.0%			
1	$9,900	$2,800	$2,545	$2,800	$1,700	$1,545	$1,700	$14,400
2	$5,186	$3,136	$2,592	$2,960	$1,802	$1,489	$1,749	$9,894
3	$3,619	$3,512	$2,639	$3,127	$1,910	$1,435	$1,797	$8,543
4	$2,839	$3,934	$2,687	$3,301	$2,025	$1,383	$1,846	$7,986
5	$2,374	$4,406	$2,736	$3,482	$2,146	$1,333	$1,895	$7,751
6	$2,066	$4,935	$2,785	$3,670	$2,275	$1,284	$1,945	$7,681
7	$1,849	$5,527	$2,836	$3,866	$2,411	$1,237	$1,994	$7,708
8	$1,687	$6,190	$2,888	$4,069	$2,556	$1,192	$2,043	$7,799
9	$1,563	$6,933	$2,940	$4,280	$2,710	$1,149	$2,092	$7,935
10	$1,465	$7,765	$2,994	$4,499	$2,872	$1,107	$2,141	$8,104

Table 9.7: EUAC results for an asset first cost of $9,000 over an N=10 year horizon with a nominal pre-tax MARR of 10% and no salvage value at any time.

For the example shown in Table 9.7, if we assume a baseline cost of $2,800 and operating expense growth rate of 12%, then for year $n = 2$

$$OC_2 \;=\; OC_1(1+0.12) \;=\; \$2,800(1.12) \;=\; \$3,136$$

Be sure to notice that Column B provides the estimated operating cost for *that year only*, and the required EUAC calculation involves the annual recovery of the *aggregate* operating expense up to and including the year the asset is retired. The pattern for EUAC should be fairly clear from our previous discussion in Section 4.3, leading to Table 4.7. Note that if the annual operating cost were a constant value (for all years), then the EUAC for any given year would simply be that uniform baseline value. However, situations like the sample problem (where OE is non-uniform) require us to convert all OC values to the equivalent PV calculation (at t=0) as shown Column C; the calculation is based on Equation 3.4:

$$P_8 = \frac{F}{(1+i)^N} \;=\; \frac{\$6,190}{(1+.10)^8} \;=\; \$2,888$$

And as previously demonstrated, we work out the EUAC from the *cumulative* PV_n as shown in Column D:

$$\begin{aligned}
\texttt{EUAC}_{OE,3} &= \sum_{n=1}^{3} PV_n \left(\frac{A}{P}, 10\%, n\right) \\
&= (\$2,545 \;+\$2,592+\$2,639)\,(A/P,10\%,3) \\
&= \$7,776\;(0.4021) \;=\; \$3,127
\end{aligned}$$

Again, this should look very familiar to the calculation sequence and example presented on page 109.

EUAC of maintenance costs

In principle, typical maintenance costs of predictable growth can be treated in much the way that operating expenses were treated in the previous section. For our sample problem we are assuming we have baseline maintenance expenses of $1,700 that grow by 6% annually. The EUAC for year 4 is easily found:

$$\begin{aligned}
\texttt{EUAC}_{ME,4} &= \sum_{n=1}^{4} PV_n\left(\frac{A}{P}, 10\%, n\right) \\
&= (\$1,545\ + \$1,489 + \$1,435 + \$1,383)\,(A/P, 10\%, 4) \\
&= \$5,852\ (0.3155)\ =\ \$1,846
\end{aligned}$$

Of special interest is the case when maintenance requirements exceed normal periodic expenses, such as with an overhaul. Treating such situations is quite easy with a spreadsheet. For the sample problem we have been examining, consider that, in years 3, 6, and 9, the system must have a thorough failure analysis stress test which will add $2,500 to maintenance those years. Table 9.8 illustrates how the EUAC is impacted. Spreadsheets greatly simplify

E	F	G	H
Maintenance Expenses			Total
Cost	PV	EUAC	EUAC
6.0%			
$1,700	$1,545	$1,700	$14,400
$1,802	$1,489	$1,749	$9,894
$4,410	$3,313	$2,553	$9,299
$2,025	$1,383	$2,439	$8,579
$2,147	$1,333	$2,391	$8,247
$4,776	$2,696	$2,700	$8,437
$2,413	$1,238	$2,670	$8,384
$2,557	$1,193	$2,660	$8,416
$5,210	$2,210	$2,848	$8,690
$2,873	$1,108	$2,849	$8,813

Table 9.8: Impact on EUAC results from Table 9.7, but with additional $2,500 overhaul expenses in years 3, 6, and 9.

the ease with which we can accommodate non-uniform, irregular, or one-time expenses. The workbook that accompanies this textbook provides the spreadsheets discussed here and enables you to test changes and ideas of your own to better see cause and effect and to help validate if your hand calculations are correct.

9.3.3 EUAC and economic life

Preceding sections illustrated how the EUAC trends vary with the length of the project horizon. Some trends are common sense, as in the case of the capital recovery (Column "A" of Table 9.7) that has a steep initial decrease that tapers off over time. Summing individual EUAC components to produce an single overall system EUAC often results in a concave curve that exhibits a minimum EUAC value. The minimum value will most typically reflect the decrease in capital recovery being offset by increases resulting from rising maintenance and operating costs. The year corresponding to the minimum EUAC marks an elapsed period defining the *economic life of the asset.*

An example of aggregate EUAC shown in Figure 9.3 is based on data from Table 9.7. The curve marked `Total EUAC` is a plot of the data from Column H of Table 9.7. Year 6 marks the *minimum* EUAC and thus the asset's economic life is 6 years, even though the project horizon is 10 years. Note the shape of the curves. Some aggregate EUAC curves have a relatively flat out-year profile reflecting relative insensitivity of the asset EUAC to changes in operating and maintenance costs.

EUAC curves can help in the examination of investments where a higher initial asset cost might be acceptable if that minimizes downstream maintenance costs. From a systems perspective, unexpected production interruption and "rush" maintenance might have more of an impact on overall production line cost variance (and is less preferable) than initial capital cost of one subsystem (asset). The shape of the EUAC curve can be as important and revealing as the economic life it defines. UPS has been known to keep its delivery trucks in service for up to 25 years – attention to maintenance is legendary to ensure reliability as part of the brand. (Compare with the allowable MACRS depreciation schedules indicating a 5 to 10 year life for typical service trucks!). What do you think the UPS aggregate EUAC curve looks like? A majority of the assets will have the composition and shape that are similar to the aggregate plot of Figure 9.3, and the economic life of the asset will normally be fairly easy to pick out.

Our sample problem outlined a project life of 10 years, a life much longer than the asset economic life of 6 years. The difference in life is important to know for any given asset, be it new, old, challenger, or defender. However, it is essential to *have or be able to estimate the data* available to work through the required calculations. Information that is absent creates a void, and we have to decide how to handle the situation. In the sections ahead there are three different techniques whose choice is dependent on the nature of the data included in the problem statement. Of course, just because we may know or can find economic life does not mean it will always be used in a given problem analysis. Sometimes the horizon is fixed and minimum EUAC is less of consequence in decision-making. An asset can be economically inferior yet

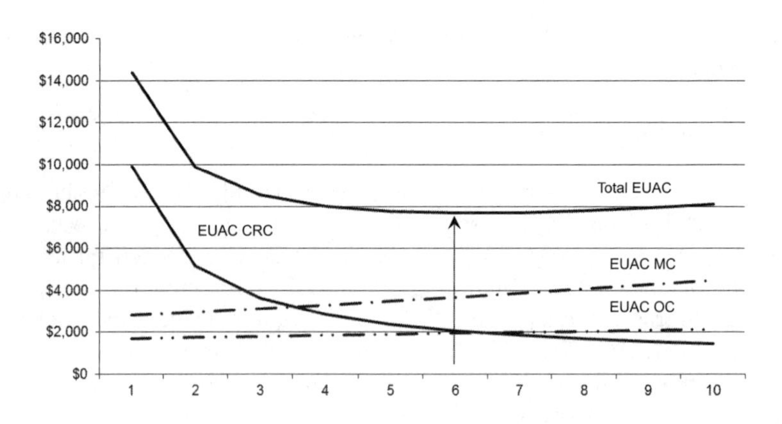

Figure 9.3: Aggregate EUAC for the sample problem to establish *economic life* of the asset.

still be physically satisfactory for the task.

9.3.4 Objectivity in analysis

Replacement analysis requires objectivity about asset information, but obtaining a third-party viewpoint can sometimes be hard to achieve. There might be a rich history surrounding the defender that has polarized corporate stakeholder opinions. A major effort and debate may have occurred in the advocacy, installation, and operation of the original project associated with the asset. As well, it may be slightly difficult to accept that (in hindsight) there were some errors made in the process (of learning something new). If we misread technology changes in the market, maintenance requirements, or asset features or specifications, then it is entirely possible that the book value of the asset is dramatically different than market value. Not replacing the asset is a way to avoid realizing such an error. Although we cannot live in the past, we can certainly learn from the past and look forward to future success. Regardless of prior decisions, objectivity produces value for the enterprise in the long run.

Earlier in Section 2.3.1 we mentioned the value of the "outside view," and this perspective has benefit in avoiding the trap of internal optimism and a can-do attitude that can bias judgment; the list from Mauboussin[8] on p. 23 helps break with the past on project analysis. It is helpful for engineering managers to visualize themselves as outside consultants for this type of decision-making.

Objectivity will assist in proper financial analysis by avoiding the *sunk cost trap*, *wishful thinking about prior decisions*, and *failure to realize new opportunities*:

1. Despite large sums of money involved in capital asset acquisition, whatever has been *spent in the past* is the so-called *sunk cost* that can be problematic for inexperienced managers. Sunk costs are funds already

spent that cannot be recovered; the fallacy of staying with something to the bitter end often leads to waste of another kind! If an asset is underperforming it is prudent to examine the situation regardless of how much money has already been spent.

2. Since few prior capital investments ever predicted the future accurately, there is bound to be a difference between the current market value of the asset under consideration for removal (defender) and what the depreciated book value states. There are income statement implications in such differences (favorable or unfavorable); adjusting entries must be anticipated and included in the decision-making process; and plans for implementation are required depending on the outcome of the decision.
3. Opportunities might be hidden to someone too close to the project and thus opportunity costs understated. For instance, if a new system consumes significantly less space in a confined area, then there is the opportunity to extract value by reconfiguration of complementary systems or in the addition of other assets. Floor space is not free and certainly if special utilities are already present (compressed air, 240V supply, overhead cranes) this can have a huge impact in reducing project first cost.

Stated another way, objectivity helps us in three ways:

1. Prevents sunk costs for being included in replacement analysis.
2. Recognizes and includes market value impact on financial statements.
3. Assures opportunity costs are included in analysis.

Consider the following situation:

> A dicing saw for microelectronic wafers (to convert wafers into chips) was purchased 2 years ago at a bargain price of $17,000 (they often sell for $25,000) and was anticipated to have a life of 7 years. Generally, the used equipment market for wafer processing is pretty weak since the technology changes very rapidly; thus the salvage value of such equipment is 2 cents on a dollar (salvage value estimated at under $500). The current book value for the saw is about $3,500, annual operating costs are running about $6,000 per year, and it has an estimated remaining useful life of 4 years. Unfortunately the saw was quite a financial deal, but the unforeseen down-time of a less expensive saw during production lot changeover is causing the company to rethink the situation and consider a replacement saw. A new vendor wants to get the sale of this important player in the market, so they offer a new system that normally sells for $29,495 at the sale price of $24,999 *and* a trade-in of $5,000 for the old machine. "You can't go wrong," says the vendor. "You're getting a quality dicing saw for $19,999 - that's a 30% discount off list!" Operating costs are expected to drop by 15%, and salvage value is probably around $10,000 with a 6-year life.

Our consultant would summarize "just the facts" as follows

This example brings up several important issues related to replacement problems.

Project Parameter	Defender	Challenger
First cost, P	\$5,000	\$24,999
Operating costs, OE	\$6,000	\$5,100
Service period, n	4 years	6 years
Salvage value, SV	\$500	\$10,000

1. **Defender market value**. Our forward-looking perspective means that the original purchase price and current book value are irrelevant. The *market value* is the trade-in value offered by the challenger's salesperson. It is clear that different trade-in deals will affect market price, but for this problem the \$5,000 trade-in value represents the market price of the defender.
2. **Sunk cost**. The difference between the defender's book value and market price is the *sunk cost* of the defender. In this situation the sunk cost is *negative* by \$1,500 (= \$3,500 - \$5,000) and is therefore a *capital gain* to the company. Of course, there is no surprise to the idea that in order to make the sale the salesperson will offer an attractive trade-in price.
3. **Challenger first cost**. Despite the list price of \$29,495, the anticipated transaction cost of \$24,999 is the relevant challenger first cost. Please note that in this example the trade-in price is *NOT* deducted from the challenger first cost since we have already recited the trade-in value as the equivalent of the defender first cost.
4. **Service period**. Although the defender had an original expected service period of $n = 7$ years, the period $n = 4$ years is based on the consultant's current estimates for remaining useful life, not on the historical expectation of $5 = 7 - 2$. This fore-shortening of expected duration is typical for high-technology systems.

Trade-ins: Cash flow versus opportunity cost approaches

The games that are played with trade-ins can lead to confusion, but the situation can be sorted out by asking, "What transaction did or will occur?" Two basic methods of dealing with trade-ins are often used in analysis. One is referred to as the *cash flow approach* and the other is the *opportunity cost approach*; use tends to be a matter of personal opinion and problem complexity (one or more challenges being considered at the same time, each with a different trade-in value).

Cash flow approach

The trade-in value of the defender is part of the down-payment on the challenger. This works best (easiest) when the *expected life of the challenger and defender are equal* or are for a prescribed planning period. Essentially, the defender's first cost is set to \$0; no money is exchanged if the defender is selected.

Opportunity cost approach

In this case the transaction price for the challenger remains as stated (*not* offset by a trade-in), and the trade-in value of the defender is the market value of the defender, thus becoming the "first cost" investment in the defender. Although this streamlines the analysis if different expected or remaining lives are in play, the situation becomes challenging if different challengers offer different trade-in values for the defender.

Large trade-in values and abnormally abbreviated service life are common ways to manipulate decision-making. Large trade-in values can set a precedent for inflated list price and special quarter-end "deals" that can create an unfavorable urgency to make a deal. As well, the premature end to service life creates a bias in favor of the defender.

9.4 Replacement analysis for fixed project period

Over time, a variety of tools and techniques can be developed to optimize asset economics, but in some situations "optimizing" an asset to the "nth degree" is immaterial. For instance, suppose the wafer-dicing system discussed back on page 292 had a "flat" total EUAC curve as illustrated in Figure 9.3. The difference in EUAC may be under $1,000 per year over a 4 to 5 year span. If the wafer-dicing system is interlocked as part of a much larger production system, there may be *system-wide* considerations that overshadow *incremental sub-system replacement* concerns. As well, plant operations logistics may only permit replacement on a specific budget cycle; planned plant shut-downs may impose limitations and the asset may have to wait for its turn. Another way to say this is that the EUAC for a given project horizon may be satisfactory from the larger production perspective, even if the asset is beyond the optimal EUAC economic life. Analysis is simplified when the assumption of a prescribed project life or horizon can be established. Let's consider two examples to demonstrate the analysis procedure.

Problem 9.4.1: Wafer-dicing replacement

Consider the wafer-dicing problem just discussed, but let's modify the problem so that the expected service life of both units is 4 years and assume a nominal WACC of 12%. EUAC calculations follow that

$$\begin{aligned} EUAC_D &= P\ (A/P, i\%, n)\ +\ OC\ -SV\ (A/F, i\%, n) \\ &= \$5,000\ (A/P, 12\%, 4)\ +\ \$6,000\ -\$500\ (A/F, 12\%, 4) \\ &= \$5,000\ (0.3292)\ +\ \$6,000\ -\$500\ (0.2092) \\ &= \$7,541.40 \end{aligned}$$

$$\begin{aligned} EUAC_C &= P\ (A/P, i\%, n)\ +\ OC\ -SV\ (A/F, i\%, n) \\ &= \$24,999\ (A/P, 12\%, 4)\ +\ \$5,100\ -\$10,000\ (A/F, 12\%, 4) \\ &= \$11,238.19 \end{aligned}$$

And so despite the aggressive trade-in and list price discount, the situation here is that the defender should *still* be kept in service. If, however, we were to level the playing field a bit and let the capital cost of the challenger to extend over a longer project horizon, the situation is much different. It is easy to show that for $n = 6$ the defender EUAC of \$7,154.52 is still lower than the challenger EUAC of \$9,948.14, though the gap has narrowed to under \$2,800. It would actually take the dealer to reduce the challenger price by another \$5,000 to \$19,999 and for the OC to drop to \$3,500 for the challenger to compete (on a financial basis only).

To continue, let's assume the challenger to the wafer-dicing system is a lease. The salesperson suggests that leasing for 4 years with an option to buy is "the modern way to do business." The suggested lease is \$4,000 a year, but if the company agrees to an unknown but "mutually beneficial" lease-end buy-out, the annual lease could be lowered to \$2,000. The situation is as follows

Project Parameter	Defender	Challenger
First cost, P	\$5,000	lease=\$2,000
Operating costs, OE	\$6,000	\$5,100
Service period, n	4 years	4 years
Salvage value, SV	\$500	N/A

Of course, the defender's EUAC is still \$7,541.40 and the lease option is \$7,100, which places the challenger in a favorable position; miraculously the challenger EUAC is *just under* the EUAC of the defender. Increasingly equipment leases are a preferred alternative and challenge the economics of existing assets. As discussed in Section 5.8, leases can offer lower payments, tax benefits, access to the latest equipment, and conservation of *working capital*, but many individuals forget a lease is a legally binding contract that over the long run may result in a greater asset cost than simply buying the asset outright. Current data about leases and leasing trends can be found in Wescott and Karson [9].

Our example calculations have compared EUAC values, though certainly equivalent results are obtained if PV or even FV are used - all of the TVM equations are, in principle, equivalent. Differences arise in the convenience of calculation. So, for instance, we could evaluate the wafer-dicing machine for the 4-year life scenario using PV and find that:

$$\begin{aligned} PV_D &= P + OC\ (P/A, i\%, n) - SV\ (P/F, i\%, n) \\ &= \$5,000 + \$6,000(P/A, 12\%, 4) - \$500\ (P/F, 12\%, 4) \\ &= \$5,000 + \$6,000(3.0373) - \$500\ (0.6355) \\ &= \$23,260.50 \end{aligned}$$

$$PV_C = \$34,133.75$$

Since we are computing the present value of the *costs* of each system, then given $PV_C > PV_D$, the outcome of keeping the challenger still represents the preferred course of action.

The fixed project horizon approach for assets of unequal lives is convenient under EUAC if a repeatability asset value assumption is reasonable. This might imply that a back-up machine or a machine of identical attributes is available to replace the existing asset to fulfill the project horizon needs. This is a fairly naive view of the world, however, since having such a back-up would represent a highly underutilized asset when not in service (> 50% of the time).

Problem 9.4.2: HEC maintenance guarantee A "harsh environment chamber" (HEC) is used for life-testing of circuit boards, and over the past few years the salt-spray version of the testing chamber has been used to quality-check products being used in commercial marine applications. These testing chambers are not that expensive (some labs will have as many as eight systems), and the maintenance has turned out to be a real nuisance because of the salt-spray. The company anticipates strong sales for the next 5 years. The current chamber OEM is Harsh Life, Inc., but a newcomer in the field Test Prep, LLC offers a low-maintenance test chamber for which they absorb all maintenance costs for the first 4 years. The engineering manager cannot resist taking a look at that product to determine if they should replace the current system. Assume that MARR = 17%.

	Old HEC		New HEC	
Year	MV	OE_n	MV	OE_n
0	$5,250		$8,000	
1	4,300	$950	6,000	$ 0
2	3,350	950	5,250	0
3	2,400	950	4,750	0
4	1,450	950	4,250	0
5	500	950	4,000	$1,200

We can use the same spreadsheet template as in Section 9.3.2 (p. 288) to find the results for the defender and the challenger:

Calculation results shown in Table 9.9 yield that the defender's minimum EUAC = $2,158 in year 3, and the challenger has a minimum of EUAC =

Defender			Loss					
Year	MV(n-1)	MV(n)	MV	Z	OE	MAC	PV of TC	EUAC
0	**$ 5,250**				32%			
1	$ 5,250	$ 4,300	$ 950	$ 893	**$ 350**	$ 2,193	$ 1,874	$ 2,193
2	$ 4,300	$ 3,350	$ 950	$ 731	$ 462	$ 2,143	$ 1,565	$ 2,170
3	$ 3,350	$ 2,400	$ 950	$ 570	$ 610	$ 2,129	$ 1,329	**$ 2,158**
4	$ 2,400	$ 1,450	$ 950	$ 408	$ 805	$ 2,163	$ 1,154	$ 2,159
5	$ 1,450	**$ 500**	$ 950	$ 247	$ 1,063	$ 2,259	$ 1,030	$ 2,173

Challenger			Loss					
Year	MV(n-1)	MV(n)	MV	Z	OE	MAC	PV of TC	EUAC
0	**$ 8,000**							
1	$ 8,000	**$ 6,000**	$ 2,000	$ 1,360	**$ -**	$ 3,360	$ 2,872	$ 3,360
2	$ 6,000	**$ 5,250**	$ 750	$ 1,020	**$ -**	$ 1,770	$ 1,293	$ 2,627
3	$ 5,250	**$ 4,750**	$ 500	$ 893	**$ -**	$ 1,393	$ 869	$ 2,278
4	$ 4,750	**$ 4,250**	$ 500	$ 808	**$ -**	$ 1,308	$ 698	**$ 2,089**
5	$ 4,250	**$ 4,000**	$ 250	$ 723	**$ 1,200**	$ 2,173	$ 991	$ 2,101

Table 9.9: Fixed-period EUAC for Harsh Environment Chambers.

$2,089 in year 4. Whether you compare the EUAC at the end of each project horizon or compare minimum EUAC, you note the challenger has a lower EUAC and thus the defender should be replaced.

But is the case for replacement *compelling*? Maybe not. The difference between the maximum and minimum EUAC for the defender is $35 over a 5-year period – less than 2% of the lowest EUAC value. So, certainly maintenance *is increasing at 30% per year*, but the HEC market value is declining at a moderately slow rate and this has a balancing effect on the marginal annual cost. The EUAC only changes by $1.00 (yes, *one* dollar) between year 3 and year 4! In this situation, being preoccupied with one aspect of system cost can lead to an non-optimal use of resources from a system perspective for several reasons:

1. The economics of the defender and the challenger are actually quite similar. If the challenger is to displace the defender, there must be something special about the equipment that enables the challenger to make money. Possibly the quality is actually low and problems do not show up within the first few years, thus the special offer by the dealer. The manager should be careful and conduct due diligence.
2. Even if the equipment is a qualified *technical* replacement, does the differential in savings cover removal of the old system and installation of the new one? Indeed, the end-period EUAC for both systems is only off by $72, and this would not cover even 2 hours of managerial time!

Suppose the salesperson for the challenger decides to lower the price of the new system by $1,000 to make the conversion more appealing. Table 9.10 provides calculation outcomes that do have noteworthy differences, but this is still not exceptionally compelling.

In the end, the idea of having a machine with no maintenance costs (to you) for a period of 4 years is appealing at first glance, but the economics are not so dramatically different that you would recover the switching costs.

9.5 Replacement analysis for an incremental period

Comparing assets with equivalent, fixed project horizons was a simple start to the general topic of replacement analysis; in Section 9.4 we demonstrated the convenience of the approach. However, a fixed horizon overlooks two "asset dynamics" that are often at play in daily industrial life. First, Section 9.4 avoids using the *economic life* of the asset and assumes the EUAC over the prescribed project horizon is satisfactory - this is not always the case. In many companies the annual capital budgeting exercise will call upon the engineering manager to take a closer look at all expenses and identify areas of production where expenses are rising. If an asset is near the end of its life, the engineering manager might begin to wonder, "Are we getting close to replacing the incumbent asset or should we keep it another year?" It is a fair question to ask if an asset should be replaced in advance of the original asset end-date.

New Challenger

Year	MV(n-1)	MV(n)	Loss in MV	Z	OE	MAC	PV of TC	EUAC
0	**$ 7,000**							
1	$ 7,000	**$ 6,000**	$ 1,000	$ 1,190	**$ -**	$ 2,190	$ 1,872	$ 2,190
2	$ 6,000	**$ 5,250**	$ 750	$ 1,020	**$ -**	$ 1,770	$ 1,293	$ 1,996
3	$ 5,250	**$ 4,750**	$ 500	$ 893	**$ -**	$ 1,393	$ 869	$ 1,826
4	$ 4,750	**$ 4,250**	$ 500	$ 808	**$ -**	$ 1,308	$ 698	**$ 1,725**
5	$ 4,250	**$ 4,000**	$ 250	$ 723	**$ 1,200**	$ 2,173	$ 991	$ 1,789

Table 9.10: Fixed period EUAC for new challenger.

Secondly, the fixed project horizon method "assumes away" the obvious condition that the incumbent is near the end of its life. There is no question we *prefer* that the challenger have a much longer life at a lower EUAC! Project lives are different by definition. Overall, the moderate increase in *incremental project period analysis* complexity is outweighed by the improved decision-making that can potentially result; it is the topic of study in this section.

A quick look at the many engineering economics textbooks available today reveals several ways to treat incremental project replacement study. In the present work we draw on the total EUAC curve in Figure 9.3, observing that to the left of the minimum, EUAC declines as each year passes (Type I trend), and to the right the EUAC rises incrementally with each additional year the asset is retained (Type II trend). The differences in ways to study a Type I or II trend for the defender is what leads different books to present different forms of analysis. Further, the relative position of the challenger total EUAC curve will play a part in the decision, too. These four aspects of an incremental analysis can be challenging to describe clearly. A graphical view of the situation is portrayed in Figure 9.4 and Figure 9.5.

In Figure 9.4, point X represents the EUAC of the asset if we were to replace *prior* to reaching the end of its economic life; point Y follows EUAC as it is climbing. As mentioned earlier there might be reasons we find a point X or Y EUAC satisfactory even if the study period is not positioned exactly at the minimum EUAC. In the incremental EUAC analysis, we compute the marginal annual cost (MAC) for the *next year* of the defending asset to compare with a computed EUAC for the challenger. *The specific nature of the comparison depends on which side of the EUAC minimum the defender is positioned on the total EUAC curve.* This is a frequent source of confusion, and Figure 9.4 and Figure 9.5 are worthy of study, reflection, and discussion in teams prior to moving on to calculation details.

9.5.1 Procedure for incremental asset cost calculations

In this section we explore the general procedure for *incremental asset cost* procedure to set the stage for understanding how we compute and interpret EUAC outcomes. Sometimes availability of market value and maintenance data is potentially an issue for *incremental asset cost*, but we proceed as if estimates are available.

Marginal annual cost

Marginal annual cost (MAC) of retaining an asset draws on a detailed year-to-year *incremental* analysis. MAC generally has three components (see Section 9.3.2).

Erosion in asset market value
As each year passes, the market value of the asset decreases. In some cases we are provided or can estimate the decline in market

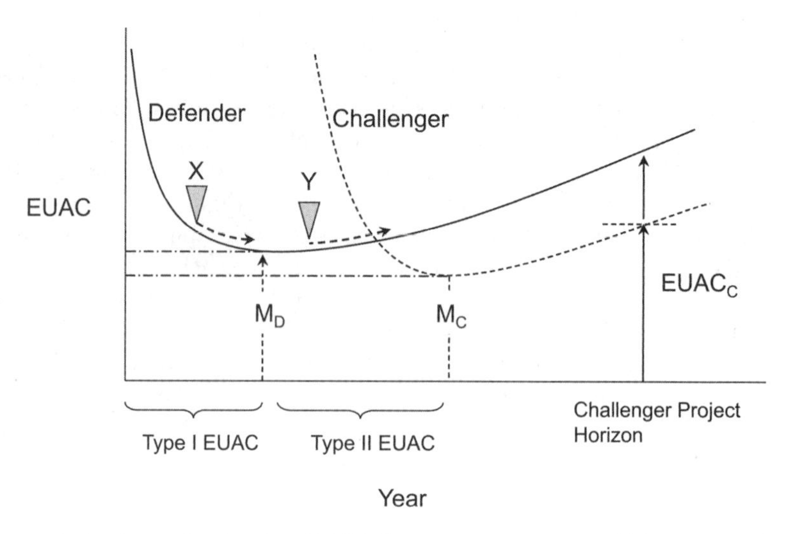

Figure 9.4: Scenario A: Total EUAC curves for Min $EUAC_D < EUAC_C$.

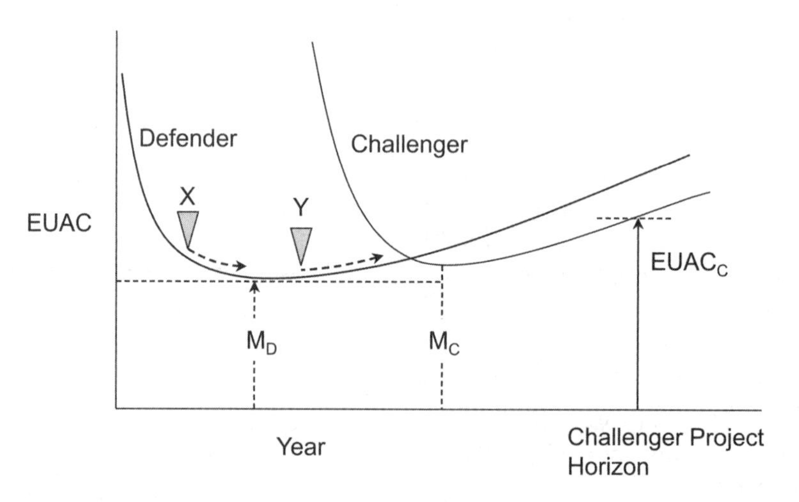

Figure 9.5: Scenario B: Total EUAC curves for Min $EUAC_D > EUAC_C$.

value. It may also be close enough to simply use book value.

$$\Delta MV_n = MV_{n-1} - MV_n \tag{9.3}$$

Opportunity cost
Objectivity asks we consider that investment in the asset for the year carries with it the cost of capital that could be used in other pursuits.

$$Z_n = i\% \; MV_{n-1} \tag{9.4}$$

Operating cost
All of the maintenance, normal repair, utilities, taxes, training, operating labor, etc., are rolled up into the operating cost. These costs might be fixed, set as a percentage of asset value, or itemized in a table

$$OC_n = \sum_{j=1}^{J} E_{n,j} \tag{9.5}$$

Now, given the definitions above, *if we are provided with cost data or can assume* the year-to-year data, then the total Marginal Annual Cost is

$$MAC_n = \Delta MV_n + Z_n + OC_n \tag{9.6}$$

At this point we are indifferent as to whether the analysis is for the defender or the challenger. This is just about the calculation procedure. An example will clarify the steps involved for a typical calculation.

Problem 9.5.1: Nickel deposition chamber replacement
A nickel deposition chamber requires an investment of $125,000 and has the expected asset MV and operational data shown below. If the corporate MARR is 15% for high-tech projects, compute the marginal annual cost for the system over a 7-year period.

Year,n	MV	ΔMV_n	OC_n	Z_n	MAC_n
0	$125,000				
1	$95,000	$30,000	$25,000	$18,750	$73,750
2	$70,000	$25,000	$32,500	$14,250	$71,750
3	$50,000	$20,000	$40,000	$10,500	$70,500
4	$40,000	$10,000	$47,500	$7,500	$65,000
5	$30,000	$10,000	$55,000	$6,000	$71,000
6	$20,000	$10,000	$62,500	$4,500	$77,000
7	$15,000	$50,000	$70,000	$3,000	$78,000

Table 9.11: Marginal annual costs over a 7-year asset project horizon.

Calculations for each year n are the same; selecting $n = 3$ in Table 9.11:

$$\begin{aligned} \Delta MV_3 &= MV_2 - MV_3 \;=\; \$70,000 - \$50,000 \;=\; \$20,000 \\ Z_3 &= i\%\; MV_2 \;=\; 15\%\;(\$70,000) \;=\; \$10,500 \\ OC_3 &= \$40,000 \quad \texttt{(Given)} \end{aligned}$$

$$MAC_3 = \$20,000 + \$10,500 + \$40,000 \;=\; \$70,500$$

Maintenance costs are on a gradient increasing by \$7,500 per year from the baseline of \$25,000, and OC_3 is given as \$40,000.

As is common with many replacement problems, organizing the problem information is more challenging than the simple mathematics involved in the problem.

Some analysts prefer a slightly different form of Equation 9.6. An alternate description of the incremental cost would use the TVM over the 1-year incremental period, and Equation 9.7 results:

$$MAC_n \;=\; MV_{n-1}(1+i)MV_n \;+\; OC_n \tag{9.7}$$

Calculation outcomes are the same, only the problem set-up is slightly different. Notice that we *did not compute EUAC* in this problem; we only needed the incremental cost of asset retention. We'll take up the issue of computing EUAC versus MAC next.

EUAC versus MAC

The use of EUAC calculations must be clear relative to the role of the MAC. Generally speaking, comparing *defender MAC* and *challenger minimum EUAC* can be of interest. However, the EUAC *of* the defender MAC has limited relevance.

Consider the calculation of the EUAC *of* the MAC just to build out the discussion of the sample problem further. Recall that determining the EUAC involves a step-wise process outlined in Section 4.3:

1. Identify the marginal cost data for year n.
2. Compute the PV of the stated annual costs for year n.
3. Compute the EUAC of the aggregate PV costs to establish the EUAC *through* year n.

This method and its outcome were presented in Section 9.3.3 (p. 291). The result for our sample problem is shown in Table 9.12. From table 9.12 we see there is a difference between EUAC *of* the MAC and the MAC value itself. This point is exaggerated somewhat in Figure 9.6.

Table 9.12 and Figure 9.6 bring out several important points. Notice how the `MAC` and the `EUAC` *of* the `MAC` start out at the same point "A" in Figure 9.6, eventually departing, and sometime thereafter the `MAC` curve rises to cross the `MAC` curve at year Y_n. MAC increases and has a terminal value at point "C" that exceeds the terminal value of EUAC at point "B". This typical

Year	MV(n-1)	MV(n)	MV Loss	Oper Cost	OE	MAC	PV of MAC	Project EUAC
0	**$ 125,000**				$ 7,500			
1	$ 125,000	$ 95,000	$ 30,000	$ 18,750	$ 25,000	$ 73,750	$ 64,130	$ 73,750
2	$ 95,000	$ 70,000	$ 25,000	$ 14,250	$ 32,500	$ 71,750	$ 54,253	$ 72,820
3	$ 70,000	$ 50,000	$ 20,000	$ 10,500	$ 40,000	$ 70,500	$ 46,355	$ 72,152
4	$ 50,000	$ 40,000	$ 10,000	$ 7,500	$ 47,500	$ 65,000	$ 37,164	$ 70,719
5	$ 40,000	$ 30,000	$ 10,000	$ 6,000	$ 55,000	$ 71,000	$ 35,300	$ 70,761
6	$ 30,000	$ 20,000	$ 10,000	$ 4,500	$ 62,500	$ 77,000	$ 33,289	$ 71,474
7	$ 20,000	$ 15,000	$ 5,000	$ 3,000	$ 70,000	$ 78,000	$ 29,323	$ 72,064
8	$ 15,000	$ 10,000	$ 5,000	$ 2,250	$ 77,500	$ 84,750	$ 27,705	$ 72,988
9	$ 10,000	$ 7,500	$ 2,500	$ 1,500	$ 85,000	$ 89,000	$ 25,299	$ 73,942
10	$ 7,500	$ 5,000	$ 2,500	$ 1,125	$ 92,500	$ 96,125	$ 23,761	$ 75,034

Table 9.12: EUAC calculation for marginal annual cost.

scenario underscores that the EUAC provides an *average* for a given period and as time unfolds the trend for increasing or decreasing marginal cost is eventually reflected in an increase or decrease in EUAC.

We might forget the reasoning for computing EUAC in the whirlwind of data and calculations involved in typical replacement problem analysis. Recall that the EUAC was to facilitate asset comparisons by providing a convenient representation of asset investment on an equivalent annual basis. This did not mean that the EUAC was "the" costs incurred for a specific year, only that on an equivalent basis the average annual costs could be estimated. For example, $\mathtt{EUAC}_7$ is the EUAC for specific project horizon $n = 7$ and provides the average annual costs over a 7-year period, not the costs at year 7.

The appropriate incremental cost for a specific year comparison is the marginal annual cost. A fundamental reason for computing EUAC is to es-

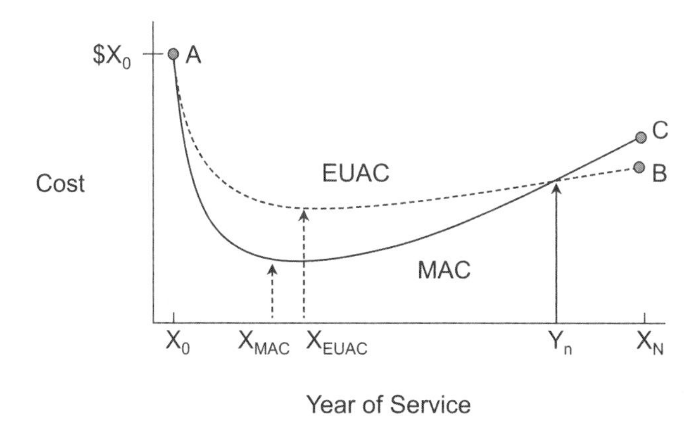

Figure 9.6: Comparison of marginal annual cost (MAC) and the equivalent uniform annual cost EUAC (of only the MAC) for a typical replacement scenario.

tablish a "normalized" asset value amenable to discussion and analysis of replacement options. A few pages ago, in Figure 9.4 (p. 302), we marked the different EUAC curve trends as representing either a Type I or Type II trend. *How* to make the replacement decision is *a function of the estimated* `MAC` *trend.*

Analysis for Type I trends (decreasing incremental MAC)

In a Type I problem the `MAC` values decrease with increasing n, up to the minimum EUAC point, X_{EUAC} on Figure 9.6. If we restrict our attention to this type of trend then let's suppose the defender incremental MAC for year X_n is higher than a challenger EUAC, which some might argue warrants the need to replace the defender. In the decreasing incremental defender MAC situation, a comparison of the defender MAC against the challenger EUAC does not provide an optimal outcome. There are two reasons why computing EUAC is a disadvantage in this situation:

Defender market value not suitable as the defender first cost Comparison of incremental EUAC for the defender against the best challenger EUAC and replacing after 1 or 2 years presumes a suitable replacement exists for the defender; it may not, and we must ask if replacement is *suitable at all*! Changes in market value may still be small enough where the opportunity has not been provided for adequate capital recovery, to the disadvantage of the defender. This elevated value of EUAC, in practice, has no comparable challenger EUAC. A critical period of time would be at least to the point where X_{EUAC} has been reached.

Asset swaps are not financially frictionless No estimate of installation, removal, or other changeover costs has been included in the analysis. Essentially, a "sterile" EUAC calculation assumes a frictionless financial transaction. These values affect the asset disposal and acquisition price, and the EUAC of the challenger must fully incorporate these marginal costs in practice.

The analysis of the Type I problem is therefore quite simple. Begin with the computation of the incremental MAC and confirm that the value is decreasing. If so, compute the defender's minimum EUAC, and if it is less than the challenger's minimum EUAC, then keep the challenger; otherwise replace with the challenger. The defender will be replaced eventually, just not this year, and certainly not until $X_n > X_{EUAC}$.

Analysis for Type II trends (increasing incremental MAC)

In a Type II problem the `MAC` values increase with increasing n. Analysis of the Type II problem begins with the computation of the incremental MAC (and confirms that the value is increasing). If so, compare the defender's MAC, and if it is less than the challenger's minimum EUAC, then keep the challenger; otherwise replace with the challenger. The defender will be replaced eventually, just not this year, but certainly when Y_n is reached.

We can easily imagine that incremental replacement analysis could become quite unwieldy if there were several challengers under consideration, particularly if we were to run through the Type I or Type II problem check for each challenger option. Fortunately, as a practical matter, Type II problems tend to dominate. Since techniques for replacement analysis have been around since 1940, it is not much of a reach to assume the competence of the engineering manager who chose the defender in the last life cycle analysis (the defender was at one time a challenger). From strictly a financial perspective the defender must have been the best choice at the time. If after one year a challenger emerges that seems a viable contender to replace the incumbent, what does this imply? Probably a more realistic scenario is that there was a disruptive technology, significant market shift, or major economic change that has caused the paradigm of the original decision-making process to no longer be valid. Surely this happens, but for the typical on-going concern we want to understand Type I problems, but find that Type II dominate manufacturing process analyses.

9.5.2 After-tax analysis

Like many aspects of replacement analysis, after-tax analysis of replacement problems is cumbersome more than it is difficult. There are essentially three main elements of after-tax cash flow (ATCF) we need to examine for a typical replacement problem:

1. Tax on capital gain or loss on asset disposal or replacement
2. Tax benefit on depreciation expense
3. Tax on ordinary income

Capital gain or loss on asset disposal

At the time of asset disposal, the book value (depreciated value) of the asset may be higher or lower than the market value. If the asset disposal transaction results in a capital gain, then a capital gain tax must be recognized and recorded. In a similar fashion, if the asset sold at a price lower than the book value, then a tax credit is allowed. The tax on the capital gain or loss is based on the prevailing corporate tax rate

$$I_{CG} = T\ (MV_n - BV_n) \tag{9.8}$$

Where

$$\begin{aligned} I_{CG} &= \text{Income tax associated with capital gain (or loss)} \\ MV_n &= \text{Asset market value in year } n \\ BV_n &= \text{Asset book value in year } n \\ T &= \text{prevailing corporate tax rate} \end{aligned}$$

Tax benefit on depreciation expense

Asset depreciation has an impact on the income statement; this non-cash expense is not shown on the statement of cash flows. Including depreciation expenses will typically reduce net income and thus reduce the tax liability to the corporation. This expense adjustment is simply the product of the prevailing corporate tax rate T and the depreciation D_n

$$I_{D_n} = T\ (D_n) \qquad (9.9)$$

As indicated earlier in Equation 9.1 (p. 274), the prevailing corporate tax rate T is the sum of the federal and state corporate tax rates, and is often in the range of 33% to 40%.

Tax on ordinary expenses

In a replacement problem there is usually associated an annual cost of operations. As we have seen in prior examples, the reason for replacing an asset might be that age and wear have resulted in increasing asset operating expenses, and we believe that by replacing the asset we will benefit from more favorable economics.

Annual expenses in *before-tax-cash-flow* (BTCF) problems are normally absent the tax benefit associated with the operations expense that reduces net income and thus reduces overall corporate tax liability. This expense adjustment is simply the product of the prevailing corporate tax rate T and the operating cost OC_n

$$I_{OC} = T\ (OC_n) \qquad (9.10)$$

We see that the inclusion of taxes has the potential to significantly change the EUAC, and the depreciation scheduled used in calculations can profoundly influence economics. This is one of the reasons depreciation schedule are of interest to regulatory and tax professionals.

The EUAC reflects the annualized capital cost of the asset. In our *after-tax-cash-flow* (ATCF) analysis we must remember to adjust the asset cost basis used in the EUAC calculation. Said another way, the first-cost of the asset investment described on page 295 must be adjusted for the tax savings associated with the expected depreciation of the asset.

$$P_{adj} = P - T\ (P - BV_n) \qquad (9.11)$$

Equations 9.8 to 9.12 are not mathematically complex at all, but do present an information management issue for many problems of practical interest. There are several concepts we have to track:

1. Depreciation
2. Before-tax cash flow
3. Taxable expenses
4. Taxable income changes

Spreadsheets are quite commonly used in practice, and a worked problem on page 313 will demonstrate a typical after-tax replacement analysis.

Tax problems are often perceived as quite difficult to the novice, but we emphasized that while state and federal tax codes may take volumes of books to explain, the net impact of taxes on a replacement decision is much less so. *And* it is essential to differentiate between normal consumer sales taxes on the order of 5% to 7%, and corporate income taxes that can vary between 14% and 40%. Sales taxes therefore have less of an effect on replacement decisions than corporate income taxes do. Knowing that many replacement problem calculations involve numerous assumptions and expense projections, what is the likelihood we are off by 4% versus 40%? Refinement of analysis is a matter of magnitude. The tax impact of a large depreciation expense for a moderately new piece of equipment is conceptually and significantly more important than the impact of sales tax on service contracts (which are likely exempt in a business-to-business transaction anyway).

It helps to view the impact of taxes on a replacement problem as simply a form of income and expenses. For instance, depreciation is an expense on the income statement, and our recognition of a reduction in corporate tax liability is equivalent to reducing the cost of owning the equipment. Similarly, annual operating expenses can actually grow to be quite large. In a worked problem to follow, the out-year maintenance expense of a defender is about 30% of the price of a new system. In this case, computing the reduction in tax liability because of the non-trivial maintenance expense is like a reduction in the expense itself.

If manufacturing operations are a significant portion of the balance sheet, taxes in localities that assess property and plant taxes based on the market value of the assets may be significant. It will matter, of course, if the replacement analysis is being done for a municipality versus a private entity, as the former is tax-exempt.

Economic policy may also be worth examining, as often local, regional, or state agencies may offer capital investment tax incentives to foster economic development. There may be allowed a 30% tax credit for capital investments exceeding a certain threshold, say $1 million, knowing that capital investment is a leading indicator of economic growth and job creation. The fundamental idea is that, over the long run, the value of higher employment in a region exceeds the tax allowance that the government provided for the private firm. The level of tax knowledge the engineering manager must acquire will be different in a small firm than in a large one. It is more likely that special economic incentives, R&D tax credits, valuation levies, and the like are more profound in the large company. However, it can be generally assumed extensive tax issues warrant tax professionals on staff at a larger firm, and less so for the start-up or a small firm.

9.5.3 Illustrative worked problems

Problem 9.5.2: Maintenance costs rapidly rising

A pick-and-place assembly system with an expected life of 9 years was purchased 4 years ago for $182,000 and currently has a market value of $24,000. Management believes the operational costs are "getting out of hand" for the next 5-year cycle, as shown in the table below. For the current budget year ($n = 1$), the market value of the system is equal to the operating costs, and the engineering manager has been told, "Surely you can do better than this!"

After speaking with a regional sales rep, there appears to be a new system that can be purchased for only $95,000, with a life of 12 years and a salvage value of $10,000. "You can't beat that" suggests the salesman, and further, "Our system's annual maintenance cost of $17,000 will be at least half the cost of what you'll pay for at the end of life of the existing system." If the corporate MARR is 12%, is this replacement a "go"?

Year, n	MV	OC_n
1	$20,000	$20,000
2	$17,000	$25,000
3	$15,000	$30,000
4	$14,000	$33,000
5	$13,000	$35,000

Solution

We begin by computing the EUAC of the proposed new system (the challenger). Following the procedure from Section 9.4,

$$\begin{aligned} EUAC_C &= P\,(A/P, i\%, n) + OC - SV\,(A/F, i\%, n) \\ &= \$95,000\,(A/P, 12\%, 12) + \$17,000 - \$10,000\,(A/F, 12\%, 4) \\ &= \$95,000\,(0.1614) + \$17,000 - \$10,000\,(0.0414) \\ &= \$15,333 + \$17,000 - \$414 = \$31,919 \end{aligned}$$

Calculations for the defender are shown in the table below, including both the MAC and the EUAC.

Year, n	MV	ΔMV_n	OC_n	Z_n	MAC_n	$EUAC_n$
0	$24,000					
1	$20,000	$4,000	$20,000	$2,880	$26,880	$26,880
2	$17,000	$3,000	$25,000	$2,400	$30,400	$28,540
3	$15,000	$2,000	$30,000	$2,040	$34,040	$30,170
4	$14,000	$1,000	$33,000	$1,800	$35,800	$31,348
5	$13,500	$500	$35,000	$1,680	$37,180	$32,226

We find the defender competes with the challenger, but the comparison must be done carefully. The challenger has an equivalent uniform annual cost of $EUAC_C$ = $31,919, which is greater than the marginal annual cost MAC_2 = $30,400 and less than MAC_3 = $34,040 for the defender. Since the marginal annual cost of the defender for year 3 exceeds the challenger's EUAC, the defender should be replaced in year 3.

Note that if we were to have used the EUAC of the marginal cost, then $EUAC_C < EUAC_{D,5}$, and the defender replacement would have been delayed an additional 2 years.

Problem 9.5.3: Keeping pace with regulatory changes

Cloud Computing Corporation has successfully provided data storage and access systems for commercial clients for about 5 years. Anticipating a rebound in the economy, they chose to launch an in-state data "server farm" capability to take advantage of investment tax credits the state was offering on a one-time basis. Key conditions for the investment tax credit were the hiring of 10 additional employees and to keep the center operational for a period of 6 years. Given the growth in data storage services this was not anticipated to be a problem. About a year after the installation "went live," the company was pleased they were rapidly acquiring customers in the health-care space. Unfortunately, a security audit revealed that the data center was not compliant with all health-care IT (HIT) concerns, especially with the growth in mobile HIT. Essentially they had a choice of (a) overhauling the current system at a lower initial cost but with higher long-term software maintenance costs, or (b) replacing the system outright with lower long-term maintenance costs.

Project Parameter	Defender	Upgrade	Replacement
First cost, P	$30,000	$15,000	$70,000
Operating costs, OE	$2,000	$10,000	$3,500
OE Annual Escalation	30%	30%	10%
Service period, n	5 years	5 years	5 years
Salvage value, SV	$0	$0	$35,000

Several of the steps involved in solving this problem should be familiar from Section 9.3.2, page 304. A twist, though, is that the defender is inadequate to provide the required level of service and the system must undergo an upgrade, shown in the table above. In this case we combine the defender and the upgrade expense, provided with matching lives and similar OE escalation rates. Let's assume that management is seeking an aggressive return (because of past errors) and sets MARR = 25%. As well, assume that due to the rapidly changing technology environment, the equipment has no salvage at the end of the 5th year. Straight-line depreciation is used for simplicity.

The calculation methods and sequence are the same for both the defender and the challenger and follow the same procedure that led to Table 9.6 on page 305:

1. Identify the marginal cost data for year n.
2. Compute the PV of the stated annual costs for year n.
3. Compute the EUAC of the aggregate PV costs to establish the EUAC *through* year n.

From Table 9.13 it is easy to show that for a fixed project horizon $n = 5$, the defender EUAC of $36,068 is higher than the challenger EUAC of $25,862; the challenger is thus the preferred option. More importantly, the decision under consideration is for $n = 1$, in which case the defender EUAC of $32,250 is *still* higher than the challenger EUAC of $28,000; in fact, *none* of the defender EUAC values are lower than the challenger, so the replacement should take place right away.

This result really is not much of a surprise, since the defender is hampered with the need for a software upgrade that adds $10,000 a year to the operational costs. We leave it for the reader to show that if the defender maintenance expense was equal to that of the challenger, the preference would be to keep the defender (defender EUAC = $23,750; challenger EUAC remains at $28,000).

Defender Year	MV(n-1)	MV(n)	Loss MV	Z	OE	MAC	TC	EUAC
0	**$ 45,000**				30%			
1	$ 45,000	$ 36,000	$ 9,000	$ 11,250	**$ 12,000**	$ 32,250	$ 25,800	$ 32,250
2	$ 36,000	$ 27,000	$ 9,000	$ 9,000	$ 15,600	$ 33,600	$ 21,504	$ 32,850
3	$ 27,000	$ 18,000	$ 9,000	$ 6,750	$ 20,280	$ 36,030	$ 18,447	**$ 33,684**
4	$ 18,000	$ 9,000	$ 9,000	$ 4,500	$ 26,364	$ 39,864	$ 16,328	$ 34,756
5	$ 9,000	**$ -**	$ 9,000	$ 2,250	$ 34,273	$ 45,523	$ 14,917	$ 36,068

Challenger Year	MV(n-1)	MV(n)	Loss MV	Z	OE	MAC	TC	EUAC
0	**$ 70,000**				10%			
1	$ 70,000	$ 63,000	$ 7,000	$ 17,500	**$ 3,500**	$ 28,000	$ 22,400	$ 28,000
2	$ 63,000	$ 56,000	$ 7,000	$ 15,750	$ 3,850	$ 26,600	$ 17,024	$ 27,378
3	$ 56,000	$ 49,000	$ 7,000	$ 14,000	$ 4,235	$ 25,235	$ 12,920	$ 26,816
4	$ 49,000	$ 42,000	$ 7,000	$ 12,250	$ 4,659	$ 23,909	$ 9,793	**$ 26,312**
5	$ 42,000	**$ 35,000**	$ 7,000	$ 10,500	$ 5,124	$ 22,624	$ 7,414	$ 25,862

Table 9.13: Healthcare IT server farm upgrade or replacement.

An interesting assumption to change is from straight-line depreciation to a 5-year MACRS schedule. The results of this calculation are shown in Table 9.14.

Defender Year	MV(n-1)	MV(n)	Loss MACRS	Z	OE	MAC	TC	EUAC
0	**$ 45,000**				30%			
1	$ 45,000	$ 36,000	$ 9,000	$ 11,250	**$ 12,000**	$ 32,250	$ 25,800	$ 32,250
2	$ 36,000	$ 21,600	$ 14,400	$ 9,000	$ 15,600	$ 39,000	$ 24,960	$ 35,250
3	$ 21,600	$ 12,960	$ 8,640	$ 5,400	$ 20,280	$ 34,320	$ 17,572	**$ 35,006**
4	$ 12,960	$ 7,776	$ 5,184	$ 3,240	$ 26,364	$ 34,788	$ 14,249	$ 34,968
5	$ 7,776	$ 2,592	$ 5,184	$ 1,944	$ 34,273	$ 41,401	$ 13,566	$ 35,752

Challenger Year	MV(n-1)	MV(n)	Loss MACRS	Z	OE	MAC	TC	EUAC
0	**$ 70,000**				10%			
1	$ 70,000	$ 56,000	$ 14,000	$ 17,500	**$ 3,500**	$ 35,000	$ 28,000	$ 35,000
2	$ 56,000	$ 33,600	$ 22,400	$ 14,000	$ 3,850	$ 40,250	$ 25,760	$ 37,333
3	$ 33,600	$ 20,160	$ 13,440	$ 8,400	$ 4,235	$ 26,075	$ 13,350	$ 34,380
4	$ 20,160	$ 12,096	$ 8,064	$ 5,040	$ 4,659	$ 17,763	$ 7,276	**$ 31,498**
5	$ 12,096	$ 4,032	$ 8,064	$ 3,024	$ 5,124	$ 16,212	$ 5,312	$ 29,636

Table 9.14: Healthcare IT server farm upgrade based on MACRS depreciation schedule.

In the MACRS case, the year 5 salvage value is non-zero, and the difference in asset depreciation leads to the situation where the defender is not as

Year	MACRS Recovery Rates MACRS Class Life					
	3	**5**	**7**	**10**	**15**	**20**
1	0.3333	**0.2000**	0.1429	0.1000	0.0500	0.0375
2	0.4445	**0.3200**	0.2449	0.1800	0.0950	0.0722
3	0.1481	**0.1920**	0.1749	0.1440	0.0855	0.0668
4	0.0741	**0.1152**	0.1249	0.1152	0.0770	0.0618
5		**0.1152**	0.0893	0.0922	0.0693	0.0571
6		**0.0576**	0.0892	0.0737	0.0623	0.0528
7			0.0893	0.0655	0.0590	0.0489
8			0.0446	0.0655	0.0590	0.0452
9				0.0656	0.0591	0.0447
10				0.0655	0.0590	0.0447
11				0.0328	0.0591	0.0446

Table 9.15: MACRS depreciation schedule.

immediately poor an option as before. The depreciation difference between the defender $EUAC_1 = \$32,250$ and challenger $EUAC_{min} = \$31,498$ that was \$4,250 (straight line) is now \$2,750. This makes since the purpose of the MACRS schedule is to accelerate the rate of depreciation, which will increase the marginal annual cost, thus causing the EUAC curve to rise.

Possibly more important than the difference between depreciation schedules is that way in which outcomes can be influenced by assumptions. Some degree of consistency must be maintained for results to be meaningful, and this is where corporate policy is important. For an on-going concern, a specific policy on replacement will adjust over time so that the prediction technique approaches the reality that unfolds. For a specific industry, taxes, property valuations, incentives, and tweaks to the depreciation schedule should play out in the long haul.

Problem 9.5.4: After-tax analysis Bethesda Imaging purchased a pick-and-place electronics assembly system with a built-in UV paralyne coating system for \$87,000 4 years ago. Originally the system was expected to last 10 years and have minimal operational support costs, about \$2,700 annually. To respond to changes in the market calling for greater variety of UV coating options for circuit boards, Bethesda is considering replacing the system with one that does not cost *much* more than they originally spent, as shown in Table 9.16. Management believes they could make the switch easily since the new system is more flexible, less expensive to operate, and has a surprisingly high salvage value. The vendor trying to sell the new system is willing to pay \$55,000 for the old system and sell it to another OEM for \$55,000.

Despite the attractive nature of the new equipment deal, the before-tax analysis was not very favorable. The existing system had an EUAC of \$17,626 that was \$5,500 less than the challenger system EUAC of \$23,140. Man-

Project Parameter	Defender	Challenger
First cost, P	$87,000	$90,000
Operating costs, OE	$2,700	$1,500
Service period, n	10 years	6 years
Salvage value, SV	$0	$25,000

Table 9.16: Pick-and-place assembly system options for Bethesda Imaging.

agement feels that the nature of this project warrants a look at after-tax calculations, since it is believed that "too much was left out" of the original calculation.

An after-tax calculation using the process outlined on page 307 provides the results shown in Table 9.17. Although the difference in EUAV has shrunk by $2,900, the existing system has an EUAC of $10,515 that is *still* less than the challenger system EUAC of $13,109. In this worked problem, the calculation objective is to work through three common elements of after-tax cash flow (ATCF):

1. Tax on capital gain or loss on asset disposal or replacement
2. Tax benefit on depreciation expense
3. Tax on ordinary income

where the capital gain (or loss) component is given by Equation 9.8:

$$I_{CG} = T\ (MV_n - BV_n),$$

given

$$\begin{aligned} I_{CG} &= \text{Income tax associated with capital gain (or loss)} \\ MV_n &= \text{Asset market value in year } n \\ BV_n &= \text{Asset book value in year } n \\ T &= \text{prevailing corporate tax rate} \end{aligned}$$

The second component is the tax benefit on depreciation expense, Equation 9.9:

$$I_{D_n} = T\ (D_n)$$

And lastly, the tax shield from operating costs, Equation 9.10 :

$$I_{OC} = T\ (OC_n)$$

And, of course, we have to adjust the "first-cost", with Equation 9.12

$$P_{adj} = P - T\ (P - BV_n) \tag{9.12}$$

The detailed calculations and results are presented in Table 9.17. It is quite helpful to reference the workbook that accompanies this chapter to trace the detailed calculations.

Present Machine (Defender)

Before Tax			After Tax	
Original Cost:	$87,000		Annual Depreciation:	$8,700.00
Original Life	10	years	Book Value:	$52,200.00
Investment:	$55,000		Rec'd depr (RD) on Trade-In:	$2,800.00
Upgrades:	$0		RD tax savings:	$952.00
Net Investment:	$55,000		After tax investment	$54,048.00
Remaining life:	6	years	Annual tax savings:	$3,876.00
Annual Expenses:	$2,700		Actual after tax expenses:	-$1,176.00
MV (Salvage):	$0			
EUAC: Annual Expense		$2,700.00	EUAC: Annual Expense	-$1,176.00
+ Investment Annualized		$14,926.44	+ Investment Annualized	$11,691.41
- Salvage Annualized		$0.00	- Salvage Annualized	$0.00
		$17,626.44		**$10,515.41**
			Difference (savings) of Tax-Free and ATCF:	$7,111.03

Challenger

Before Tax			After Tax	
Investment:	$90,000		Depreciation:	$15,000.00
Life:	6	years	Annual tax savings:	$5,610.00
Annual Expenses:	$1,500		Actual after-tax expenses:	-$4,110.00
MV (Salvage):	$25,000		RD for salvage:	$8,500.00
EUAC: Annual Expense		$1,500.00	Annual Expense	-$4,110.00
+ Investment Annualized		$24,425.09	+ Investment Annualized	$19,468.38
- Salvage Annualized		-$2,784.75	- Salvage Annualized	-$2,249.20
		$23,140.34		**$13,109.18**
Delta:		**($5,513.90)**	**Delta:**	**($2,593.77)**
			Difference (savings) of Tax-Free and ATCF:	($2,920.13)
MARR =	**16%**			
i =	**8%**	After Tax		
t =	**34%**	Tax Rate		

Table 9.17: After-tax replacement problem.

9.6 Summary

The ability to understand the issues, sort through the chaff and work through capital budgeting and after-tax replacement problems places the engineer at a competency level that lends credibility to the individual at the board level. It should be evident that these types of calculations are not mathematically difficult as much as they can be hard to get organized. Knowing what is relevant is an issue, too.

Several facets of career growth come with success at capital budgeting and replacement analysis:

1. Capital budgeting impacts operations for a very long time, and playing a role in steering investments in the future can be rewarding and provide visibility to the engineering manager. It is not very often that capital equipment is replaced after 1 year of use, but if it is, there will be a lot of meetings about what went wrong. Keep your resume polished if the pendulum swings the wrong way.
2. Capital equipment projects provide an opportunity for creativity and future scenarios to be envisioned. It can be fun to see the variety of proposals people can generate in response to a request for proposals. You can learn as much from the unfundable ideas as you can the funded ones.
3. Working through a variety of project options in advance of budget committee meetings gives the engineering manager a chance to network throughout the company, as often you'll be interacting with sales, marketing, operations, and finance in shaping the story.
4. For some engineering managers, the work with vendors might be their first real experience in negotiating terms and conditions. Many engineers go though their undergraduate education without any practical (or even simulated) negotiation experience and this new facet of business has its own tricks to learn. Working from small projects to large ones can even potentially entrain the engineer in merger and acquisition discussions.

The problem sets to follow most certainly offer the reader the chance to practice calculations, but we've added quite a few discussion questions, too. This is intended to spark discussion in small groups, giving the opportunity to move beyond just getting the right answer and digging in to see if basic concepts can be rewarding. To be able to argue a point of view will set you apart, and this chapter is rich with all sorts of tax, ethics, and financial issues that are very much a part of engineering management today.

9.7 Problems to work

Problem 9.1
There are three general categories of capital budget scenarios: replacement, expansion, and investment in a NewCo. Describe the overall decision-making context for each. How do they draw on similar skills? Dissimilar skills?

Problem 9.2
The overall process of creating a capital budget proposal has a lot of similarities to writing a business plan for a start-up company. Describe three aspects of the similarities between a budget proposal and a business plan and three ways in which they are distinctively different.

Problem 9.3
In analysis, some focus seems to be on the need for NPV equations to be applied to projects that are mutually exclusive. But in practice we find that the lines are blurred in capital budgeting. Provide a few reasons this situation might exist.

Problem 9.4
In capital budgeting, a distinction has been made regarding the process of "assessment" as being distinctively different than "evaluation." What is the difference and why is it important?

Problem 9.5
If bond money tends to be the least expensive cost of capital, why do we bother with the WACC in project analysis? It seems like the cost of equity is hard to estimate, and maybe we are just guessing anyway. Why is WACC so popular?

Problem 9.6
In developing a capital budgeting portfolio, there are times when a project with a lower return is squeezed in to the budget but a higher-return project is excluded. How can this be fair? Wouldn't this result in project managers inflating expected returns?

Problem 9.7
In ranking projects with the NPV criterion, there are situations where a short-term high cost of capital project is more favorable than a long-term low cost of capital project. Does this make sense? How might this lead to short-term thinking?

Problem 9.8
Assigning an MARR seems more like an art than a science. How, then, is the assignment of MARR meaningful?

Problem 9.9
In evaluating projects of unequal lives, what are some of the limitations of the "repeatability analysis" that might be cause for concern?

Problem 9.10
What is the difference in the reinvestment assumption of NPV and IRR, and what might be the limitation of those strategies in selection of capital equipment?

Problem 9.11
Suppose that changes in the economy are making investors more risk-adverse. How would this affect the WACC (through the cost of debt and the cost of equity)?

Problem 9.12
As discussed in Section 9.2.4, WACC is popular and routinely used in a variety of capital budgeting situations. It seems, though, that the factor is not without controversy. What are some of the controversies and what cautions would you offer to avoid mis-use?

Problem 9.13
Examine the MARR values listed in Table 9.2, p.276 and provide some reasonable explanations for at least 2 of the project types and corresponding MARR values. Try to find an article in the current business press that reinforces your explanations.

Problem 9.14
Select an article from the current press related to the after-tax cost of debt and explain what effect this might have on company financing goals.

Problem 9.15
Szechwan Gardens, Inc. florist chain has a tax rate of 40%, and they have some bonds outstanding (from an earlier store expansion) with a yield of 8%. The CFO determined that the "best" capital structure for their business is 60% debt. What is the cost of equity capital if the WACC is 8.75%.

Problem 9.16
In the capital budgeting discussion of Section 9.2.7, the scoring of project criteria in Table 9.3, p. 278, seems somewhat arbitrary. Wouldn't it be easy to manipulate the scoring to "fix" the scale so that certain projects are favored? What can be done to avoid this?

Problem 9.17
Does it seem odd that ROI can have several limitations and it is the basis for prioritizing projects in Figure 9.2? Why or why not?

Problem 9.18
The project criteria outlined in Table 9.5 was provided for a sample problem. What other criteria might be a useful part of the more general case of project portfolio management?

Problem 9.19
To build out a complete set of services, a lab manager would like to add a small e-beam deposition machine that costs about $60,000 to be more full-service to the lab customers as well as adding about $15,000 in annual revenue to the lab. Free-cash flow is tight for the lab right now, so he was approached about leasing the system for 1/3 the sale price of the machine for each of three years. Since the machine would have a market value of $10,000 in three years, the leasing agent tells the manager he'll save money. The lab is part of a community college and is tax-free, but they do have an MARR of 12%. Should they lease or buy?

Problem 9.20
For Problem 9.19 a special "Tiger Team" has identified a different use for the $60K that involves an incremental profit and cash flow (before tax and depreciation allowance) of $20,000. They are proud to have beat the other team with higher revenue. If this project were subject to net tax of 24%, what would be the after-tax cash flow?

Problem 9.21
If a firm always uses retained earning to fund capital equipment, then is the cost of capital irrelevant? Why or why not?

Problem 9.22
In replacement problems the "life" of an asset is defined three ways. What are the three types, and why is the difference important?

Problem 9.23
Szechwan Gardens, Inc. has been thinking about further expansion of the business. They have $40,000 available to spend. If the required rate of return is 9%, which of the following should be selected for investment? Assume that all options have a 7-year life and a PW analysis is preferred.

Project	Investment	Cash Flow
A. New product coolers	$12,000	$2,000
B. On-line order entry	$15,000	$8,000
C. New delivery truck	$22,000	$3,000
D. New mall kiosk store	$41,000	$9,000

Problem 9.24
Early Stage Venture Partners has been presented with four investment proposals and they have $12,000 left in their venture fund. If the required rate of return is 25%, which of the following should be selected for investment? Assume that there is more than one project that is viable but the investment is $12,000 in each case.

Project	PW	Horizon
A. Teddy-Fan cookie cutter kits	$450	5
B. LNG motorcycle conversions	$15,000	10
C. Catalytic converter recycling	- $2,000	12
D. Cell phone on-line sales	$1,900	7

Problem 9.25
Projects A and B are being considered by a firm with an MARR of 18%; the VP has emphasized that the projects are mutually exclusive (the other 2 VPs don't seem to think it matters) and that he prefers the analysis be done with PW calculations as "today matters more than tomorrow" (and the engineering manager thinks there's a problem with that, too). Which of the two alternatives should the engineering manager recommend?

Project	Y0	Y1	Y2	Y3
A	-$10,000	$4,000	$5,000	$6,000
B	-$15,000	$7,000	$7,000	$7,000

Problem 9.26

Projects A, B, and C are being considered by a firm with a MARR of 18%. The company has lagged competitors for years, and changes in the FDA regulations means something must be done soon. Management does not want to deal with the problem again for the foreseeable future and asked for cash-flow estimates for the next 10-year period.

	Fix A	Fix B	Fix C
Initial investment	$300,000	$400,000	$500,000
Salvage value	$50,000	$60,000	$70,000
Expected gross income	$200,000	$300,000	$400,000
Expected expenses	$100,000	$150,000	$175,000

With a maximum capital budget of $900K, what combination of projects makes sense to invest in?

Problem 9.27

A decision must be made between two very high-technology projects, Project X and Project Y, both with a very short life of 3 years and each having the same cost of $9,499. The firm's normal MARR is 10%. The difference is in the risk of each project, as shown below:

Project X		Project Y	
Probability	Net Cash Flow	Probability	Net Cash Flow
0.25	$5,000	0.25	$0
0.50	9,000	0.50	8,000
0.25	12,000	0.25	20,000

1. Identify which of the two projects is more risky.
2. If the riskier project has a MARR 2% higher than the company norm, what is the NPV for these projects?

Problem 9.28

Three mutually exclusive projects all have the same life of 5 years and the MARR is 10%. Is there a difference in selection outcome if rate of return or present worth is used in the analysis? Which alternative is the preferred investment?

Project X		Project Y	
Probability	Net Cash Flow	Probability	Net Cash Flow
Project investment	$12,000	$16,000	$20,000
Expected annual income	3,000	5,000	8,000
Return on investment	8%	10%	12%

Problem 9.29

Let's assume Bethesda Imaging, Inc. has $30 million cash on the balance sheet. Executives at the annual board meeting passed on any of the investments from the prior year, and correctly anticipated lower profitability for the past year. The new strategy is to set aside $7.5 million for future investments. In response, the Chief Operating Officer (COO) went back to her direct reports for investment ideas that (a) would meet the minimum MARR set forth in Table 9.2 and (b) would not each exceed a $4 million investment cap. The four division managers under the COO came up with just 4 projects; managers have tried to be persuasive with the value of their proposals with very long documents and supporting studies, but in the end the COO simply extracted the data shown below:

Project	Project Type	Investment	ROI
A.	Replace coil-winding machine to maintain production reliability	$1.1M	10%
B.	Expand new plant locally (with investment tax breaks)	$4.8M	16%
C.	New imaging accessory product line in existing market	$4.5M	22%
D.	Quality system to improve software regulatory compliance	$2.5M	30%

What is the investment portfolio that you would recommend?

Problem 9.30

Which of the alternatives described below is more economical if (a) the assets are depreciated on a straight-line basis, (b) the tax rate is 40%, and (c) the after-tax MARR is 6%?

	Investment	Cash Flow
A. Asset investment	$12,000	$18,500
B. Salvage value	$750	$3,000
C. Life (years)	8	8
D. Operating cost	$2,000	$1,000

1. How does the outcome change if this was a before-tax analysis?
2. What is the after-tax result if the life had to be shortened by 3 years?
3. Describe the difference between a before-tax and after-tax MARR.
4. Perform a parametric analysis on the corporate tax rate and explore what impact this might have on decision-making.

Problem 9.31

Tax credits enabled and persuaded a company to invest in a new single-material 3-D prototyping machine. A year after the system was up and running, two-material 3-D prototyping became the rage, and anyone who was not at the leading edge was losing customers. The company had a choice of (a) overhauling the current system at a lower initial cost but with higher long-term maintenance costs, or (b) replacing the system outright with lower long-term maintenance costs.

Project Parameter	Defender	Upgrade	Replacement
First cost, P	\$25,000	\$15,000	\$80,000
Operating costs, OE	\$2,000	\$7,000	\$4,000
OE Annual Escalation	25%	25%	10%
Service period, n	5 years	5 years	5 years
Salvage value, SV	\$0	\$0	\$30,000

Assume that management is seeking an aggressive return (because of past errors) and sets MARR = 22%. As well, assume that due to the rapidly changing technology environment the equipment has no salvage value at the end of the 5th year. Straight-line depreciation is used for simplicity. Which alternative is most favorable?

Problem 9.32

A deep reactive ion etching (DRIE) system with an expected life of 10 years was purchased 5 years ago for \$175,000 and currently has a market value of \$92,500. Predicted operational costs the next 5-year cycle are shown in the table below. A state grant incentive program has recently been launched that encourages capital investment for industrial modernization. A new \$120,000 DRIE has a life of 10 years and a salvage value of \$12,000. If annual operating costs are \$13,222 and the corporate MARR is 12%, is this replacement a "go"?

Year, n	MV	OC_n
1	\$71,000	\$15,000
2	\$60,000	\$18,000
3	\$50,000	\$21,000
4	\$23,000	\$24,000
5	\$9,000	\$28,000

References

[1] ABET. *Criteria for accrediting engineering programs.* ABET, 2010.

[2] D.H. Marshall, W.W. McManus, and D.F. Viele. *Accounting: What the numbers mean.* McGraw-Hill, 7th edition, 2007.

[3] Eugene F. Brigham and Michael C. Ehrhardt. *Financial management.* South-Western CENGAGE Learning, 2008.

[4] James A. Tompkins, John A. White, Yavuz A. Bozer, and J.M.A. Tanchoco. *Facilities planning.* Wiley, 2010.

[5] Kerry Patterson. *Crucial conversations: Tools for talking when the stakes are high.* McGraw-Hill, 2002.

[6] Barbara S. Romzek and Melvin J. Dubnick. *Accountability in the public sector: Lessons from the Challenger tragedy*, volume 3. 7 1987.

[7] George W. Terborgh. *Dynamic Equipment Policy.* McGraw Hill, 1949.

[8] Michael J. Mauboussin. *Think twice: Harnessing the power of counter-intuition.* Harvard University Press, 2009.

[9] Robert Wescott and Adam Karson. *2012 Equipment leasing and finance: US outlook.* Equipment Leasing Foundation, 2012.

Chapter 10

Leadership

10.1 Learning objectives

Leadership is important in any organization, and poor decisions are often remembered more vividly than good ones. Simply put, decision-making quality is considered synonymous with leadership competency. Even for a well-liked leader, engineers will attribute faulty decisions to managerial incompetence, inexperience, and (in cynical cases) a lack of intelligence. In this summary chapter we examine financial decision-making at three levels: individual, group, and organizational. We provide case studies to draw on and integrate the knowledge from earlier chapters. Case studies are intended to (a) draw on knowledge that should have been gained from your investment in the subjects of this book, (b) motivate your further development of competencies to assist in dealing with ambiguous situations, and (c) inspire a desire for a broader view of financial decision-making whereby this book is just a springboard for self-directed learning.

After careful consideration of the material in this chapter and a reasoned approach to addressing key issues in each of the case studies, the reader should be able to:

1. Extract and articulate critical-path issues from ambiguous business settings, leading to identification of problems and opportunities.
2. Demonstrate an ability to think critically about solutions to problems that have a more complex context than typical end-of-chapter problems.
3. Understand a few of the challenges faced by decision-makers who may have limited information available on which to assess alternatives.
4. Exhibit self-directed activity by identifying questions and strategic pathways that draw on the need to acquire new knowledge beyond what is provided in the case.

10.2 Leadership and decision-making

A great deal has been written about decision-making because this is one of the most important skills a leader can develop and refine over the course of a career. The truth is that one or two flawed high-stakes decisions can define a reputation faster than a dozen excellent ones; engineering management activity might involve superb daily efforts, but it might seem "you only get noticed when something goes wrong". We have spoken about decision-making extensively in Section 1.2.3 (p. 7) and wish to provide a greater context in the current discussion through several case study exercises.

Of all the approaches to learning about the vast topic of leadership, case studies are quite useful in two fundamental ways. First, they tend to be quite interdisciplinary and reflect working situations in which there is limited time to work all the relevant issues for which there never seems to be enough data that we need. Second, case studies require the reader to become an actor in the process, one who is not permitted the safety of standing on the sidelines. The latter is critical since we often fail to recognize that there are fundamental attribution errors in the way we judge others' decisions in comparison to our own. There seem to be three myths about managerial decision-making that become most vivid when judging others:

1. *Decisions are primarily an intellectual exercise that involves linear, rational processes and analyses of adequate data in which a decision follows and action is taken.* This myth suggests a tidy formulaic situation, but even a brief tenure in industry suggests that decisions on meaningful issues have few such attributes.
2. *The manager makes the decision.* Actually, even the idea that the manager is in control is mythical, as many people in hallways, break-rooms, and a variety of other informal situations have great bearing and influence on any decisions made; surely, some decisions *are* made exclusively by an engineering manager, but team activity often mitigates. In large companies there are many teams, political alliances, and subcultures that will influence outcomes.
3. *Decisions are followed by implementation.* Actually, the myth here derives from the idea that a decision is an event when it is more accurate to think of it as a process. There are many branches and options as the team iterates on available information and begins to see new possibilities for action.

It is hard to break away from this frame of thinking unless (a) decision-making is experiential, (b) situations include information that is not numerical data, and (c) implementation expectations are clear that might affect strategic formulation of options. Rare is the engineer comfortable with selling, debate, and negotiation, and team case study exercises foster a more realistic sense of day-to-day decision-making behavior.

Learning how to make decisions as a leader takes time. Recall from Section 1.3, Figure 1.7, p. 10. That a novice decision-maker might be characterized as having a slow, methodical, deliberate style in which the manager carefully and consciously steps through a decision based on an well-defined issue, the presentation of an examination of the data, and the selection of an alternative. This style is afforded for the novice only in limited situations. More commonly the situation emerges over time and an overall system view is needed to frame the context properly. For managers who have "been around the block." the ability to filter and prioritize important and relevant information comes quickly, and with practice some steps executed unconsciously provide the appearance of expertise. As long as the domain of the problem remains similar to prior experiences, then decision-making comes faster and easier to the manager, but this is not to say decision-making is any *better* over time, as the work of Nutt [1] clearly has illustrated that most managers are wrong at least half the time! That might be an easy source by which to erode managerial credibility, and we address that through four case studies.

This chapter will feature "scenario analysis" to expand the discussion, recognizing that the references are incomplete models of entrepreneurial decision-making (one is external, the other internal). In this spirit the work of Tingling and Brydon helps to build a more complete picture in an organizational context; this is combined with some discussion of Tingling and Brydon to round out an understanding of how decisions are (or could be) made.

10.2.1 Why decisions fail

As mentioned above, the work of Nutt [1] reported on the results of a multi-year study of decisions made by managers. A key aspect of the work was that after a two-year period had passed, Professor Nutt went back and interviewed the managers who made the decisiosn, only to discover that the decisions were abandoned or not followed by implementation. What went wrong? In essence, the assumption that technical implementation or incorrect framing of the problem was not a root cause was faulty; instead, social processes dominated.

Key findings of the study are:

1. Rush to judgment -
   (a) Locking in on the first solution
   (b) Narrow look at motivations and remedies
2. Misuse of resources -
   (a) Failing to take a systems perspective
   (b) Inadequate scenario analysis

   Failure-prone tactics -
   (a) Inadequate participation in decision-making
   (b) Failure to uncover all concerns
   (c) Limiting search for remedies
   (d) Pressure to "have an answer"

With the issues above in mind, some ideas and approaches to consider in analyzing the case studies are:

1. Establishing a positive decision-making culture
2. Generating potential solutions

3. Evaluating alternatives
    (a) Frame and manage risk; most decisions involve some risk
    (b) Anticipate unintended consequences since it is hard to predict implications of a decision
    (c) Explore "feasibility" and ask if the choice is actually implementable (realistic)
    (d) Follow up and follow through: checking the decision and communication

Benjamin Franklin is claimed to have stated "Well done is better than well said." While this textbook has presented many topics in finance, accounting, and engineering economics and offered insight on decision-making, the ability to put the tools *in action* is hard work. Case studies allow you to work though many financial-related issues in a more complicated and richer context than typical end-of-chapter problems. All involve finance but sometimes in subtle ways, and this reflects everyday management challenges. The focus on ethics will be very evident as it seems more important today than ever before.

10.2.2 Individual decision-making skills

At the individual level, there are several aspects of individual decision-making to work to refine:

1. Understanding cognitive biases and decision-making traps
2. Managing risk and recognizing opportunity
3. Recognizing intuition and pattern recognition risks
4. Reasoning by precedent versus first principles
5. Asking the right questions

The work of Christopher Chabris and Daniel Simons [2] is a remarkably clear book with many examples of the concepts itemized above. Spending even a small amount of time exploring the notion of intuition and pattern recognition risks is quite valuable. Several "illusions" that influence financial as well as general management issues are illuminated in the book.

The chapters of this book have focused on a decision-making process that *in practice* requires the collection of a lot of data. We seek this information from many sources within the company, and despite "best efforts," we end up being cognitively limited, that is, we simply are unable to collect all of the data we need for the "right" decision. A limitation here is that the manager can fall into the mode of finding what is *satisfactory* rather than what is *optimal.* Critical thinking would ask us to frame what data are needed and articulate the procedure to follow or compromise involved if certain types of information are missing, compromised, or potentially mis-represented. The "expert" is known for prowess in knowing what to do, but this comes with the assumption of a specific set of problem patterns that are implied to be accurate; upper-level executives and consultants alike are sometimes not "asked the question" as on the one hand this might be viewed as "challenging authority" but on the other hand, questions are at the root of starting necessary discussions.

Sometimes short-cuts are good ways to avoid a sunk-cost trap of "time invested" by an analyst, who may have so much personal investment and belief in a solution strategy or final decision that the person cannot step back and see the broader context of a decision. A brief "back of the envelope" discussion can help cut to some more general issues that finance alone cannot resolve. Off-balance-sheet financing and its impediment to transparency is a great example. The longer someone has taken to work and arrive at a solution, the more likely that person is to continue to invest in working through the solution framework he or she crafted. It becomes clear the team must work to examine a problem using a variety of techniques, as in the before-tax and after-tax calculations. Spreadsheets can be difficult to question, and the expansive workbook structure can sometimes lead to flawed assumptions about what solution architecture frames the workbook solutions. How often does anyone *really* check all the equations?

Even with a small amount of experience, analysts may get a bit overconfident in their judgments, and this leads to bias. It is healthy to have a very positive view of oneself, but when critical high-stakes decisions are involved, the risk is one of being "delusional" to some extent and simply being too systematically overconfident in decisions. That is why a good team is critical when the stakes are high.

And while expertise can be a drawback when the context of the problem is changed or new, extensive prior working knowledge can be an asset in overcoming the so-called "recency effect" where the most recent trends are viewed as *the* trend. In such cases, despite the potential downer of "we've tried that before," it helps to take time to probe and find out why a particular decision or approach did not work. The set of questions that eventually are generated is bound to be worth testing for continued relevance.

10.2.3 Group decision-making skills

Engineers training to become technology leaders must be able to lead groups and navigate situations where the stakes are high. The work of Patterson [3] is an excellent reference for the development of personal competence in this area; some of the important topics covered are:

1. Finding a pathway through ambiguous situations
2. Stimulating debate, but keeping conflict constructive
3. Achieving closure in incremental steps
4. Managing transition

For instance, in Figure 10.1, the "stories" we have in our mind about people places and events might make us very efficient about making decisions ("I've seen this before"), but there may be overlooked critical pieces of data that would affect outcome. Allowing these prior experiential reference points may lead to quicker decisions but can also help find the wrong answer the fastest, too.

Once again I emphasize the value that groups *can* play in providing improved decision-making, beyond what a single analyst or engineering manager might conclude alone. For important decisions, the saying "without debate

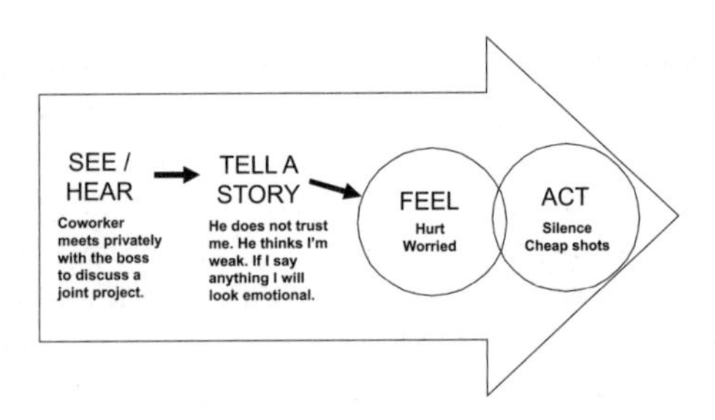

Figure 10.1: Crucial conversations: issues of "anchoring" and "illusory correlation" in negotiation.

there is no real decision" holds true. Leading healthy debate can be one of the most valuable tools an engineering manager can develop when becoming involved in strategic corporate activity.

We will not work through the book of Patterson [3] but leave Figure 10.1 as a "tickler" to find the time to explore the book – especially Chapter 6 "Master My Stories" – as a personal development item to put on your career development to-do list.

10.2.4 Organizational-level attributes

As the reader begins to work through the cases, it will be worthwhile to pause at some point and draw upon the work of Tingling and Brydon [4]. They ask the questions "What is the relationship between evidence and the decision process that an organization actually uses? Why is evidence collected in the first place?" While this may seem a bit irrelevant or tangential, the authors bring out the idea of the illusion of data-driven decisions:

> Clearly, the ideal evidence-based decision process was subverted in this case by the perceived requirement to marshal facts and analysis to support a decision that had already been made elsewhere in the organization. We call this practice decision-based evidence making and argue that it is more widespread than many managers acknowledge.

This is not a contrarian view to the preference for decisions that "supported by hard facts and sound analysis are likely to be better than decisions made on the basis of instinct, folklore or informal anecdotal evidence.[4]" But exercise caution about using data to support decisions already made in advance. This is an organizational-level issue that is not often obvious.

Individuals who embrace the basic concepts we will explore are more likely to network, find opportunity, develop a pathway forward, and manage risk better than their counterparts. Industry today demands managers who understand the importance of interdisciplinary approaches complemented by critical thinking skills, enabling them to solve the problems of tomorrow that we don't even know exist today.

10.3 Case studies to explore in teams

You must engage in the topics just presented: leadership is a contact sport and you will not be successful watching from the sidelines. Be interactive, working with colleagues to provide insight and analysis results. Crafting and sharing meaningful "elevator pitches" is a critical part of a daily review process.

Case studies and scenario analysis provide the opportunity for the structured analysis (guided by targeted questions) of moderately realistic business situations that still have lots of ambiguous and uncertain data surrounding them. An ability to state problems objectively, understand the role of management theories, look at situations from various perspectives, extract what is meaningful in a situation, and place your own experiences in context are outcomes that mark the development of an entrepreneurial mindset. The ability to properly write and analyze scenarios is an important part of this course of study. Four cases are presented here to augment the development of decision-making skills. A few aspects of the cases are important to note:

1. Each of the cases are based on actual events, but the names, companies, and products are all fictitious. Some cases are combinations of events that occurred at different times and different places to ensure that real people and companies can not be presumed.
2. Every case presents issues that could have a profound impact on the financial success of the organization, but some cases may not contain financial data to analyze. This can be a bit frustrating to the student, but is intentional to broaden perspective.
3. Situations are presented that should require discussion of the role of ethics in some cases and the ways in which problems arise - or the potential for problems to be created - as a result of managerial behavior or attitudes. By now it should be clear that even with extensive assistance from GAAP there still is room for misrepresentation, abuse, and fraud; again, we need only read the business press to see how rules can be rendered useless by those with the wrong intent.
4. We have chosen to embellish some of the situations for effect, and these cases should not be assumed as "normal" managerial behavior within companies.
5. Several of the situations involve dilemmas that new manufacturing engineers might encounter as they work to understand and eventually become a manager. Life at the "interface" of technology and management is not simple, and the cases should be cause for reflection about the challenges of technology management.

The four cases span managerial issues from team formation to the financial repercussions of laying off employees. The cases are:

- Case A: The team That Wasn't
- Case B: Disruptive Innovation at Tonowanda
- Case C: Die Cast Testing
- Case D: Welcome to FR4

Generally, the first three cases are designed to be solved by teams of 3 to

5 people within a three-hour time period. This is long enough to begin to understand the issues and think about the models (some presented within this book and others not!) that could be used to propose a solution.

The so-called "Grand Challenge" for each case often provides alternatives to choose from. In such a situation, the desired solution may not be listed, but it is important to pick a pathway and then articulate the compromise in that choice. In a few cases it is highly recommended that analogous situations or related stories be identified in the press and brought forward and included in the team activity. In this way the team can customize the case and understand the strong relevance of the case situation to everyday life. The ability to choose one alternative – even if it is not a perfect solution – provides the opportunity to reflect on your own decision-making processes and what is important to you. In essence, what compromises are *you* willing to make to "get the job done"?

The fourth case, Welcome to FR4, is suitable as a term project. Of all the cases, this one has the most evident link to many of the financial topics of prior chapters, and so the ability to work with data and equations will provide some assurance to some engineers. But this case involves choices that are not trivial, and again, teams would do well to research situations in the current press to examine the business mindset around labor issues today and how they play strongly into the financial and social health of the organization.

10.4 Case A: The team that wasn't

10.4.1 Background

Joyce's promotion to manager of Marketing and New Business Development was an exciting move - and welcome, too - after "paying her dues" working 2 years in the field. Now, she'd be able to spend more time working at headquarters on projects that could better utilize her recent MBA degree. The promotion came at a time of change for Zimmer Bakelite Technologies; what was once a moderately entrepreneurial firm of 400 employees had become more bureaucratic when it doubled in size over 5 years, and top management felt it was time to "shake things up" and become "innovative again." Younger employees were seen as a pathway to rejuvenate the company. ZBT specialized in large-oven capital equipment projects, and management felt something was needed to help cut costs and improve EPS. Again, plenty of *HBR* articles suggested innovation was the way out!

Marketing and New Business Development was a mixed bag of 5 marketing department employees who had generally been employed at ZBT from its inception 10 years ago, and a relatively new set of 4 employees who had been working custom systems for customers that management felt could become "hot" new products. The new hybrid organization was the brainchild of Sam Davis, Senior VP of Sales and Marketing. Anyway, Sam was a little tired of the constant "whining" by the old marketing department about their workload and felt the injection of a new product responsibility would energize the teams to innovatively "do more with less," thus freeing up cash to hire more salespeople. The thinking was that a seasoned marketing team would be a key asset in the team effort to develop exciting, innovative new products. Joyce was thrilled to have the chance to lead a team that could have an impact on ZBT's future; at the same time, Sam was glad the "whiners" were no longer direct reports to him.

It wasn't long before Joyce concluded that maybe the "mixed bag" that Sam described could more accurately be described as "total drag." Prior to the new department's inaugural monthly meeting, Joyce made the rounds to her relatively small team to visit 1-on-1 to hear of their concerns, interests,and personal goals for the new department. Joyce's first stop was Donna, as Donna had responsibility for all phases of the spare parts and product catalogues. Donna had a reputation for complaining frequently and loudly about every aspect of her job, but this was generally tolerated since the customers seemed to like her and she always seemed to bring the catalogues out on time and under budget. Joyce saw the opportunity to not only try to make Donna's job easier, but in doing so it might help Joyce herself politically since Donna was married to one of the "old-boy" service managers - a win with Donna could be a win for the reputation of the department.

Another key visit was to Jasper Mastick, who was well-known to have wanted (and bid on) Joyce's current management position. Jasper had very limited field experience and seemed to prefer talking to customers on the phone; being in the office was important to him so that he could hang out with, Roger Crennel VP of Sales. Jasper always worked to appeal to Roger's ego in the hope that if he could get "the nod" by a VP, then he could get the same promotions (and bonuses) the sales guys got. A challenge here was the outright hostility of Jasper. Barely 5 minutes into the conversation and Jasper made it clear that the new organization left him without anyone reporting to him and this was not only an embarrassment, but a demotion since Joyce had the job "I am more qualified for anyway."

The idea of developing a strategic vision for the new department seemed "a world away" after these initial visits, but the good news was that discussions with the other members of the team were easier. For instance, Chester was a designer who was brought onto the team since he not only knew the existing product line well but also had been a valuable contributor to some recent new products. He added a clear multi-disciplinary flavor to the team. Only odd thing was his strong ties to Donna and their incessant fixation with office gossip.

In comparison, Theresa was fairly young and had once been a contractor supporting the spare parts catalog but was now a full-time employee. While she got along great with Donna, she had an odd way of disagreeing and getting away with it. The truth be known, Theresa was actually at the root of many traditional marketing innovations, but often was fine letting Donna get credit. For instance, it was Theresa who figured out how the cost of printing could be lowered and customers served better by producing literature in PDF format. Donna had argued incorrectly that "our customers don't have email and don't want anything in electronic form anyway." (ZBT literature and user manuals were free-of-charge and easily obtained by calling customer service whenever you wanted some.) Additionally, Theresa had come up with some ideas to make a new database-driven catalog system work and had shared these with Joyce, but Theresa was cautious about discussing this since she was concerned that Jasper and Joyce might not like the change in workflow that would result.

At the inaugural meeting of the department, Sam, the Senior VP, was to make welcoming remarks, but was called away by Roger, the VP of Sales, so Joyce had to self-introduce. The meeting started somewhat civilly but soon you could have cut the tension with a knife. It did not take long for Donna to announce that "she had been lied to" about getting a promotion and this was going to be a problem because the new digital media strategy "simply was not going to work" and anyone who "had been with the company long enough would know that." Jasper let it be known that he was excited about some of the product ideas for the new department to consider, but also shared that "there is a certain way things get done around here and soon you'll figure that out." Joyce noted that "she, too, had worked in the field," and a quick call to several customers suggested that both the modernization of marketing materials and the teams' work on branding and product support materials for the new product ideas were actually welcomed. Despite the moderately positive and supportive comments by the other 7 members of the department, the overall tone of the meeting was set by Donna's and Jasper's sarcasm.

Later in the day Joyce was still searching for a positive twist to the department's first meeting and felt she was close to creating a positive summary memo to share with senior management. This, of course, prior to Roger, VP of Sales, stopping by; he seemed eager to share that "Jasper thinks you have your hands full, so good luck using your MBA to bail you out this time." He said he was joking, but Joyce was not so sure.

10.4.2 Grand challenge

Investigate the following issues:

1. Identify the three top leadership skills that Joyce needs to have going forward.
2. What are her first and second "action steps" for the execution of department strategy?
3. Describe the adaptive learning cycle for ZBT.

10.5 Case B: Disruptive innovation at Tonowanda

10.5.1 Background

Ralph Chase, VP of Sales for Tonawanda Industries, was thrilled with the order for 10,000 custom parts from Caliber Press Manufacturing. Tonawanda and Caliber had a successful supplier-customer relationship going back more than two decades, but recently competition from imported parts was depressing profit margins on sales. This most recent order from Caliber involved a protracted negotiation, and Ralph felt he was lucky getting a $75 unit price, knowing this drove down the product unit margin to just under $25. "Not close to a 50% margin," he thought, "but we'll take it." An argument made by Tonawanda to Caliber was that the cost of raw materials, not labor, drove price.

As a fresh new project manager at Tonawanda, Julia Haverstock was assigned to the Caliber project and given the lead to ensure project execution. The Caliber parts were not a typical design for Tonawanda, but close enough that none of the managers were concerned. About two weeks after the Caliber agreement was signed, Julia discovered that a slightly different alloy composition would cut Tonawanda's material and manufacturing costs by $12.50 per unit. She did note that the change in properties was likely to compromise integrity of the final parts, but not so much that the "average bear" would notice. It would have to be a fairly remote use condition for Caliber to ever detect a difference. Her engineering manager, Jason, was not at all familiar with the technology, and offered the advice, "If you say it works, then it works." We need the business and I like good news."

Julie decides to "seize the moment" and presents her disruptive innovation at the weekly project team meeting. Ralph Chase and Todd Hopkins (recently hired from VOT), VP of Marketing, are at the meeting. Todd is curious about the "average bear" and if "anyone would really know the difference." Julie becomes slightly concerned and indicates, "It seems unlikely the performance difference would be noticed unless Caliber was looking closely and performed independent tests, which we know they never do. I believe a majority of the parts will perform fine, but a few might have shorter lives. I don't really know." By now, Todd and Ralph are joyful. Todd says, "What a great hire you were, Julie. You've just increased our margins close to our favorite 50% mark and everybody is going to win. This disruptive technology thing really does pay off! Let's make sure we require an NDA to block the need to share details."

Later, Julie starts to become more concerned. "Ralph, as the sales person who signed the agreement, shouldn't you tell Caliber about the substitute material?" "Excuse me? Hello?" Ralph asks. "Look, we've got a long-term relationship with Caliber, and they expect innovation from us. If we satisfy Caliper's needs with good quality parts – and you've just said we will – what exactly is the problem?" "Wow," Julie thinks to herself, "I feel like something is missing. Are we providing what was promised? Contractually? And so even if Caliber is satisfied with the improved product, shouldn't they be given the opportunity to decide if the change is acceptable? Maybe benefit from lowered cost?" Julie decides to explore these questions further with Todd. He replies, "I don't see a problem, Julie. This is business, not engineering. The NDA will protect us. We're not in the business of giving away money, you know. Besides, as a new employee, don't you want to be known as an engineer who understands business?" Julie decides there is nothing further for her to do. The new parts based on her disruptive innovation are produced.

As the first shipment is prepared to be sent to Caliber, Julie hesitates when asked to sign a report verifying that the specifications for the part have been met; she is sensitive to the usage of the new materials and notices that the original composition of the metal is listed rather than the new material. She tells Todd that she needs some time to think about the situation. When she is out for lunch Todd persuades Shannon, another project manager, to sign. Julie is upset about Shannon's certification, but Todd indicates, "I've known Shannon a long time. We worked together at VOT, and she reviewed the situation and saw no problems. We needed to get the order on the trucks and you were out of the office; there was nothing else I could do. The NDA kept us from updating documents."

10.5.2 Grand challenge

Investigate the following issues:

1. Identify at least three issues and write a one-sentence summary of each issue.
2. Pick one of the issues and expand in more detail; use 3 to 4 brief points. Consider how tables and figures from prior chapters can be used to reinforce or explain your points (will potentially save a lot of writing).
3. Select the one most urgent issue that needs action immediately.

Be sure to identify who, when, where, and why.

10.6 Case C: Die Cast Testing

10.6.1 Background

Ordinarily, Wei Chang enjoyed the occasional mid-afternoon walk across the university-like campus of Die Cast Testing, Inc. (DCT), to visit the R&D lab. His father, Ming "Bud" Chang had co-founded DCT over 30 years ago with his colleague Harley LeRose (now, the VP of R&D at DCT) from Duke University and DCT had gained notoriety as a local success story - steady growth in instruments for casting integrity test equipment with virtually no debt. Numerous students from Duke had taken jobs with DCT's R&D lab. Wei prided himself in not only knowing each of the 170 company "associates," but also having maintained enough technical expertise to make trips to the R&D lab a fun part of the job. "Too bad none of the other executives like hanging out here," he often though. "They can't fully appreciate our industry if they don't spend at least a little time visiting R&D."

Today was slightly different. Wei was bringing unfavorable financial news to Harley LeRose, and one of the options to discuss at the next board meeting was the need to possibly terminate some new project development activity. Harley had been pretty content with a career in the lab, and this had made it easy for Wei to step in to the role of president when Wei's father retired. But now, DCT was at a fork in the road. Incremental product development had worked for a long time, and it was Wei who encouraged R&D to move into more radical measurement systems associated with microsystems sensors for high-speed multi-axis inspection - geometric control of complex part design was back in vogue in the casting industry. Specifically, the automotive segment was beginning to look like capital investments were on the horizon. The R&D initiative was a fairly big investment requiring DCT to suffer a drop of cash reserves. Up to the last fiscal quarter the board had been supportive, but now had been pushing Wei to consider either acquiring more debt or trimming R&D dramatically. They were at the point where jobs might actually have to be cut.

Working capital was an increasing issue. Customer Days Sales Outstanding (DSO) continued to skyrocket (from 89 to 132 days over the last year) at the same time product development was 9 months behind on the launch of a new microsystem electronics analysis product line (reliability test equipment). Sales was having the company "be the bank," with extended payment terms to support prices; it seemed to Wei that in many cases the combination of price/terms to get product out the door was just plain unfavorable. EPS trends were downward, and the last quarter drop to a loss of $0.25/share meant Wei would be presenting to the board an estimated year-end EPS of $1.15; if it tanked any further and went below $1.10, all managerial bonuses would be suspended. Still, they were not doing badly; with a market capitalization of around $132M, at least the stock price of $10.25 was double digit. More of concern to Wei was that even in a down economy, DCT's competitors seemed to be doing OK. As a private company, this was not too much of an issue, since as old-time employees like to say "Private money means patient money." It was a source of management pride that the company was always in control and thus "investing in the future" when competitors had to lay off employees to restore profitability. Still, debt was not part of the "company DNA," and Wei was nervous about costs. The last thing he wanted was poor press if they had to borrow more money.

One aspect of DCT's business that Wei was proud of since his becoming president a decade ago was "going global." He had managed to preserve jobs by sending all manufacturing offshore, keeping the "strategic front end" (R&D, product development, marketing) local. While this had many appealing attributes, there seemed to be grumbling among the local managers that they just "didn't have a handle" on overseas project status. In fact, it seemed that DCT's own plants were using temp labor as a substitute for DCT employees, and the word on the street was that part of the 9-month delay was the constant retraining of temporary employees. The Hefei plant was a particular challenge, as it was somewhat inland and not a popular destination for the managers back in the US. So far the idea was to set profitability levels and let the locals do things the way the local culture dictates. "If it is not about managing profit, then why go offshore in the first place" thought Wei. "Maybe I'll go to Hefei in a month or two and see what's up."

Along with the global strategy, Wei had also pushed for a more diverse board. Rather than a majority of the board coming from engineering backgrounds, 6 of the 11 current board members were executives whom Wei had brought in from the outside over the last decade. This "fresh" external oversight had been central to the bank approving a credit line increase a few years ago. It troubled Wei that the board seemed comfortable with debt - some board members actually joked he was preoccupied with avoiding debt and needed to be more "modern," but high debt levels was not a company situation his father would have fostered.

Traditionally, DCT had a debt/equity ratio of 7.5% that was normally 10% below their industrial peer group. Today, problems with existing product sales and new product launches not only drove D/E to 11%, but the line of credit was nearly exhausted. Wei was hoping Harley would help him with ideas to avoid acquiring more debt to continue funding R&D. Maybe Harley would even offer some R&D projects that could be terminated.

10.6.2 Grand challenge

Wei Chang was hopeful that Harley LeRose could help him decide on one of three basic "go-forward" options. Actually this might be pretty easy work, particularly since Harley was a co-founder and knew DCT better than anyone else. Wei's three ideas were the following:

1. Do nothing at all and work through this in real-time with the board at the next meeting. They were, after all, there to help shape the company, and he needed them now!
2. Cut the new microsystems test product line developments and focus on

the core products that had made DCT successful. This was painful and almost unthinkable to Wei since it would mean a 20% reduction in workforce at the main campus. This would look terrible in the press.
3. Acquire more debt to fund the current level of activity, continuing to "wait out the storm" of the current economy. This option was almost unthinkable to Wei, since it would drive D/E to a historic high of 18%,and Wei felt this would leave the company in a position the co-founders had always wanted to avoid.

10.7 Case D: Welcome to FR4

10.7.1 Background

Rick Treadway, VP of Operations for FR4 Technologies, LLC, was becoming increasingly concerned about the DuraBoard product line. The so-called ruggedization process involving spraying off-the-shelf circuit boards with paralyne had created a popular product line. As a supplier to OEMs, FR4 Technologies salesmen had been able to rapidly penetrate the aerospace and military markets by talking up the unique value proposition of advanced technology at commodity pricing. Rick was not immediately concerned about the typical day-to-day problems any technology company would have; Rick was actually thinking more and more about the risk of tin whiskers to his product line. He wasnt exactly sure they really had a problem on their hands, but the DuraBoard product line was carrying most of the sales these days and those sales were increasing rapidly. He was concerned that if DuraBoard sales were to be compromised there were no other products in the pipeline to pick up the slack. If FR4 lost DuraBoard the current workforce of 55 people would have to be halved; it would be like the early '90s all over again. Unfortunately, everyone at the company was clueless about Tin Whiskers. It was only at the most recent board meeting that Al Cedar, one of FR4s longest standing board members, mentioned the lead-free solder failures back in the Apollo days when he was a contractor. Thats when the CEO got the idea for the Ranger Squad to take care of the problem. It was a hit with the board and they told the CEO to act on it immediately. That was 9 months ago at the February board meeting.

Tin Whiskers was an old problem that was new again. A few years ago many vendors in the EU had moved to lead-free solders in the production and assembly of components and subassemblies used in circuit boards. Vendors had performed a variety of functional and short-term quality checks that enabled (what seemed like, anyway) a seamless transition for FR4 to purchase EU-compliant subassemblies in aerospace circuit boards. There was a lot of excitement at FR4 Technologies when they realized that the previous effort of their European suppliers had simplified FR4s international expansion goals since EU vendors had mastered lead-free production. Then the VP of Sales for FR4 didnt believe they had to do anything to their domestic product lines to be compliant domestically and internationally. Well, OK, Rick thought, they really were not 100% compliant, so FR4 did have to do something to address the prospect that Tin Whiskers could be an issue. Still, it was easy and comforting enough to talk about EU vendor compliance, and this created a good enough story for sales to keep pushing the product in the market. Besides, anyone who wanted to dig a little deeper about the impact of lead-free solders was simply told that between the EU supplier monitoring and FR4s newly formed Special Applications Grou – now nicknamed the Ranger Squad – from any perspective everything seemed under control.

The Ranger Squad was talked about at a high level in the company, and even with the recent hiring freeze at FR4, an exception was made for the new lead manager position. The CEO and VP of Sales like to talk about the investment in the Ranger Squad as their commitment to quality. Sales continued to grow at the company, so management at the company seemed happy – most importantly the CEO. Thats the way it was at FR4: when you make the boss happy, everyone is happy.

It had been surprisingly difficult to form the Ranger Squad. Even with the technology employment market looking a bit dismal, finding a leader for the team had not gone very well. At the outset, the idea of having just the VP of Sales, HR, and the CEO interview the candidates seemed like the right thing to do, but for a variety of reasons it was tough to make a hire. Two good candidates turned down offers. At first, it seemed like many of the candidates were under the impression they would just be sales support jockeys, so after about 6 months they included the VP of Operations on the interview team. Interviews seemed to get a little better after that, and within 2 months they were able to hire Alexis Chambers for the position. Almost 3 full quarters had been lost in the hiring of the lead for the Ranger Squad, so there was some concern among top management that there might not be enough time for the new team to have an impact before the next board meeting. It was already November, and typically by December, staff were already thinking about the February board presentations.

Digging In

Rick was getting feedback from staff that returns were beginning to take a lot of space in the warehouse. He thought, "This is strange, I did'nt think wed have Tin Whiskers returns so soon. Rick realized it had already been about 3 months since Alexis had joined the company; maybe it was time to bring in Alexis. Within a few days they met in his office, and immediately it was evident there was a lot to talk about. Hardly 15 minutes had passed when Alexis said, "I thought I was being given the chance to run my own show. My expectation was that I would be hiring my own staff, but sales pre-selected all the staff for my group due to the recent hiring freeze. They claimed they had the best knowledge of staff members that were customer-service oriented and that I should interview each of them only to indicate those I could not work with. Also, the budget has been dramatically cut. When I had my orientation with the CFO on the companys financials, she told me that to make room for my salary they had to eliminate most of the business development funds

I had to work with. Any funds I have are supposed to come from technology improvements the Ranger Squad creates that pay their own way.

After Alexis paused for a moment, Rick began thinking out loud: I think we have two problems. First, you were hired to look at the Tin Whiskers issue, and Im not sure how much progress youve made. Its a complicated problem and it sounds like youre not off and running. Your lack of progress is a problem. Second, there may be other issues with the DuraBoard product line since our team is predicting well hit a 30% return rate by the end of the fiscal year. Alexis was a bit startled by the comments but agreed. OK, I really have not made as much progress at all on Tin Whiskers as I would have liked, but I have started to work through and analyze three options. And, as far as the DuraBoard product returns increasing, I would also agree with you. I have some specific thoughts on why sales numbers are high but profitability is low. If I read the number right, then it also appears as if cash flow has been, well, negative, for two years running. Why doesnt anyone at the top notice?

Rick seemed friendly enough to Alexis, and Rick ended the discussion with the advice that "You should not ask questions you really do not want to know the answer. I would recommend you keep focused on how your Tin Whiskers team is going to improve the top line, especially if youre meeting with the CEO or anyone from sales. Look, Alexis, this discussion has been really informative, and although I am sure youve got ideas on Tin Whiskers, it seems clear to me they are not going to help me solve the warehouse problem. Lets find some time in another month or two to meet again.

Alexis also figured the warehouse returns were not her problem, and decided she should focus on the reason she was hired – to lead the Ranger Squad and provide a solution to the Tin Whiskers problem. Three options had been floating around her team: (a) do nothing at all, (b) create an internal process for corrective action, or (c) partner with an outside firm that had specific experience in this area.

Option A had some merit. Although Alexis was not originally sure about this pathway, it started to make some sense once she had her orientation with Sonya Peterson, the CFO. Sonya was the fourth CFO for FR4 within the last 6 years and seemed to Alexis to have great insight on the way decisions were made at FR4. Dont expect to spend a lot of money if you do not have a convincing story about the benefit Sonya warned. "Cash is tight, and I am not sure we could get a bank loan if we really needed one. The CEO and VP of Sales will want to hear that you can bootstrap the project and not impact profitability.

Although taking a wait-and-see approach had merit, the idea of recommending doing nothing did not seem to be the central theme Alexis would want to present at her first board meeting. She thought "Surely I can come up with something more exciting than that in the next month. The other major option was to create an internal manufacturing cell to rapidly prototype a variety of Tin Whisker mitigation solutions. Initial estimates were that this would require about $750,000 in capital equipment purchases and might have a reasonably high impact on product development. But there were issues taking people off other projects to launch this project and project management impacts to be considered. Since the VP of Sales was also the only source for marketing data at FR4, it was a little tough to prioritize internal development projects. The VP was always fond of saying I know our current customers well enough to know whats going on in the market.

The other option was to partner with Jackson Engineering, an outside vendor that had worked with FR4 on several other occasions. Jackson Engineering was a competent but expensive firm to work with, and the most recent disagreement with FR4 involved a hold-back on a payment from FR4 to Jackson. This type of disagreement was common between FR4 and vendors. As the VP of Sales put it you have to let the vendors know who runs the show. Although the work with Jackson was viewed as having a lower probability of success to Alexis two product development team members Sam and Ajit, there was something about Jacksons technical insight that was appealing. Alexis has been informed by Sam that he had already talked to the VP of Sales about this option, and the VP was concerned that customers might find out FR4 was investigating a potential DuraBoard quality problem. Ajit knew of Jackson Engineerings reputation and disagreed with that general conclusion on customer risk, but Sam said at the last team meeting, Look, sales and I golf in the same Saturday league, and Im telling you the $500,000 over two years to launch has too much risk and a lower probability of success than keeping the issue close to home.

Meeting the CEO

Sam Thomas had been the CEO and president of FR4 Technologies since its founding after his retirement from the military in 1985. Although business was a little rough during the recession during the early 1990s, they had managed to secure a few sole-source contracts to stabilize sales. More recently, Sam himself led the company drive to expand beyond being a government contractor, and over the last 2 years had increased the size of the operations team to keep up increasing sales. In fact, the size of the sales team had not expanded at all for the past 5 years, but year-over-year growth in the top line sales volume was nothing short of phenomenal. This further underscored the recent decision for the CEO to take more of an arm's reach on day-to-day decisions; in the spirit of letting sales run sales Sam was pleased to have Ken Willow, his long-time friend and FR4s first executive hire back in the '80s, be the VP in complete charge of the sales, marketing, and business development aspects of the company.

Alexis was excited about the idea of meeting with the CEO. This was the first real sit-down since the time of her interview, and she had worked hard to understand as many aspects of the business as possible. As well, she had worked hard on understanding some of the issues associated with potential solutions to the Tin Whisker problem. Alexis was hoping this early meeting would give her a chance to get some insight on which options might be favored by the CEO. Some validation of anything would be useful.

After Sams secretary brought him coffee, he moved from his desk and sat down with Alexis at his conference table. She did not recall his office being so large during the interview process. Sam began, So by now youve met with all FR4s leadership and from what Ken Willow tells me, the Ranger Squad is starting to make progress. Since youve had some time to apply your years in academia to the real world of manufacturing, what do you think?

Alexis welcomed the invitation to jump right in, especially with her insight on the business growth and profitability. "FR4 seems to be doing a great job in penetrating the market with a product that solves a customer need, but from my review of the financials, it seems like there could be some problems with our sales growth. I am thinking FR4 might want to consider basing sales commissions on net income, not the top line. We may also want to watch the gross sales number as it is increasingly based on product sold on account, which seems like an increasingly risky situation.

Sam sat back in his chair. "Well, he said sarcastically, I am happy to see youve so quickly figured out how everyone else should be doing their job. This makes me even more anxious to hear how you are going to solve the Tin Whiskers problem, too.

Alexis was caught a little off guard by the tone of Sams remarks, and it was tough to know how to immediately respond. She decided to continue by outlining the solution options to the Tin Whiskers problem. While Alexis felt confident in what she was describing, she sensed some uneasiness with Sams lack of engagement on the technical discussion and brought her presentation to a close more quickly than she had planned.

So, let me see, youve pretty much figured out whats wrong with everything else in the organization, but regarding the performance of your own job, you only have a few options to discuss, one of which is to do nothing. Youre sitting with the CEO, presenting him with more problems, and not solutions. Sam paused, leaned forward in his chair, then continued Do you see anything wrong with this situation? How about this: lets meet on the anniversary of your 180-day tenure with the Company and review what specifically it is youre going to do to fix the Tin Whisker problem and keep sales growing. This will give you 2 weeks to help me understand that maybe I did not make a mistake in hiring you.

Alexis returned to her desk, and Ajit stopped by. "I heard that Ken has already talked to Rick about your meeting with the CEO and it did not go well. So, youve got just a few weeks to work out a plan. What are you going to do?

10.7.2 Grand challenge

After the meeting with CEO Sam Thomas, Alexis Chambers decided her first meeting with the CEO was not going to dampen her spirit. In a follow-up meeting with Sonya Peterson, the CFO, Alexis was pleased to discover there had been some work by the CFOs predecessor looking at an innovative test cell that had some of the attributes of the internal manufacturing cell that could potentially rapidly prototype a variety of Tin Whisker mitigation solutions. Alexis liked the idea that two existing process machines for the DuraBoard line (coating, flowed by curing) could be replaced with a single more efficient machine that had soldering process module options. The old report was not well documented, but it seemed that enough data were there to apply to the current situation:

1. The two existing process machines could be sold for $70,000 in the secondary market. Their depreciated book value is $120,000 with a remaining useful and depreciable life of 8 years. Straight-line depreciation was used on these machines.
2. The new integrated system machine can be purchased and installed for $480,000. It has a useful life of 8 years, at the end of which a salvage value of $40,000 is expected. The machine falls into the 5-year property class for accelerated cost recovery (depreciation) purposes.
3. Due to its greater combined efficiency, the new process machine was expected to result in incremental annual savings of $120,000.
4. The companys corporate tax rate is 34%; if any loss occurs in any year on the project, the company will receive a tax credit of 34% of such loss.

Alexis thought this could be a great part of the story for the in-house effort. Besides, it had a lot of fun MBA-type stuff to talk about, and the VP of Sales might like this as a way to showcase the issue of Tin Whiskers as a process improvement project.

The CFO suggested Alexis consider that the CEO might really be more interested in the idea that the new process will reduce the technician workforce by 50%; Sam was more inclined to give bonuses to managers who provided labor cost reduction news than machine production efficiency. Rick Treadway, VP of operations, agreed. Look, do the analysis with the data youve got from the old report and when youre done I need you to consider a few more things. Of the 10 technicians in production, currently 8 are working the production line. The word on the street from Ajits work is that if there are any layoffs, the remaining workers will probably slow things down in production. Sam continued, I do not know how much production would be affected, but I suspect theres a 80/20 chance the annual savings will drop by 30% and 60% of the projected savings in each of the next two years. The idea is the guys on the floor may want to prove that the old machines with the old crew are better than a smaller crew on a new machine. This is what may have killed the project last time, but I really dont know. And, of course, I know you want to prototype new process steps to fix the Tin Whiskers problem, but youll be a real hero if the new process also pays its own way.

The CFO suggests the average line worker is making $38/hour (with benefits), and because of cash flow this quarter, production is just one shift a day, 5 days a week. When Alexis asked for suggestions on what should be documented, Sonya simply replied, Work this up so Sam will give the other VPs a bonus this year. The company might be sold before we have to deal with any long-term production issues.

References

[1] Paul C. Nutt. *Why decisions fail: Avoiding the blunders and traps that lead to debacles.* Berrett-Koehler Publisher, 2002.

[2] Christopher Chabris and Daniel Simons. *The invisible gorilla; How our intuitions deceive us.* MJF Books: New York, 201.

[3] Kerry Patterson. *Crucial conversations: Tools for talking when the stakes are high.* McGraw-Hill, 2002.

[4] Peter M. Tingling and Michael J. Brydon. *Is decision-based evidence making necessarily bad?* Sloan Management Review, 2010.

Appendix A

Problems and Problem-Solving

Decision-making is, indeed, a *very* broad field, and this section can only elucidate a few obvious points. Some commentary on the subsection discussions to follow will clarify very pertinent and subtle concepts to keep in mind and guide us as we move forward through this Appendix.

- We explain more about the difference between a *problem* and the *problem-solving process* with the design process analogy, as this helps underscore that the problem-solving process may cause the problem to be re-defined, and, as we comment later, highlights that problem outcomes are often reached as the result of an *iterative* process. This is no real surprise in general, except there is a cohort of engineers who tend to trivialize financial analysis and may be perplexed by this idea.
- We very briefly explore two basic problem categories – stochastic and deterministic – as a way to introduce and then differentiate between risk and uncertainty in the parameters associated with problem solving. Students often find stochastic problems *much* more difficult to solve as these problems require significantly more judgment of an interdisciplinary nature. This is not a skill usually practiced during typical engineering education. For engineering managers, the subject of risk and uncertainty are routinely encountered, and the key is to learn to *manage*, not *avoid* the issues.
- We spend just a few minutes examining *type of outcomes* of decision-making, since that, too, seems to be the source of frustration for those new to financial decision-making. Frequently there is more than one acceptable (and certainly no single perfect) selection among financial solution options. Students accustomed to finding and featuring an "answer" by drawing a box around a number computed to three significant digits are puzzled (or just plain surprised) by the presence of choices.
- *Finally*, we consider the impact of organizational form, emphasizing how

this influences decision-making, problem definition, and the problem-solving process; it is *emphasis*, not *exclusion*, that is important to keep in mind.

A.1 Design process analogy

The broad scope of the engineering discipline and the need to explore subjects in some depth have driven undergraduate engineering educators to draw heavily on textbook problems, which the students learn to solve. More often than not, the problem is *given to* and not *crafted by* the student. In contrast, business problems within which technical problems are nested present a managerial, team-based process. No simple matter at all. In fact, some problems are not problems at all – many are simply not worth solving, despite the fact that in industry, many employees are consumed with tasks and problem solution efforts that are unimportant and lead to organizational inefficiency and frustration and that common question "Why can't we get anything done around here?" (Lefton and Loeb[1], p. 30). In the words of Lefton and Loeb, an ideal situation is that where "capable employees find fulfillment and enjoyment from [important work]" (Lefton and Loeb[1], p. 33) – working on the right problem can be fun! Within this book we attend to some strategies and topics in problem identification and prioritization, but most often problems are presented that are extracted from either real-world engineering management situations or are characteristic of "classic" problems that a typical engineering manager will face in daily practice.

The art of asking questions and developing problem statements is a fascinating course of study in itself, but I'm sorry to say that we can't squeeze as much as we'd like about that subject into this short appendix. Just to give a little insight into this very broad subject, the work of Benner[2] provides a delightful look at the transformation in thinking and tactics that must emerge as the novice develops expertise in the skill of framing the correct problem in nursing; another facet explored by Zenios and colleagues at Stanford[3] is a methodology drawing on a "needs analysis" to fine-tune a systemic approach to identify the "right" problem, avoiding a self-limiting investment in solving the narrow problem right in front of us. And the work of Nutt[4] is an enjoyable read also pointing to the importance of framing the problem properly. It is a limitation of the present work that we treat problem identification somewhat casually and make it appear more obvious than exists in reality.

We've hinted that a *problem* and the *problem-solving process* are two different things. They are. Just because this book does not emphasize problem definition does not mean that we're left with the easy part of learning about financial decision-making within the problem-solving process. Managerial problem-solving is the "stuff MBAs are made of," but just as in problem definition, we can fall into many traps. Russell Ackoff [5] provides an entertaining, practitioner's viewpoint conveyed through several "fables"; take a weekend to read his book!

How rewarding, then, that with an engineering background, the "scientific method," introduced in the early years of college, now comes to the rescue! At the risk of oversimplifying, problem-solving begins with crafting what the problem is, developing options to resolve the problem, choosing a course of action from the options, and then following up and following through to see if the course of action had the desired outcome. In the professional arena, problem-solving not only involves decision making, but setting priorities, exercising judgment, and establishing accountability. Problem-solving has a lot of similarities to the engineering design process. Table A.1, adapted from the works of Dym and Little [6] and Marshall [7] summarizes this process.

	Problem-solving		Engineering design process
1.	Encounter problem	1.	Define problem
2.	Collect data	2.	Collect data
3.	Analyze data to specify problem	3.	Formulate hypothesis
4.	Determine plan of action	4.	Design plan to test hypothesis
5.	Execute action plan	5.	Test hypothesis
6.	Evaluate plan for effectiveness	6.	Interpret results
		7.	Evaluate for study conclusion

Table A.1: Similarity between problem-solving and the engineering design process (adapted from Dym and Little[6] and Marshall [7].)

Table A.1 underscores *decision-making as a central part of the problem-solving process.* It is making a choice between options. Critical thinking plays a large role in effective decision-making. Although not *all* decision-making is driven by the identification of a problem, the general categories of problems addressed in this book are. Deciding what color to paint a living room involves choices, but generally involves *preferences* and does not require one to go through the problem-solving process.

A.2 Two basic categories of problems

Decision-making related to, for instance, inventory, supply chain, and workflow management typically distinguishes between *deterministic* and *stochastic* problems. The distinction is not academic, as fundamentally different solution strategies must be engaged:

- The more difficult category of **stochastic problems** involves uncertainty, primarily in that some of the information necessary to solve the problem behaves in a probabilistic fashion.
- The (generally) much easier **deterministic problem** is more familiar to undergraduate engineers. Here, all the information needed to solve the problem is known with a great deal of certainty.

Relevant to this volume is that stochastic problem solution methodologies build on many concepts introduced in deterministic problem solution technique; it is quite difficult – if not impossible – to appreciate the nuances of solving stochastic problems if mastery of deterministic problems is weak or absent. Among other things, the impact of an interdisciplinary problem context escapes meaningful attention if solution of deterministic problems is not routine.

Central to the discussion is the introduction of *risk* and *uncertainty.* Though definitions vary from textbook to textbook, let's begin by referencing *The American Heritage Dictionary*[8] and toss in a description of *regret*, too:

- Risk: "... the possibility of suffering harm or loss; danger ... probability of loss ..."
- Uncertainty: "... not knowing or established; questionable; doubtful ... vague ..."
- Regret: "... distress over desire unfilled or an action [not] performed ..."

For the layperson, risk often implies that something "bad" or negative will happen with some degree of probability. In contrast, uncertainty is a bit more innocuous, absent "value judgment," and implies a simple indefinite or incalculable nature of events. Michael Mouboussin[9] discusses that a good manager simply provides the best insurance possible for situations which seem the most likely to happen:

> *How should we think about risk and uncertainty? A logical starting place is Frank Knight's distinction: Risk has an unknown outcome, but we know what the underlying outcome distribution looks like. Uncertainty also implies an unknown outcome, but we don't know what the underlying distribution looks like. So games of chance like roulette or blackjack are risky, while the outcome of a war is uncertain. Knight said that objective probability is the basis for risk, while subjective probability underlies uncertainty.*

An example folds all these ideas together. If I decide that today is the day that I need to upgrade a web server to expand an online business for a new product and market entry, then I can go out and get three or four quotes *today* for the exact price of the service. Based on specific criteria (an objective function) I can undertake a fairly rigorous "defender versus challenger" analysis and have a service in place by the end of the week. This deterministic problem involving a choice of vendors has an outcome that has *certainty* since we are confident that the lease will be signed and the web server will be made available to build our business (well, OK, this is predicated on knowing exactly what the costs are). When outcomes are unknown they are *uncertain.* Certainty comes from events that are guaranteed to occur or are highly reliable. If, however, the lease process had an outcome that was unknown or variable (a "low-bid" service from a company on the verge of bankruptcy), then there is *uncertainty* about securing the service and *risk* that the decision might result in a *loss* of our payment for the service. If we contemplate making the decision three months from now we are dealing with some uncertainty about price and service provider, and now have compounded uncertainty.

Thus, in the presence of *uncertainty* and *risk*, deterministic problems with a single outcome morph into stochastic problems with a variety of outcomes.

Both types of these problems are addressed in the present work since the solution procedures and decision approaches are fundamentally different.

We could argue that the average engineer spends most of his or her time at the university learning to perfect the solution to deterministic problems when, in reality, a majority of problems for the front-line engineering manager are actually probabilistic. And, indeed, whether it is the "New Century Engineer" (Sheppard [10], p. 3), or the *Olin Triangle* to "educate the whole person" (NAP [11], p. 103), developing comfort with an interdisciplinary mind-set helps the engineering manager navigate uncertainty and manage risk; often these solution methodologies are iterative.

The Accreditation Board for Engineering and Technology (ABET) underscores the need to emphasize decision-making (ABET, 2010, p 4., Criterion 5b) and the likelihood of iterative solutions:

> Engineering design is the process of devising a system, component, or process to meet desired needs. It is a decision-making process (often iterative), in which the basic sciences, mathematics, and the engineering sciences are applied to convert resources optimally to meet these stated needs.

A.3 Organizational form

Project management decisions are influenced in part by the structure of the organization. Distinguishing the form of the business has project financial and accounting issues regardless of whether you have been in business 10 days or 10 years. It is unfortunate when the form of the organization does not align with the nature of the business and operational risks involved. Three facets come to mind:

1. The formalities and expense required to maintain a specific organizational form.
2. Income and securities reporting situations are quite different.
3. Pathways and mechanisms for project investments are influenced.

The last step is often a stumbling block for start-ups. For example, an LLC does not have issued stock that can be the source of investment revenue (but an "Inc." can). Overall, the best form of the organization is a function of (a) the risk level of the products and services provided to customers, and (b) the manner in which you will capitalize projects. Figure A.1 illustrates the four basic ownership structures in use today.

Sole proprietorship

Many small businesses (retail, service, farm) and professional practices (consulting, law, medicine) owned by a single person are sole proprietorships. This works well if you are starting the business on a shoestring, since there are no specific legal formalities involved and no special operating rules. However, a sole proprietorship is considered "inseparable" from its owner, so all losses are his or her to absorb and the proprietor is personally liable for lawsuits and court judgments. The owner pays taxes on income through a Schedule C attached to one's personal tax return.

This organization form will also limit access to capital for projects since the ownership cannot be transferred or sold. No real succession planning make sense here. If you have a risky business (e.g., equipment rental or roofing), you may not be able to obtain business insurance. A complaint sole proprietorships sometimes have is that they feel isolated without co-owners or partners, and this seems to be most prevalent when difficult decisions (such as acquisition of debt) have to be made.

Partnerships

Much like a sole proprietorship, a partnership does not necessarily require any paperwork to be filed with the state, and the business is off and running the moment you start business with your partner. While it is prudent that a partnership agreement be drafted, often that does not take place. A partnership is easy to start, and the right set of partners might complement each others' skills, which could lead to improved business success. There are taxation benefits as income and losses are pass-through to each of the partners (no double taxation); the one tax document required of a partnership is the IRS Form 1065 Partnership Return. There is some limited opportunity to attract capital for projects through ownership benefits (defined in an operating agreement). Like the sole proprietorship, the partnership works well when product or service liability is a minimum and you are not likely to be sued. The structure of limited versus general partners must be carefully thought out, as this structure has a high potential for authority conflicts and exiting is not always easy.

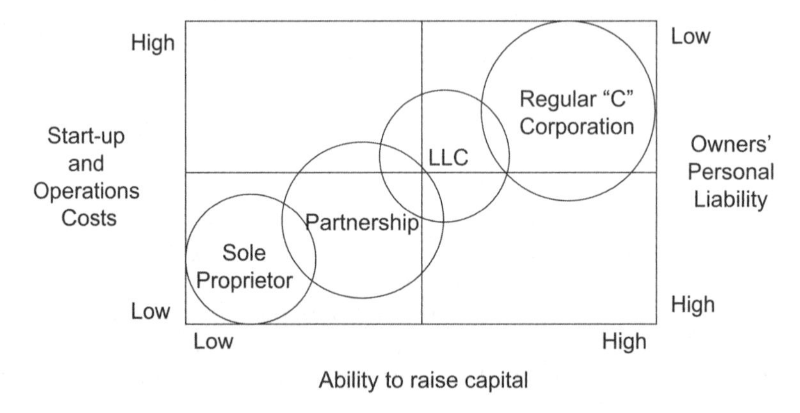

Figure A.1: Categorization of organizations.

Limited Liability Company (LLC)

As shown in Figure A.1, the limited liability company (LLC) is only slightly more costly to set up and run than a partnership, but the organizational form provides legal liability protection afforded a corporation. In most states the process for establishing the LLC is fairly straight-forward (and several steps can sometimes be done online), so the combination of limited personal liability and ease of creation makes this one of the most popular organizational structures today. Further, there are the pass-through tax benefits of a partnership, and also like a partnership, minimal external and internal state and federal filing requirements. LLCs are preferred by those organizations that have somewhat risky business operations and for which they wish to limit liability for LLC debts and claims.

Although an LLC may not always require filing articles of incorporation with the Secretary of State, the process is fairly straightforward, and some states have online databases of company records so that you can always look up examples of how to complete the forms. As with the partnership, it is highly advisable to develop an operating agreement, a useful document when you are attempting to raise money and potential investors can obtain an "interest" in the LLC.

C Corporation

This organizational form derives its name from "26 USC Chapter 1, Subchapter C - Corporate Distributions and Adjustments" of the IRS Code. The "C" designation simply used to differentiate it from a Subchapter "S" corporation (which itself is a corporation that is treated in some ways like a partnership). The C Corporation organizational form is what comes to mind when we simply say "corporation."

The C Corporation offers limited liability of shareholders, the shielding of management from personal liability for business liabilities, and the potential to attract capital easily. A corporation is an entirely independent, legal entity, with a entirely separate identity from the individuals employed to manage and control corporate affairs. C Corporation status means the "entity" pays taxes just as an individual would, but this leads to the notion of "double taxation" when the owners of the corporation (shareholders) subsequently have to pay taxes on salary and bonuses they draw from the corporation.

As a separate entity, the corporation does have considerable internal and external filing requirements at the local, state, and federal levels; the bureaucracy that can ensue might also create the impression of a slow organization with a lot of red tape and potential for diminished managerial incentives for those with ambition. Below a certain size, the need to conduct and document regular meetings of the officers of the corporation can seem burdensome.

The way transactions are treated is often linked to the type of corporate entity chosen when starting a business. In the U.S., the filing of initial tax forms requires a declaration of the intended accounting method. Many small businesses and professional practices find the cash accounting system simple to implement and suitable for much of their business activity. However, the IRS mandates an accrual method of accounting if any of the following three conditions apply:

1. The company is a C Corporation.
2. Gross sales revenue is normally (or expected to be) greater than $5 million.
3. Your company has inventory.

As with anything, there are exceptions to these guidelines, and it would be best to call in tax experts if there was a compelling reason to not follow the guidelines!

A.4 Problem solution outcomes

Managers, then, have sharpened acuity over desired outcomes, undertaking problem analysis that involves

- The need to have a *desired state* that frames the gap between "what should be" and "what is" in the mind of the decision-maker.
- Having some idea of a quantifiable outcome from the problem solution that you can control (through decision-making), also known as the *controllable variable.*
- Knowing some (or any) of the problem *constraints* that may exist, such as upper and lower bounds to acceptable controllable variable outcomes.
- Identifying "everything else needed to solve the problem," that is, the *problem data* not under your control and often known as *uncontrollable parameters.*

The problem-solving process is iterative and educational since it frequently results in the discovery of new data pertinent to the problem, but unidentified at the outset. An "outcome," of course, can also be a *course of action*, but generally, for a decision to be made, the problem analysis must entail (a) more than one course of action, and (b) possible outcomes of *un*equal value.

Different courses of action have different levels of effectiveness:

- A problem is considered *solved* if the outcome optimizes the resources associated with the decision variables. In medicine this is *efficacy* in that the solution did what it was intended to do.
- If a selected outcome is "good enough," then the problem is simply *resolved.* Why would you settle? Perhaps you prefer a solution that uses the least amount of resources and is therefore an *efficient* problem solution.
- Possibly the problem is *dissolved* if during the iterative discovery process there are changes to key variables rendering the problem trivial or nonexistent.

Some people with control-theory backgrounds will recognize that methods exist to optimize problem solutions (best possible outcome for the variables) and the heuristic (and simpler) approach where there is an acceptable choice or satisfactory values to the variables. Heuristic solutions are often faster and simpler than optimal solution methodologies, particularly if the model for the problem is complex.

The successful financial decision-maker is one who embraces and finds delight in the problem-solving process. Most likely the engineer who has some comfort level with ambiguity and interdisciplinary activity. Many times the *cause of the problem* emerges from working through many iterations of the problem definition, in which there is a pattern to relevant changes. And it is most gratifying to find the outcome that explains all the facts. It's sort of like a mystery novel! The reality is that such gratification is hard to come by, so we have to be content with success in the *art of decision-making.* We really do not understand the problem completely, yet still manage to make a decision that satisfies most of our criteria.

References

[1] Robert E. Lefton and Jerome T. Loeb. *Why can't we get anything done around here: The smart manager's guide to executing the work that delivers results.* McGraw-Hill, 2004.

[2] Patricia Benner. *From novice to expert: Excellence and power in clinical nursing practice.* Prentice Hall, 2004.

[3] Stefanos Zenios, Josh Makower, and Paul Yock. *BioDesign: The process of innovating medical technologies.* Cambridge University Press, 2010.

[4] Paul C. Nutt. *Why decisions fail: Avoiding the blunders and traps that lead to debacles.* Berrett-Koehler Publisher, 2002.

[5] Russell L. Ackoff. *The art of problem solving: Accompanied by Ackoff's Fables.* John Wiley and Sons, 1978.

[6] Clive L. Dym and Patrick Little. *Engineering design: A project-based introduction.* John Wiley and Sons, 2009.

[7] D.H. Marshall, W.W. McManus, and D.F. Viele. *Accounting: What the numbers mean.* McGraw-Hill, 7th edition, 2007.

[8] William Morris. *The American Heritage Dictionary.* Houghton Mifflin, 1978.

[9] Michael Mauboussin. *More than you know: Finding financial wisdom in unconventional places.* Columbia University Press, revised and expanded edition, 2008.

[10] Sheri D. Sheppard, Kelly Macatangay, Anne Colby, and William M. Sullivan. *Educating engineers: Designing for the future of the field.* Jossey Bass, 2009.

[11] National Academies. *Educating the engineer of 2020.* National Academies Press, 2005.

Appendix B

Mechanics of Accounting

B.1 Learning objectives

The world of accounting, finance, and engineering economics is at the core of business operations and is central to the process of gathering, classifying, analyzing, and reporting financial and economic information. Although the novice might initially sense accounting is more of an art than a science, recent industry scandals have led to an increase in the provision of guidelines and recommended practices. Overall, much actually goes "according to plan," and the latitude in systems and processes provides the needed flexibility for the broad range of entities they must serve.

There is a new vocabulary for engineers that goes with this general subject - knowledge of the basic terms is essential. This appendix explores this fascinating industry and sets the discussion framework for the chapters. It is imperative to perform the preparatory work that follows. If you fail to grasp the foundations of financial statements, your ability to ask questions as a manager is significantly compromised.

After reading and discussion sessions, the student should be able to:

1. Identify the difference between the "inside view" and "outside view."
2. Understand who uses accounting information and why this information is useful.
3. Identify the accounting tools and principles that shape financial statements.
4. Understand the ethical issues in accounting and the challenges accountants face.
5. Know the different types of financial statements and the relationship between them.
6. Explain what a company's annual report is and why it is used.

B.2 Accounting to support financial statements

Our transaction-based view of business shown earlier in Figure 1.8 illustrated the role of accounting systems to provide the tools and techniques by which we capture, analyze, record, and communicate information about transactions. The accounting system produces the data needed to create the financial statements summarized in Section 2.3.4; we now proceed to dig in a little further to clarify the essential accounting tools of the trade.

Recording and accounting for the transactions of business has a long history - fascinating, really - and reflects how the basic characteristics of *double-entry accounting*, journals and ledgers were structured to meet the emerging need of economies in the 15th and 16th centuries. Folklore varies, but many sources point to the Italian merchant Lucas Pacioli as the "father of accounting" with his systematic methods for recording business transactions in 1494 (Lauwers and Willekens[1]). Further refinements are attributed to more complex business activity (and the role of external investors in organizations) around the time of the Industrial Revolution. The *accounting equation* is at the center of the double-entry system and has remained so for centuries.

B.2.1 T-accounts

Figure B.1 presents the accounting equation in the form of the so-called "T-account" format, the name resulting from the resemblance of the chart to the letter "T". In this T-account example, several common asset, liability, and owners' equity categories are featured.

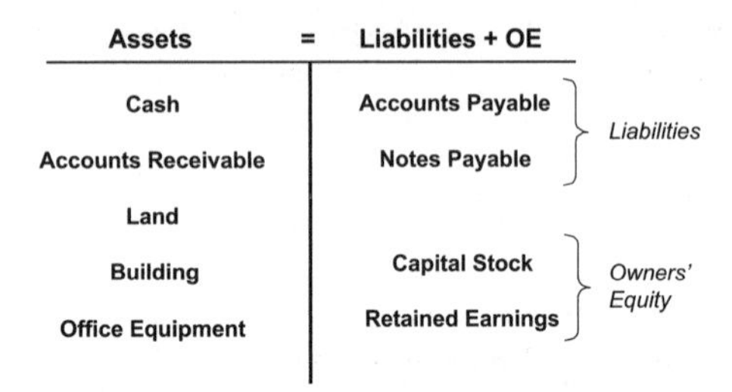

Figure B.1: The accounting equation in "T-account" format with example category listings.

By now we understand that both sides of the accounting equation must balance at any given time and therefore both sides of Figure B.1 should balance. Each of the categories (e.g., "Cash") listed in the chart is referred to as an *account*. Changes in any account category must be balanced by changes in other account categories. Since changes to at least two accounts must be made for any given transaction, this is called *double-entry accounting*. The layout of the T-account reflects the historical placement of "Assets" on the left-hand side and "Liabilities and OE" on the right-hand side. The general form of a T-account is simply

T-Account	
Left Side: **Debit** entry	Right Side: **Credit** entry

As mentioned earlier, assets are the resources used for the production of goods and services, and the claims on those assets are represented (and balanced) by the non-owner equity (liabilities) and owner equity (stock). Tradition has it that the left-hand side is called the *debit* side (short for the old-English "debitor") and the right side the *credit* side (as in "creditor"). Others have mentioned the terms come from the Latin word *debere* for "left" and *credere* for "right." Historical anecdotes notwithstanding, the custom in the 21st century is for accounts listed in Figure B.1 to be called either *debit accounts* if they are on the left-hand side and called *credit accounts* if they are on the right-hand side.

B.2.2 Chart of accounts

In practice there may be dozens of accounts needed to capture all of the information related to a company's business transactions, so it is handy to have an enumerated *chart of accounts*, also called the COA. The detailed structure of the COA is not specified by law (in the US), but recommended COA formats are provided by organizations such as FASB (`fasb.org`) and the National Center for Charitable Statistics (`http://nccs.urban.org/projects/ucoa.cfm`). Increasingly, companies use the default COA provided with accounting software such as QuickBooks (`www.quickbooks.com/`) or SAP (`www.sap.com`). The idea that the COA is simply obtained by modifying the baseline from a software package is fairly widespread.

Asset Accounts (1xx)		Liability (2xx) and OE Accounts (3xx)	
Code	Account Name	Code	Account Name
101	Cash	201	Accounts Payable
102	Bank Accounts	210	Tax Payable
110	Accounts Receivable	220	Employment Expenses Payable
120	Prepaid Business Expenses	230	Bank Loans
130	Inventory	270	Accrued Expense
140	Equipment	300	Common Stock
145	Depreciation	350	Retained Earnings

Table B.1: Example Chart of Account (COA) items

It is important to understand that the "debit" and "credit" terminology refers to the position within the T-account, and the reader should not confuse the similarity of names with their use in handling a personal checking account. A "debit" does not mean the trend of a value ("increase" or "decrease"). Figure B.1 illustrates the concept of *normal balance* to correspond with the normal presentation of the accounting equation. That is, the left-hand side traditionally is the asset (debit) side and the right-hand side is the equity (credit) side. Thus, a positive cash position is a normal debit balance, and having accounts payable outstanding represents a normal credit balance.

Each of the accounts shown in Table B.1 can itself be represented as a T-account. This additional method to track transaction detail facilitates the organization of transaction data and enables more complicated business activity to be clearly communicated when developing financial statements. Using the T-account framework just described, we can express the 101 Cash Account in the form:

101 Cash Asset Account	
Left Side is a **Debit** entry	Right Side is a **Credit** entry

So in the case where a company has a variety of cash transactions over the period of a quarter, the cash asset account would track each of the transactions. Consider a series of possible cash transactions associated with the rental business:

101 Cash Asset Account			
Starting investment	$50,000	June supply expense	$750
Collection of rent	$1,920	Investment in HVAC	$14,500
		Property taxes	$2,388
		Office supplies	$76
	$51,920		$17,714
Balance	$34,206		

Table B.2: Example *Cash Asset Account* for six typical transactions.

In this case we have 6 cash transactions, beginning with an investment of $50,000, and over the quarter the income and expenses have left a cash balance of $34,206.

The T-account in Table B.2 for cash might resemble the Statement of Cash Flow, but the Statement of Cash Flow is structured according to *categories*, and the T-account for cash is *chronological* by debit or credit column. T-accounts are used to support the generation of financial statements, but serve a different purpose in communicating information than the financial statement itself; this is explored further when, for instance, we examine the Statement of Cash Flow in Section 2.11.

B.2.3 General journal

Another observation about Table B.2 is that it provides a very limited description of the transaction; the summary statements provided are quite brief and seem more like "key word" lists than descriptions, and do not inform about possible special conditions about the transaction that could help with audit or error analysis. The solution is to create a separate document called a *general journal* (or just *journal*) that provides a complete description of each transaction. Thinking back to Figure 1.8 (p. 11) we characterized accounting as involving three major processes as shown below in Figure B.4. Journal entries are part of the second step of the three-step process.

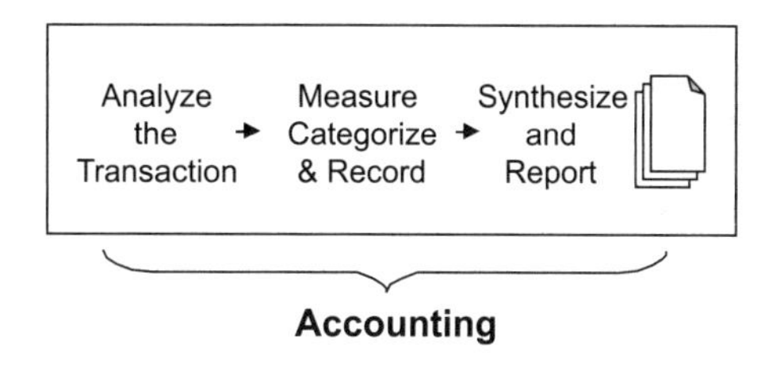

Figure B.2: Summary of key functions of the accounting process.

Although there are slight differences among journal formats in industry, overall they have the general layout shown in Table B.3.

The universal features of a journal we would expect to find are as follows:

1. A *date* for each transaction. For entries in the same year or same month, we do not need to repeat the year and month, though with computerized systems this is often added automatically.
2. A *description* of each debit and credit COA line item, along with the affected accounts. Note the tradition of indenting the credit record.
3. The *amount* of the debit and credit.
4. The chart of account *posting record* (essentially the COA identifier).
5. A brief note or explanation (italicized in Table B.3) to enable enough of the transaction to be understood. Sometimes this may refer to a *source record* like an invoice, services contract, or bank statement.

General Journal for Island Rentals, Inc. *Page 1*

Date	Description	PR	Debit	Credit
2013 June 14	Cash	101	$50,000	
	Common Stock	300		$50,000
	Investment in the company by a founder			
June 17	Office Supplies	810	$76	
	Cash	101		$76
	Purchase of office supplies from XYZ			

Table B.3: A typical set of General Journal entries.

Table B.3 includes an account number of 810 for office supplies, and the leading digit of "8" does not match any of those listed in Table B.1. There are sometimes thousands of account numbers used in business, and the leading digit often indicates the "top level" categorization of the account. Table B.1 was primarily concerned with the classic accounting equation accounts (asset 1xx, liability 2xx, and owner equity 3xx). In Table B.3 items like "office expense" are actually linked to the income statement; this different relationship is reflected in a different leading numeral. It is somewhat academic to list all the possible COA numbers, but Figure B.3 illustrates a few expense varieties. This list was extracted from a QuickBooks file for a company that had approximately 100 account numbers in use.

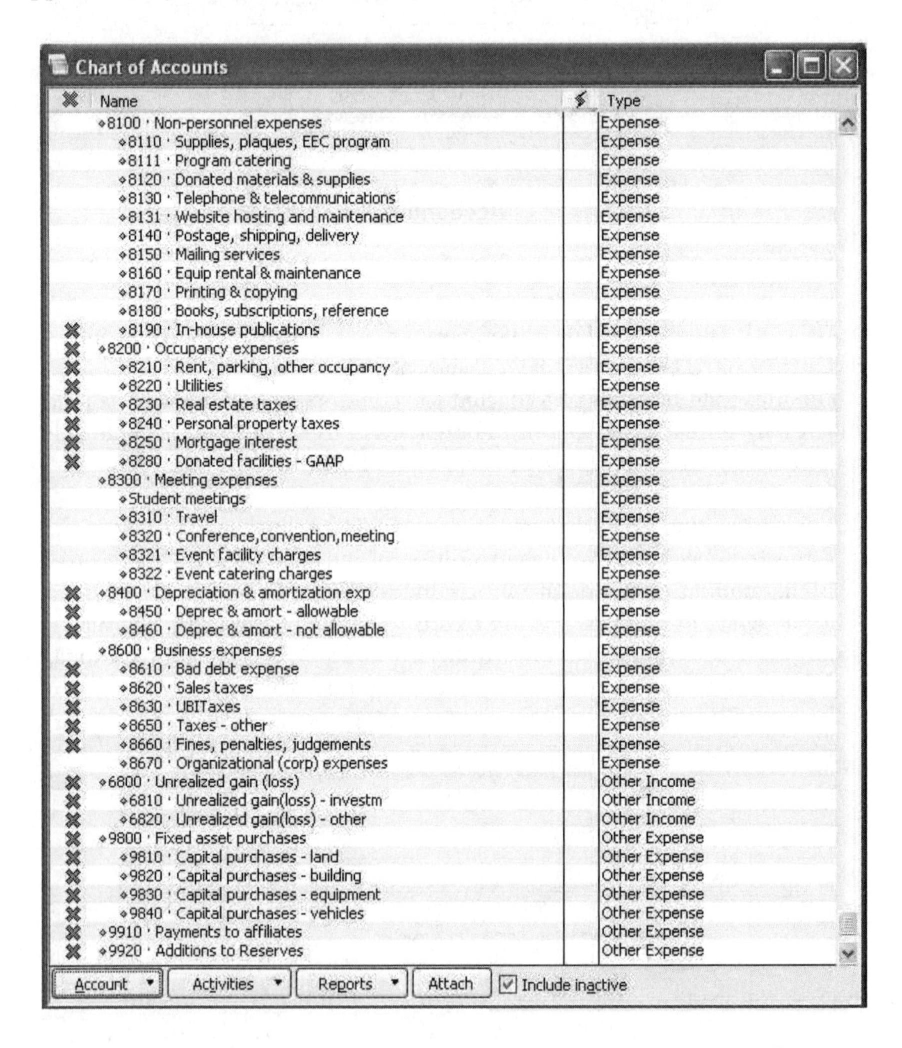
Chart of Accounts

Name	Type
8100 · Non-personnel expenses	Expense
8110 · Supplies, plaques, EEC program	Expense
8111 · Program catering	Expense
8120 · Donated materials & supplies	Expense
8130 · Telephone & telecommunications	Expense
8131 · Website hosting and maintenance	Expense
8140 · Postage, shipping, delivery	Expense
8150 · Mailing services	Expense
8160 · Equip rental & maintenance	Expense
8170 · Printing & copying	Expense
8180 · Books, subscriptions, reference	Expense
8190 · In-house publications	Expense
8200 · Occupancy expenses	Expense
8210 · Rent, parking, other occupancy	Expense
8220 · Utilities	Expense
8230 · Real estate taxes	Expense
8240 · Personal property taxes	Expense
8250 · Mortgage interest	Expense
8280 · Donated facilities - GAAP	Expense
8300 · Meeting expenses	Expense
Student meetings	Expense
8310 · Travel	Expense
8320 · Conference,convention,meeting	Expense
8321 · Event facility charges	Expense
8322 · Event catering charges	Expense
8400 · Depreciation & amortization exp	Expense
8450 · Deprec & amort - allowable	Expense
8460 · Deprec & amort - not allowable	Expense
8600 · Business expenses	Expense
8610 · Bad debt expense	Expense
8620 · Sales taxes	Expense
8630 · UBITaxes	Expense
8650 · Taxes - other	Expense
8660 · Fines, penalties, judgements	Expense
8670 · Organizational (corp) expenses	Expense
6800 · Unrealized gain (loss)	Other Income
6810 · Unrealized gain(loss) - investm	Other Income
6820 · Unrealized gain(loss) - other	Other Income
9800 · Fixed asset purchases	Other Expense
9810 · Capital purchases - land	Other Expense
9820 · Capital purchases - building	Other Expense
9830 · Capital purchases - equipment	Other Expense
9840 · Capital purchases - vehicles	Other Expense
9910 · Payments to affiliates	Other Expense
9920 · Additions to Reserves	Other Expense

Figure B.3: A subset of expense account numbers, extracted from a small company QuickBooks file.

Remark on accounting software

In 2000, Albrecht and Sack [2] were clear that accounting software would change the field of accounting; from chapter 2 (p. 6):

> While these [accounting] change drivers have significantly impacted everything we do, including the way we live, they have had two dramatic impacts on business. First, they have eliminated the old model that assumed information is expensive. Today anyone, armed with the right software, can be an "accountant" and produce financial information.

Surveys differ on what might be the most popular accounting software program, though commonly mentioned are names like QuickBooks, SAP, PeopleSoft, Oracle, ADP, Great Plains, JD Edwards, and Peachtree. Product cost and features vary widely. Some systems are designed around a limited numbers of users and other limitations, so it is an art in itself to select the right package for a business (Johnston[3])). I have used the first 4 packages in practice and found each suitable for the intended organization size, scope, and complexity of accounting activity.

Ths book uses snapshots from QuickBooks to demonstrate a few software features, but one should assume that for the purposes of this book the digital philosophy of other software systems could have equally been used.

As an example, compare the General Journal from Table B.3 (page 361) with the QuickBooks sample shown in below in Figure B.4.

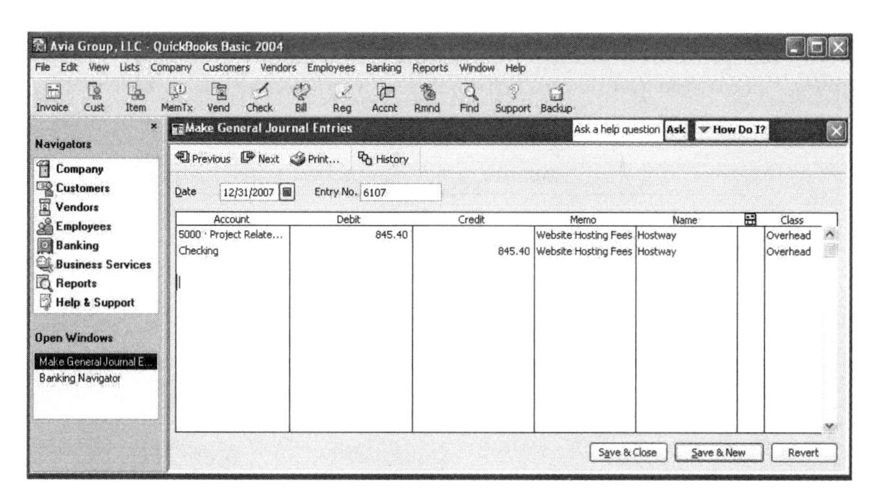

Figure B.4: A journal entry extracted from a small company QuickBooks file.

Figure B.3 and Figure B.4 have similar features with the same content available to the user, but the formatting is slightly different, as would be expected. For the present work we assume some form of software is available in practice, and the role of the current work is to elucidate essential principles for decision-making. Often you can pick up the use of, say, QuickBooks or Peachtree in your spare time.

Key points about the general journal

The general journal is a bookkeeping tool that helps us organize and record the details of each transaction. Although in paper-based systems the journal was a specially ruled book, in electronic systems the journal is simply equivalent to a spreadsheet file. Two key functional features of a journal are (a) the detailed description of the transaction, and (b) identifying links from the transaction to debit and credit accounts. In software systems the journal screen is the primary data-entry screen for the system. Knowledge of the chronological journal and its function in the accounting cycle is important and relevant to *both* paper-based and electronic accounting systems.

B.2.4 General ledger

Once general journal entries have been updated, the next critical step is to organize and aggregate data into accounts. Recall Table B.2, the Cash Asset Account, that organized data into debit and credit columns and also provided the user a running total of the debit, credit, and net account balances. It is easy to look at Table B.2 and visualize this as a prelude to the creation of financial statements.

A slight problem occurs when we try to integrate the numerous T-accounts into a single T-account sheet (recall the format of Figure B.1) since this greatly expands the table (equivalent to the accounting equation) and becomes as shown in Figure B.5.

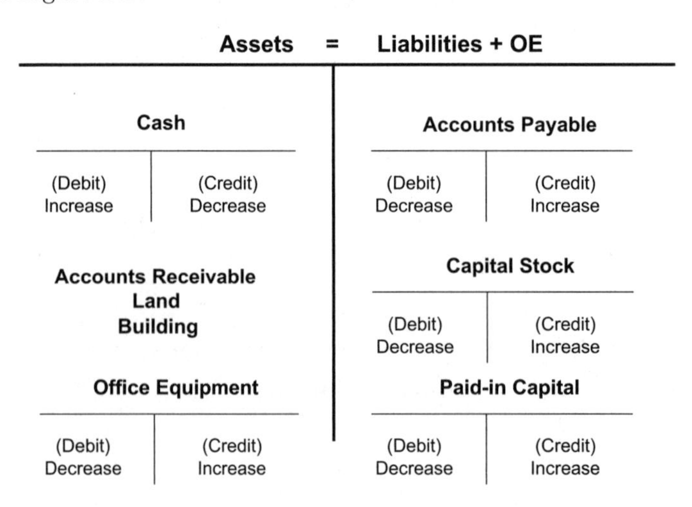

Figure B.5: The accounting equation in "T-account" format with expanded account items (several accounts have not been expanded in T-account format for brevity).

Conceptually Figure B.5 helps with the development of the accounting process but is not practical as an accounting tool. Fortunately, we can organize data in *ledger account forms* (also known as *balance column accounts*) as shown below in Table B.4.

General Journal for Island Rentals, Inc. *Page 1*

Date	Description	PR	Debit	Credit
2013 June 14	Cash	101	$50,000	
	Common Stock	300		$50,000
	Investment in the company by a founder			

Table B.4: A simple ledger account form.

The so-called "posting" of journal accounts into a complete set of ledger accounts is illustrated in Figure B.6.

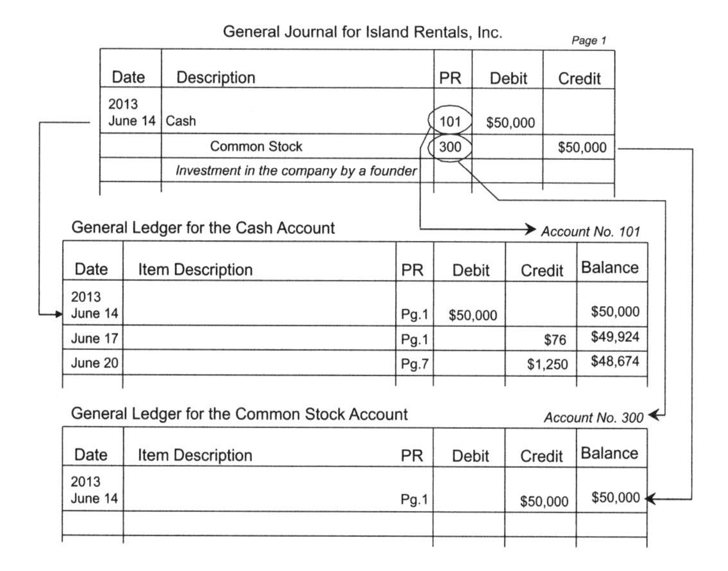

Figure B.6: Posting of journal entries to ledger account forms.

All of the information needed for the general ledger is derived from the general journal. Any errors in completing either the general journal or general ledger will propagate throughout the system.

The idea of "posting" is based in paper systems, and the process is automated in software like QuickBooks and Peachtree whereby the completed journal entry has all the data needed to automatically post. Paper-based systems would normally conduct "posting" on a periodic schedule (daily, weekly), but posting is automatic and immediate with software. The general journal is a "lilly-pad" activity that is more mechanistic than requiring any interpretive activity. Of course, the ledger accounts *do* have value in reviewing and analyzing account activity. A comparison of accounting textbooks over time suggests a diminished need to elaborate on this step, with bookkeeping descriptions giving way to the meaning of the data. Having said that, my experience is that when attempting to diagnose accounting errors (even with automated software procedures), the auditor is lost if the fundamental concepts are not clear. There seem to be many ways for incorrect entries to find their way into even disciplined organizations.

The following process summarizes the posting process.

1. Begin with the debit (left) entry of the journal account (101 Cash, in our example) and transfer to the ledger account (from the general journal) the date, journal page, and journal transaction amount. When this critical data is entered, then compute the ledger account balance.
2. Double-check or add the posting reference from the ledger account to the general journal. In the simple sample case, the Cash Ledger reference states "Pg. 1" (first page of the general journal) in the ledger column Posting Record "PR".
3. Now, move to next line in the journal account (the credit account, "right side") and transfer to the ledger account (from the general journal) the date, journal page, and journal transaction amount. Then compute the ledger account balance
4. Double-check or add the posting reference *from* the ledger account back to the general journal. Quite similar to step 2 above, the Common Stock Ledger reference states "Pg. 1" (first page of the general journal) in the ledger column Posting Record "PR."

It might seem odd at first, but for paper-based systems the "explanation" space normally echoes exactly what is in the journal entry. As long as the journal page reference is satisfactory, there is no need to re-enter it in a ledger account. Accounting software programs are not burdened by the extra step since transferring the journal description to the ledger explanation is trivial and effortless, so many software general journals do include the explanation.

The ledger does not indicate whether, for instance, the balance in Figure B.4 is a debit or credit balance. Going back to Section B.2.1, the balance is assumed to be the *normal balance* for the account. Cash accounts normally have a debit balance, so the $43,000 is assumed a debit balance.

B.2.5 Adjusting entries

We've made a big deal out of the transaction-based perspective and the need for a clear understanding of the type (investing, financing, operations) and accuracy with recording of transactions. So why would adjustments to entries be necessary if the general journal and postings to the ledger were done right? Do we always need adjusting entries? The answer is, "It depends." Let's find out why.

Accounting period

The concept of an *accounting period* is an important frame of reference for business transactions. This is sometimes referred to as the *reporting period* and is closely related to the dates we mentioned when discussing naming convention for financial statement titles (Section 2.3.4, p. 26). At the time, those statements were implied to be *annual* statements, though interim financial statements might be produced throughout the accounting period. Figure B.7 illustrates the idea of monthly, quarterly, and annual periods. There is generally no requirement that the accounting period end on December 31; many organizations use other year-end dates like April 30 or June 30.

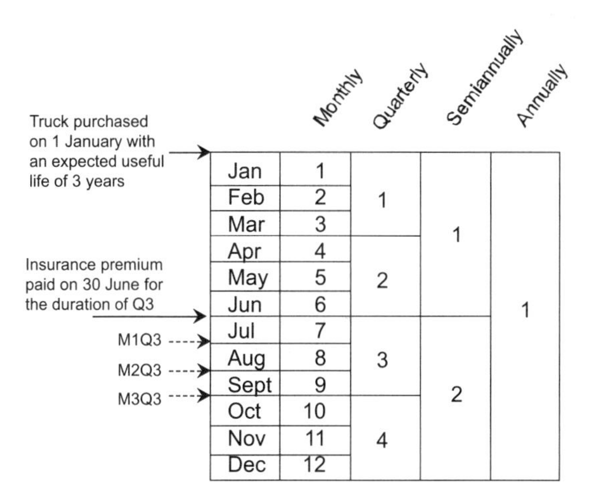

Figure B.7: Accounting periods for a typical business. Organization activity reporting can be done on a monthly, quarterly, semiannual, or annual basis.

Figure B.7 illustrates some expense examples that span multiple months. Such expenses might be cash expenditures at the beginning of the year that benefit the business throughout the year. As well, there may have been income that was earned but might not be collected for many months. Expenses could have been incurred but credit extended with no requirement for the customer to pay until the following month. In short, we are saying that income and expenses might overlap or extend over multiple accounting periods. Adjusting entries enable us to correct transaction allocations such that the expense and period are aligned.

The process by which revenue earned is matched to expenses incurred (to earn that revenue) and further adjusted to the reporting period is called *accrual accounting*. The accrual basis of accounting recognizes revenues and expenses when they accrue ("come into existence as a legally enforceable claim" [4]) and are measurable, not necessarily at the time of the cash payment. While there is some mild complexity associated with accrual accounting, the general belief is that it better represents the profitability of a company and how it performs period-to-period. Accrual accounting, adjusting entries, and the accounting period become three concepts central to an accurate portrayal of the way assets are managed to reach the goals and objectives of a company.

Nature and form of the business

The *nature and form* of the business could dictate whether adjusting entries are logical. And sometimes we do things because we are required by law. For instance, the nature of a computer repair service provided by a small business might require debit or cash payment immediately upon completion of repairs. If the computer repair service is small enough, job revenue might align with "same-day-service" and the income aligns exactly with services provided. In this case "recognizing" cash flow might easily and fairly represent service income. On the other hand, a publicly traded multinational C-Corporation with over $5 million in sales may not have a choice, and the SEC might prescribe adjusting entry guidelines to ensure transparency of activity to external stakeholders. Other private, medium-size investment banking companies might have a wide portfolio of securities activities and subject every transaction to an adjusting-entry check (and double-check) as a matter of policy to build client confidence or minimize opportunity for fraud.

As a general rule it is highly recommend that regardless of the form of the business, accrual-basis accounting should be the first choice, but understanding that in many situations a cash basis is suitable and appropriate.

Operational focus

The main activities of a company were categorized in Section 1.3 as either *investing*, *operating*, or *financing* activities. We established at the outset that our engineering manager viewpoint would tend to center attention on operational activity (Section 1.2, p. 2). The existence of historical activity, the assumption of an on-going concern, and the desire of an engineering manager to understand and improve operational productivity and performance were a priority. One of the foundational objectives for creating financial statements is to provide information about use of resources, claims on those resources, and the effect of operational activity on the changes of those assets. As such, we find adjusting entries impact the balance sheet and the income statement, *not* the statement of cash flows.

Adjusting entry quad-chart

Figure B.8 is a very concise representation of four situations where adjusting entries are required and the manner in which they impact the balance sheet and income statement.

For instance, in Figure B.7 consider the situation where a company pays a $90 quarterly invoice for Q3 insurance on June 30, for liability insurance for the months of June, July, and August. Because the insurance coverage was a future time that had not arrived, you are in quadrant DE of Figure B.8, since the prepayment was for a deferred expense that had yet to be recorded for the

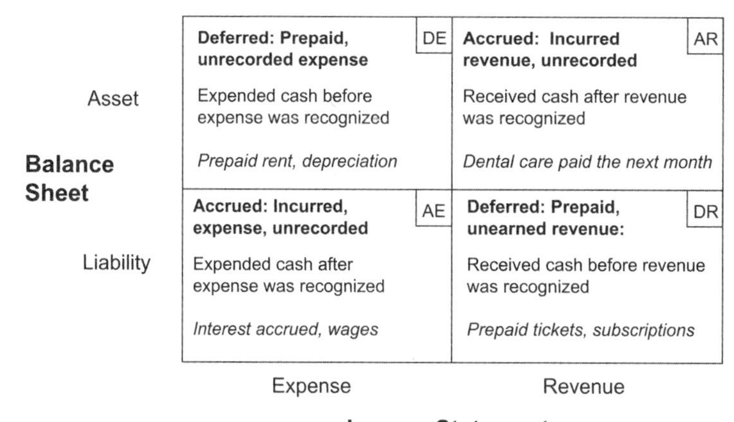

Figure B.8: Quad-chart for deferred and accrued expenses and their relation to the balance sheet and income statement.

given monthly accounting period. Here, we have an expense on the income statement that is an asset on the balance sheet.

Quadrant DE is also used to account for asset depreciation, a situation where a capital investment on the balance sheet is converted over time to a series of expenses on the income statement. This topic is covered further in Section 2.6.

If you run an airline it is common practice to ask passengers to pay for tickets in advance. In such a case you have collected cash for services yet to be rendered. This situation is identified in quadrant DR, whereby the receipt of cash affects revenue on the income statement and a liability on the balance sheet. This can easily be viewed as a good thing, but must be treated with caution. If the organization is a bit short of cash (and discipline) then "cash in the bank" may be viewed as available for other things, when in fact it is truly a liability. The idiom "robbing Peter to pay Paul" falls within this quadrant; to some extent pyramid fraud schemes emanate from here, too. On a more day-to-day level, the idea that you prepay some other firm (e.g., upfront payments for website updates) works well as long as the firm (in quadrant DR) does not become insolvent and your pre-paid expense (quadrant DE) is lost. Some airlines have gone bankrupt leaving passengers with worthless tickets.

Quadrant AR has the potential to cause trouble, since it involves accrued expenses for which cash has not been received. This is often where credit has been provided to the customer and the company had the risk of the promise of payment. A dentist might provide a cleaning service to you, for which an invoice is later sent to you and you might pay within a month. For the dentist, the promise of receipt of cash (account receivable) is an asset (albeit not free of risk) on the balance sheet and recognized as revenue on the income statement.

Some people like the idea of sitting in quadrant AE, as this is the case where you've received services or something of value, but have not yet paid for it. How fortunate you are (or are not) to be in this situation was hinted at earlier in Section 2.4.1, p. 39 in the description of the way in which ratios provide performance insight. Equation 2.9 described the idea of "Days Sales Outstanding" (DSO) that is a reflection of the efficiency with which account receivables are converted to cash (we have the cash cycle of Figure 2.3 in mind). There are pros and cons to quadrant DE. On one hand, quadrant AE can be viewed as advantageous in that you have non-owner assets provided that help you generate income. On the other hand, the peril of quadrant AE is that too large a liability with extended DSO might put you in a cash crunch.

Adjusting entries involve a wide variety of accrual and deferred expenses and revenues. Some of the complications and downsides of "transactions gone bad" might foster an attitude that "cash is king" in search of a simpler financial life, but understanding how adjusting entries work is essential for today's engineering manager. Interdependence of organizations through partnerships and collaborative activity is the norm, and project managers must know how assets move between the balance sheet and income statement in the form of adjusting entries.

This section underscores the distinctive difference between an income statement and a cash flow statement. The discussion above provided several examples where adjusting entries needed to craft an accurate income statement actually involve a variety of speculative situations. In comparison, cash flow is less speculative and is easier to produce, but unfortunately is not a good tool for describing the productivity of assets that leadership is entrusted to manage and meet the organization's performance goals.

Adjusting entry account table

In situations where a transaction impacts *both* the balance sheet and income statement, it can help to *link* T-accounts as shown in Figure B.9 to provide a framework for tracking adjusting entries. This is a convenient graphical depiction of transaction analysis when several T-accounts might be needed to be accurate, yet the presentation confusing. The adjusting entry table provides convenience and simplicity in understanding transactions, though in practice software systems make the use of an account table somewhat academic.

Industry today tends to make extensive use of accounting software, much in the same way calculators and computers are used to make engineering calculations. Figure B.9 helps convey key concepts without us getting lost in T-account mechanics. Think of this as an accounting tool for learning accounting concepts. Let's work though three simple examples.

1. **Item 1: Cash investment of $50,000**. This transaction only affects the balance sheet. We record to the cash (debit) account the sum of $50,000 and record in the owners' equity (credit) account an equal sum of $50,000. There is no impact on the income statement since this is a financing activity, not an operations activity.
2. **Item 2: Purchase office supplies for $79**. On the balance sheet, cash is credited in the sum of $79 – the dual-entry system balances the

	Balance Sheet		Income Statement	
	Assets = Liabilities + Owners' Equity	⇦	Net Income = Revenue - Expenses	
1.	Cash +$50,000	Paid-in capital +$50,000		
2.	Cash -$76	Loss -$76		Office Supplies -$76
3.	Cash +$1,920	Gain +$1,920	Rent +$1,920	
4.	Long-term asset -$2,000	Loss -$2,000		Depreciation -$2,000

Figure B.9: Transaction summary for adjusting entry table.

credit with a $79 debit to OE. Of course, OE is impacted since office supplies are normally considered an operational expense (and so debit expenses for $79) which represents a negative net income. The negative net income amount is equal to the loss in OE.

3. **Item 3: Receipt of $1,920 rent**. Rental revenue is favorable and thus a credit, just as OE normally carries a credit balance. Net income is a computed quantity (not shown on the T-account, but is shown on the income statement financial statement) and is used to debit or credit OE.

The convenience of Figure B.9 for the recording and aggregation of impact of the transactions on financial statements becomes more evident when we look at the equivalent set of T-account entries shown in Figure B.10.

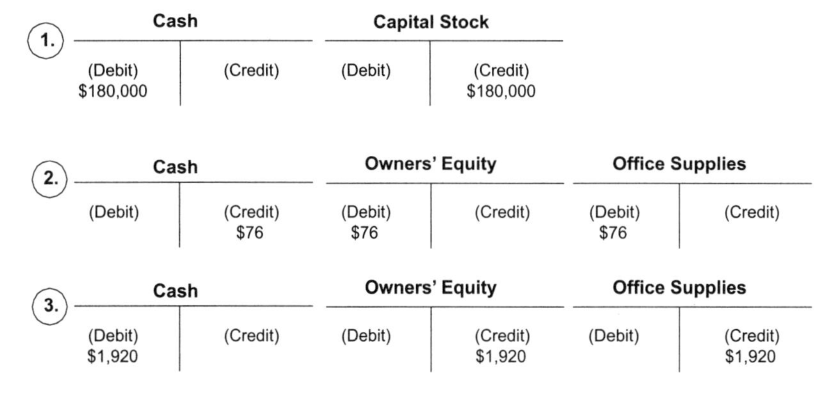

Figure B.10: T-account format for the transactions shown in Figure B.9.

Generally, it is through working specific problems that competencies improve in the classification and recording of transactions. Both the horizontal approach in Figure B.9 and the T-account approach shown in Figure B.10 are satisfactory for "learning the ropes" of accounting transactions; choice just becomes a matter of personal preference. In either case, the same general approach is needed when identifying, analyzing and recording transactions:

1. Analyze the transaction: What is going on?
2. Enter the transaction into the general journal: What accounts are affected?
3. Post the entry: How are the accounts affected?
4. Examine the accounts and check if the balance sheet balances.
5. Does the outcome make sense?

Table B.5: Five critical questions to ask when adjusting entries.

Interestingly, the last step is crucial whether you use a software system or (still) use paper ledger systems, though errors propagate into reports faster in an electronic system. Neither paper nor the computer can tell you that you, say, incorrectly credited a rent payment. Mastery of each step in Table B.5 comes with practice over time; the reader should make it a priority to gain access to a desktop accounting system and test out the entry of various transactions and their impact on financial statements.

Trial Balance

Modern accounting systems automatically produce the *trial balance*, based on data entry and related bookkeeping tasks associated with the general journal. The trial balance is simply a list of all the debit and credit balances (as posted in the general ledger) at a given point in time, possibly the end of a month, the quarter, or another reporting period. The trial balance is *not* a financial statement in the formal sense used in external decision-making (part of a 10K or event in investor conference calls), but is primarily a statement used for error checking and diagnostics. For instance, the entries might be chronological to assist with tracking down errors or questions.

We can easily imagine a set of T-account entries more complex than the simple example of Figure B.10 that could result in errors, particularly if a paper-based system were in use (and the impact of a general journal entry not immediately observable). In the "paper world," it was suggested above that the trial balance was generated as an intermediate step on the way to creating the financial statements (cash flow, income, balance, OE). It is more useful as an *internal control* to help spot journal and ledger errors. Many of the errors we can think of relate more to paper-based systems than computer-based systems. Typical errors are

1. Ensuring the aggregate ledger postings have the result DR = CR.
2. Identifying journal entries that may have been missed during posting.
3. Incorrectly posting a DR when it should be a CR.
4. Incorrectly posting a DR to account 101 and the corresponding CR to account 110.
5. Incorrectly posting the DR and CR to the same column.

Hopefully by now you've taken the time to explore a computer-based system; if so, it should be pretty clear the trial balance is a component of days gone by as few errors discussed above find their way into today's software systems, but sometimes errors do occur and a trial balance might be needed. For instance, in Quickbooks some people create trial balance reports to work from if they are making a large number of general journal entries. The Quickbooks export function is useful for generation of off-line creation of trial balance spreadsheets.

More relevant to today's engineering manager is the use and interpretation of data already entered into the system. Few managers actually complete an entry in the general journal, but many do query accounting databases to inspect various statements (trial and otherwise) and might reflect on the process provided in Figure B.10.

Despite who does the work, it is imperative to complete all adjusting entries at the end of the reporting period (often on a monthly basis) to enable financial statements to properly reflect the business activity for the reporting period. Many different internal managerial accounting and finance decisions are dependent on monthly and quarterly reports.

B.3 Problems to explore

Problem B.1
Discussion questions.

1. Provide several examples of "intangible" assets.
2. Is it possible for Working Capital to be negative? Retained Earnings? Why?
3. The liquidity of assets is important to creditors. Does the liquidity of assets affect OE? Why?
4. What might be of consideration in a transaction whereby intellectual property (IP) is exchanged for common stock.
5. Does preferred stock have a preferential position during bankruptcy proceedings?

Problem B.2
Roxanne Chester graduated from dental school in May 2013 and has decided to launch a practice called "Surf Smiles, Inc." Her family thinks she should join an existing practice since she will have to learn a lot of business skills, but that just serves to annoy Roxanne since she thinks figuring out ledger entries can't be any harder than getting adults to floss their teeth daily or endodontic surgery (root canal). Besides, she wants the company to be located near a beach. The transactions of her company in May are as follows:

May transactions	
1st	Roxanne used the last $2,250 of her college loans to form SSI
4th	Purchased office supplies and medical supplies for $1,200 in cash
18th	Purchased $18,250 in medical equipment on credit with 5% down
21st	Performed routine cleaning and consult with her first patient for $95 on credit
25th	Paid $75 for the first payment on her medical equipment (from May 18th)
30th	Paid her office electric bill of $40

1. Analyze the transactions in terms of their effect on the accounting equation.
2. Develop an income statement and balance sheet for this new business.
3. Is this business doing well?

Problem B.3
Sarah Vandenberg has started a website business since her brother Todd has opened an successful garage called Toddies Gear Guys, and now Sarah wants to be an entrepreneur, too. Below are the events that Sarah experienced in May 2013 for her firm Sarah's Web Wonderland. Assume Sarah's business started with Accounts Payable of $1,350 and $1,350 in Supplies.

	May transactions
a.	Sarah received $15K as a gift from her brother to start the web business
b.	Performed website work for $950
c.	Purchased a work truck (Kia Soul) for $12,250 credit with 2% down
d.	Purchased an annual truck insurance policy for $600
e.	Paid $75 to advertise in the local *Plain Dealer*
f.	Purchased reference books for $1.99 from Amazon.com but paid $32 for expedited shipping.
g.	Paid rent of $2,200 and utilities of $160
h.	Paid temporary help $1,700 for the month of May

1. Analyze the transactions in terms of their effect on the accounting equation.
2. Develop an income statement and balance sheet for this new business.
3. Is this business doing well?

Problem B.4

Ian McMahon opens a solar-powered moped rental business near a college campus. He is the sole owner of the proprietorship, named Campus Motor Bikes. Transactions for the first month of operations, July 2013, are shown below.

1. Analyze the following transactions in terms of their effect on the accounting equation.
2. Prepare the income statement and balance sheet of the business after recording the transactions.

	July transactions
a.	Ian invests $20,500 of personal funds to start the business.
b.	He purchases on account office supplies costing $350.
c.	Ian pays cash of $30,000 to acquire a lot next to the campus for a future business office.
d.	Ian rents scooters to clients and receives cash of $1,900.
e.	Ian pays $100 on the account payable in transaction (b).
f.	He pays $2,000 of personal funds for a vacation.
g.	He pays cash expenses for office rent, $400, and utilities, $100.
h.	Ian withdraws $1,200 cash for personal use.

References

[1] Luc Lauwers and Marleen Willekens. Five hundred years of bookkeeping: A portrait of luca pacioli. *Tijdschrift voor Economie en Management*, 39(3):289–304, 1994. ISSN 0772-7674.

[2] Steve Albrecht and Robert Sack. Accounting education: Charting the course through a perilous future. *American Accounting Association*, 16, 2000.

[3] Randolph P. Johnston. A strategy for finding the right accounting software. *Journal of Accountancy*, September 2003.

[4] William Morris. *The American Heritage Dictionary*. Houghton Mifflin, 1978.

Appendix C

Reference Tables

Z	0.00	0.01	0.02	0.03	0.04	0.05	0.06	0.07	0.08	0.09
-3.4	0.4997	0.4997	0.4997	0.4997	0.4997	0.4997	0.4997	0.4997	0.4997	0.4998
-3.3	0.4995	0.4995	0.4995	0.4996	0.4996	0.4996	0.4996	0.4996	0.4996	0.4997
-3.2	0.4993	0.4993	0.4994	0.4994	0.4994	0.4994	0.4994	0.4995	0.4995	0.4995
-3.1	0.4990	0.4991	0.4991	0.4991	0.4992	0.4992	0.4992	0.4992	0.4993	0.4993
-3.0	0.4987	0.4987	0.4987	0.4988	0.4988	0.4989	0.4989	0.4989	0.4990	0.4990
-2.9	0.4981	0.4982	0.4982	0.4983	0.4984	0.4984	0.4985	0.4985	0.4986	0.4986
-2.8	0.4974	0.4975	0.4976	0.4977	0.4977	0.4978	0.4979	0.4979	0.4980	0.4981
-2.7	0.4965	0.4966	0.4967	0.4968	0.4969	0.4970	0.4971	0.4972	0.4973	0.4974
-2.6	0.4953	0.4955	0.4956	0.4957	0.4959	0.4960	0.4961	0.4962	0.4963	0.4964
-2.5	0.4938	0.4940	0.4941	0.4943	0.4945	0.4946	0.4948	0.4949	0.4951	0.4952
-2.4	0.4918	0.4920	0.4922	0.4925	0.4927	0.4929	0.4931	0.4932	0.4934	0.4936
-2.3	0.4893	0.4896	0.4898	0.4901	0.4904	0.4906	0.4909	0.4911	0.4913	0.4916
-2.2	0.4861	0.4864	0.4868	0.4871	0.4875	0.4878	0.4881	0.4884	0.4887	0.4890
-2.1	0.4821	0.4826	0.4830	0.4834	0.4838	0.4842	0.4846	0.4850	0.4854	0.4857
-2.0	0.4772	0.4778	0.4783	0.4788	0.4793	0.4798	0.4803	0.4808	0.4812	0.4817
-1.9	0.4713	0.4719	0.4726	0.4732	0.4738	0.4744	0.4750	0.4756	0.4761	0.4767
-1.8	0.4641	0.4649	0.4656	0.4664	0.4671	0.4678	0.4686	0.4693	0.4699	0.4706
-1.7	0.4554	0.4564	0.4573	0.4582	0.4591	0.4599	0.4608	0.4616	0.4625	0.4633
-1.6	0.4452	0.4463	0.4474	0.4484	0.4495	0.4505	0.4515	0.4525	0.4535	0.4545
-1.5	0.4332	0.4345	0.4357	0.4370	0.4382	0.4394	0.4406	0.4418	0.4429	0.4441
-1.4	0.4192	0.4207	0.4222	0.4236	0.4251	0.4265	0.4279	0.4292	0.4306	0.4319
-1.3	0.4032	0.4049	0.4066	0.4082	0.4099	0.4115	0.4131	0.4147	0.4162	0.4177
-1.2	0.3849	0.3869	0.3888	0.3907	0.3925	0.3944	0.3962	0.3980	0.3997	0.4015
-1.1	0.3643	0.3665	0.3686	0.3708	0.3729	0.3749	0.3770	0.3790	0.3810	0.3830
-1.0	0.3413	0.3438	0.3461	0.3485	0.3508	0.3531	0.3554	0.3577	0.3599	0.3621
-0.9	0.3159	0.3186	0.3212	0.3238	0.3264	0.3289	0.3315	0.3340	0.3365	0.3389
-0.8	0.2881	0.2910	0.2939	0.2967	0.2995	0.3023	0.3051	0.3078	0.3106	0.3133
-0.7	0.2580	0.2611	0.2642	0.2673	0.2704	0.2734	0.2764	0.2794	0.2823	0.2852
-0.6	0.2257	0.2291	0.2324	0.2357	0.2389	0.2422	0.2454	0.2486	0.2517	0.2549
-0.5	0.1915	0.1950	0.1985	0.2019	0.2054	0.2088	0.2123	0.2157	0.2190	0.2224
-0.4	0.1554	0.1591	0.1628	0.1664	0.1700	0.1736	0.1772	0.1808	0.1844	0.1879
-0.3	0.1179	0.1217	0.1255	0.1293	0.1331	0.1368	0.1406	0.1443	0.1480	0.1517
-0.2	0.0793	0.0832	0.0871	0.0910	0.0948	0.0987	0.1026	0.1064	0.1103	0.1141
-0.1	0.0398	0.0438	0.0478	0.0517	0.0557	0.0596	0.0636	0.0675	0.0714	0.0753
0.0	0.0000	0.0040	0.0080	0.0120	0.0160	0.0199	0.0239	0.0279	0.0319	0.0359

Table C.1: Standard Probability Distribution: Part A from the mean of 0.5.

Z	0.00	0.01	0.02	0.03	0.04	0.05	0.06	0.07	0.08	0.09
0.0	0.0000	0.0040	0.0080	0.0120	0.0160	0.0199	0.0239	0.0279	0.0319	0.0359
0.1	0.0398	0.0438	0.0478	0.0517	0.0557	0.0596	0.0636	0.0675	0.0714	0.0753
0.2	0.0793	0.0832	0.0871	0.0910	0.0948	0.0987	0.1026	0.1064	0.1103	0.1141
0.3	0.1179	0.1217	0.1255	0.1293	0.1331	0.1368	0.1406	0.1443	0.1480	0.1517
0.4	0.1554	0.1591	0.1628	0.1664	0.1700	0.1736	0.1772	0.1808	0.1844	0.1879
0.5	0.1915	0.1950	0.1985	0.2019	0.2054	0.2088	0.2123	0.2157	0.2190	0.2224
0.6	0.2257	0.2291	0.2324	0.2357	0.2389	0.2422	0.2454	0.2486	0.2517	0.2549
0.7	0.2580	0.2611	0.2642	0.2673	0.2704	0.2734	0.2764	0.2794	0.2823	0.2852
0.8	0.2881	0.2910	0.2939	0.2967	0.2995	0.3023	0.3051	0.3078	0.3106	0.3133
0.9	0.3159	0.3186	0.3212	0.3238	0.3264	0.3289	0.3315	0.3340	0.3365	0.3389
1.0	0.3413	0.3438	0.3461	0.3485	0.3508	0.3531	0.3554	0.3577	0.3599	0.3621
1.1	0.3643	0.3665	0.3686	0.3708	0.3729	0.3749	0.3770	0.3790	0.3810	0.3830
1.2	0.3849	0.3869	0.3888	0.3907	0.3925	0.3944	0.3962	0.3980	0.3997	0.4015
1.3	0.4032	0.4049	0.4066	0.4082	0.4099	0.4115	0.4131	0.4147	0.4162	0.4177
1.4	0.4192	0.4207	0.4222	0.4236	0.4251	0.4265	0.4279	0.4292	0.4306	0.4319
1.5	0.4332	0.4345	0.4357	0.4370	0.4382	0.4394	0.4406	0.4418	0.4429	0.4441
1.6	0.4452	0.4463	0.4474	0.4484	0.4495	0.4505	0.4515	0.4525	0.4535	0.4545
1.7	0.4554	0.4564	0.4573	0.4582	0.4591	0.4599	0.4608	0.4616	0.4625	0.4633
1.8	0.4641	0.4649	0.4656	0.4664	0.4671	0.4678	0.4686	0.4693	0.4699	0.4706
1.9	0.4713	0.4719	0.4726	0.4732	0.4738	0.4744	0.4750	0.4756	0.4761	0.4767
2.0	0.4772	0.4778	0.4783	0.4788	0.4793	0.4798	0.4803	0.4808	0.4812	0.4817
2.1	0.4821	0.4826	0.4830	0.4834	0.4838	0.4842	0.4846	0.4850	0.4854	0.4857
2.2	0.4861	0.4864	0.4868	0.4871	0.4875	0.4878	0.4881	0.4884	0.4887	0.4890
2.3	0.4893	0.4896	0.4898	0.4901	0.4904	0.4906	0.4909	0.4911	0.4913	0.4916
2.4	0.4918	0.4920	0.4922	0.4925	0.4927	0.4929	0.4931	0.4932	0.4934	0.4936
2.5	0.4938	0.4940	0.4941	0.4943	0.4945	0.4946	0.4948	0.4949	0.4951	0.4952
2.6	0.4953	0.4955	0.4956	0.4957	0.4959	0.4960	0.4961	0.4962	0.4963	0.4964
2.7	0.4965	0.4966	0.4967	0.4968	0.4969	0.4970	0.4971	0.4972	0.4973	0.4974
2.8	0.4974	0.4975	0.4976	0.4977	0.4977	0.4978	0.4979	0.4979	0.4980	0.4981
2.9	0.4981	0.4982	0.4982	0.4983	0.4984	0.4984	0.4985	0.4985	0.4986	0.4986
3.0	0.4987	0.4987	0.4987	0.4988	0.4988	0.4989	0.4989	0.4989	0.4990	0.4990
3.1	0.4987	0.4987	0.4987	0.4987	0.4987	0.4987	0.4988	0.4988	0.4988	0.4988
3.2	0.4988	0.4988	0.4988	0.4989	0.4989	0.4989	0.4989	0.4989	0.4989	0.4989
3.3	0.4990	0.4990	0.4990	0.4990	0.4990	0.4990	0.4990	0.4991	0.4991	0.4991
3.4	0.4991	0.4991	0.4991	0.4991	0.4991	0.4991	0.4992	0.4992	0.4992	0.4992

Table C.2: Standard Probability Distribution: Part B from the mean of 0.5.

Z	0.00	0.01	0.02	0.03	0.04	0.05	0.06	0.07	0.08	0.09
-3.4	0.0003	0.0003	0.0003	0.0003	0.0003	0.0003	0.0003	0.0003	0.0003	0.0002
-3.3	0.0005	0.0005	0.0005	0.0004	0.0004	0.0004	0.0004	0.0004	0.0004	0.0003
-3.2	0.0007	0.0007	0.0006	0.0006	0.0006	0.0006	0.0006	0.0005	0.0005	0.0005
-3.1	0.0010	0.0009	0.0009	0.0009	0.0008	0.0008	0.0008	0.0008	0.0007	0.0007
-3.0	0.0013	0.0013	0.0013	0.0012	0.0012	0.0011	0.0011	0.0011	0.0010	0.0010
-2.9	0.0019	0.0018	0.0018	0.0017	0.0016	0.0016	0.0015	0.0015	0.0014	0.0014
-2.8	0.0026	0.0025	0.0024	0.0023	0.0023	0.0022	0.0021	0.0021	0.0020	0.0019
-2.7	0.0035	0.0034	0.0033	0.0032	0.0031	0.0030	0.0029	0.0028	0.0027	0.0026
-2.6	0.0047	0.0045	0.0044	0.0043	0.0041	0.0040	0.0039	0.0038	0.0037	0.0036
-2.5	0.0062	0.0060	0.0059	0.0057	0.0055	0.0054	0.0052	0.0051	0.0049	0.0048
-2.4	0.0082	0.0080	0.0078	0.0075	0.0073	0.0071	0.0069	0.0068	0.0066	0.0064
-2.3	0.0107	0.0104	0.0102	0.0099	0.0096	0.0094	0.0091	0.0089	0.0087	0.0084
-2.2	0.0139	0.0136	0.0132	0.0129	0.0125	0.0122	0.0119	0.0116	0.0113	0.0110
-2.1	0.0179	0.0174	0.0170	0.0166	0.0162	0.0158	0.0154	0.0150	0.0146	0.0143
-2.0	0.0228	0.0222	0.0217	0.0212	0.0207	0.0202	0.0197	0.0192	0.0188	0.0183
-1.9	0.0287	0.0281	0.0274	0.0268	0.0262	0.0256	0.0250	0.0244	0.0239	0.0233
-1.8	0.0359	0.0351	0.0344	0.0336	0.0329	0.0322	0.0314	0.0307	0.0301	0.0294
-1.7	0.0446	0.0436	0.0427	0.0418	0.0409	0.0401	0.0392	0.0384	0.0375	0.0367
-1.6	0.0548	0.0537	0.0526	0.0516	0.0505	0.0495	0.0485	0.0475	0.0465	0.0455
-1.5	0.0668	0.0655	0.0643	0.0630	0.0618	0.0606	0.0594	0.0582	0.0571	0.0559
-1.4	0.0808	0.0793	0.0778	0.0764	0.0749	0.0735	0.0721	0.0708	0.0694	0.0681
-1.3	0.0968	0.0951	0.0934	0.0918	0.0901	0.0885	0.0869	0.0853	0.0838	0.0823
-1.2	0.1151	0.1131	0.1112	0.1093	0.1075	0.1056	0.1038	0.1020	0.1003	0.0985
-1.1	0.1357	0.1335	0.1314	0.1292	0.1271	0.1251	0.1230	0.1210	0.1190	0.1170
-1.0	0.1587	0.1562	0.1539	0.1515	0.1492	0.1469	0.1446	0.1423	0.1401	0.1379
-0.9	0.1841	0.1814	0.1788	0.1762	0.1736	0.1711	0.1685	0.1660	0.1635	0.1611
-0.8	0.2119	0.2090	0.2061	0.2033	0.2005	0.1977	0.1949	0.1922	0.1894	0.1867
-0.7	0.2420	0.2389	0.2358	0.2327	0.2296	0.2266	0.2236	0.2206	0.2177	0.2148
-0.6	0.2743	0.2709	0.2676	0.2643	0.2611	0.2578	0.2546	0.2514	0.2483	0.2451
-0.5	0.3085	0.3050	0.3015	0.2981	0.2946	0.2912	0.2877	0.2843	0.2810	0.2776
-0.4	0.3446	0.3409	0.3372	0.3336	0.3300	0.3264	0.3228	0.3192	0.3156	0.3121
-0.3	0.3821	0.3783	0.3745	0.3707	0.3669	0.3632	0.3594	0.3557	0.3520	0.3483
-0.2	0.4207	0.4168	0.4129	0.4090	0.4052	0.4013	0.3974	0.3936	0.3897	0.3859
-0.1	0.4602	0.4562	0.4522	0.4483	0.4443	0.4404	0.4364	0.4325	0.4286	0.4247
0.0	0.5000	0.4960	0.4920	0.4880	0.4840	0.4801	0.4761	0.4721	0.4681	0.4641

Table C.3: Standard Probability Distribution: Cumulative – Part A.

Z	0.00	0.01	0.02	0.03	0.04	0.05	0.06	0.07	0.08	0.09
0.0	0.5000	0.5040	0.5080	0.5120	0.5160	0.5199	0.5239	0.5279	0.5319	0.5359
0.1	0.5398	0.5438	0.5478	0.5517	0.5557	0.5596	0.5636	0.5675	0.5714	0.5753
0.2	0.5793	0.5832	0.5871	0.5910	0.5948	0.5987	0.6026	0.6064	0.6103	0.6141
0.3	0.6179	0.6217	0.6255	0.6293	0.6331	0.6368	0.6406	0.6443	0.6480	0.6517
0.4	0.6554	0.6591	0.6628	0.6664	0.6700	0.6736	0.6772	0.6808	0.6844	0.6879
0.5	0.6915	0.6950	0.6985	0.7019	0.7054	0.7088	0.7123	0.7157	0.7190	0.7224
0.6	0.7257	0.7291	0.7324	0.7357	0.7389	0.7422	0.7454	0.7486	0.7517	0.7549
0.7	0.7580	0.7611	0.7642	0.7673	0.7704	0.7734	0.7764	0.7794	0.7823	0.7852
0.8	0.7881	0.7910	0.7939	0.7967	0.7995	0.8023	0.8051	0.8078	0.8106	0.8133
0.9	0.8159	0.8186	0.8212	0.8238	0.8264	0.8289	0.8315	0.8340	0.8365	0.8389
1.0	0.8413	0.8438	0.8461	0.8485	0.8508	0.8531	0.8554	0.8577	0.8599	0.8621
1.1	0.8643	0.8665	0.8686	0.8708	0.8729	0.8749	0.8770	0.8790	0.8810	0.8830
1.2	0.8849	0.8869	0.8888	0.8907	0.8925	0.8944	0.8962	0.8980	0.8997	0.9015
1.3	0.9032	0.9049	0.9066	0.9082	0.9099	0.9115	0.9131	0.9147	0.9162	0.9177
1.4	0.9192	0.9207	0.9222	0.9236	0.9251	0.9265	0.9279	0.9292	0.9306	0.9319
1.5	0.9332	0.9345	0.9357	0.9370	0.9382	0.9394	0.9406	0.9418	0.9429	0.9441
1.6	0.9452	0.9463	0.9474	0.9484	0.9495	0.9505	0.9515	0.9525	0.9535	0.9545
1.7	0.9554	0.9564	0.9573	0.9582	0.9591	0.9599	0.9608	0.9616	0.9625	0.9633
1.8	0.9641	0.9649	0.9656	0.9664	0.9671	0.9678	0.9686	0.9693	0.9699	0.9706
1.9	0.9713	0.9719	0.9726	0.9732	0.9738	0.9744	0.9750	0.9756	0.9761	0.9767
2.0	0.9772	0.9778	0.9783	0.9788	0.9793	0.9798	0.9803	0.9808	0.9812	0.9817
2.1	0.9821	0.9826	0.9830	0.9834	0.9838	0.9842	0.9846	0.9850	0.9854	0.9857
2.2	0.9861	0.9864	0.9868	0.9871	0.9875	0.9878	0.9881	0.9884	0.9887	0.9890
2.3	0.9893	0.9896	0.9898	0.9901	0.9904	0.9906	0.9909	0.9911	0.9913	0.9916
2.4	0.9918	0.9920	0.9922	0.9925	0.9927	0.9929	0.9931	0.9932	0.9934	0.9936
2.5	0.9938	0.9940	0.9941	0.9943	0.9945	0.9946	0.9948	0.9949	0.9951	0.9952
2.6	0.9953	0.9955	0.9956	0.9957	0.9959	0.9960	0.9961	0.9962	0.9963	0.9964
2.7	0.9965	0.9966	0.9967	0.9968	0.9969	0.9970	0.9971	0.9972	0.9973	0.9974
2.8	0.9974	0.9975	0.9976	0.9977	0.9977	0.9978	0.9979	0.9979	0.9980	0.9981
2.9	0.9981	0.9982	0.9982	0.9983	0.9984	0.9984	0.9985	0.9985	0.9986	0.9986
3.0	0.9987	0.9987	0.9987	0.9988	0.9988	0.9989	0.9989	0.9989	0.9990	0.9990
3.1	0.9990	0.9991	0.9991	0.9991	0.9992	0.9992	0.9992	0.9992	0.9993	0.9993
3.2	0.9993	0.9993	0.9994	0.9994	0.9994	0.9994	0.9994	0.9995	0.9995	0.9995
3.3	0.9995	0.9995	0.9995	0.9996	0.9996	0.9996	0.9996	0.9996	0.9996	0.9997
3.4	0.9997	0.9997	0.9997	0.9997	0.9997	0.9997	0.9997	0.9997	0.9997	0.9998

Table C.4: Standard Probability Distribution: Cumulative – Part B.

Index

Sense of independence
-direct control over priorities → talk to people

Forced to public present

Surprises
-Mental shift → used to not saying anything

Stood out
→technical talk -opportunities to learn about nuclear
→grid stability → securit
knock out interconnects needs to shut down

Series

☆ Be curious
-ask questions! →figure out how you fit in
-ask about customers

GE → energy division → large manufacturing
New York

7 years → made contacts

2012

EPRI
Non-profit
800

GE
200,000

EPRI is unique
- collaborative model
70% power generation
(interact with)

Work for public good

Touches entire industry

Project Manager
-scopes of work, budjet
Subcontract
- hands on work in lab
-research, data analysis
-Wales for conference

Plant maitenance → materials degradation

Switching Roles
-Social → upstate NY
- Opportunistic → new thing → more direct with customers
- Managerial
GE → indecision with projects
EPRI → expand horizons

Storage
Batteries
-more constrained grid

The Duck Curve - Vox article
Reverse hydroelectric
-requires geography
- expensive, limited

Solar Thermal
-huge thermal batteries
-portugal? SA large scale deployment